From Neural Networks and Biomolecular Engineering to Bioelectronics

ELECTRONICS AND BIOTECHNOLOGY ADVANCED (EL.B.A.) FORUM SERIES

Volume 1 FROM NEURAL NETWORKS AND BIOMOLECULAR ENGINEERING TO BIOELECTRONICS
Edited by Claudio Nicolini

A Continuation Order Plan is available for this series. A continuation order will bring delivery of each new volume immediately upon publication. Volumes are billed only upon actual shipment. For further information please contact the publisher.

From Neural Networks and Biomolecular Engineering to Bioelectronics

Edited by

Claudio Nicolini

Institute of Biophysics
University of Genoa
Genoa, Italy

Assistant Editor
Sergei Vakula

EL.B.A. Foundation
Portoferraio, Italy

Springer Science+Business Media, LLC

Library of Congress Cataloging-in-Publication Data

On file

Proceedings of the 1993 International Workshop on Electronics and Biotechnology Advanced, held July 13–16, 1993, on the Isle of Elba, Italy

ISBN 978-1-4899-1090-5 ISBN 978-1-4899-1088-2 (eBook)
DOI 10.1007/978-1-4899-1088-2

Originally published by Plenum Press, New York in 1995
Softcover reprint of the hardcover 1st edition 1995

10 9 8 7 6 5 4 3 2 1

PREFACE

This volume represents the first of a series of proceedings of the EL.B.A. Forum on Bioelectronics, a scientific discipline at the frontiers of Advanced Electronics and Biotechnology.

The name for these forums derives not only from the place (the Isle of Elba in Italy), where the conferences have been held every 6 months since 1991, but also from an acronym: **El**ectronics and **B**iotechnology **A**dvanced.

Bioelectronics is intended as "*the use of biological materials and biological architectures for information processing and sensing systems and devices down to molecular level*" and focuses its attention on three major areas:

I New hardware architectures borrowed from the thorough study of brain and sensory systems down to the molecular level, utilizing existing semiconductor inorganic materials (both GaAs and Si) and giga-scale integration;

II Protein Engineering, especially of systems involved in electron transfer and molecular recognition, integrated with Metabolism and Chemical Engineering, to develop new biomaterials by learning basic rules of macromolecular folding and self-assembly;

III Sensors, thin film and electronic devices utilizing organic compounds and biopolymers, and by implementing nanotechnology bottom up through manufacturing and characterization at the atomic level.

Bioelectronics, at the interface of biotechnology and electronics, has been characterized by the Italian Government, as early as 1989, as one of the fields of strategic importance for scientific and technological research and for potential industrial development. The decision, made in 1989 by the Italian Government, was revolutionary in this field in that for the first time in a National Program of Research scientific activity was combined with training-by-research of young scientists. For the first time, provisions were made for an eight-year plan divided into two periods, one of three and the subsequent of five years duration.

As a result, in 1990 several companies (SGS Thomson, Pharmacia-Farmitalia Carlo Erba, Elsag Bailey, Sorin Biomedica-FIAT, ENICHEM) and an industrial consortium (Technobiochip, recently transformed in a new company operating in the field of Bioelectronics, and a major sponsor of the EL.B.A. Forum), all leaders in their fields at the international level, became involved in the research activity sponsored by the National Program of Research "Technologies for Bioelectronics," under the general cover of CIREF.

Extremely significant was the formation, in 1990, of Polo Nazionale Bioelettronica, a consortium with headquarters in the Isle of Elba, that is comprised of several leading universities (Genoa, Pisa, Siena, Naples, Catania), national research institutions (ENEA, INBB, National Institute of Health), industrial consortia TECHNOBIOCHIP, CIREF), and financial holding companies (Raggio di Sole Biotecnologie, IRI-SPI).

One of the milestones in the development of the international cooperation in the fields relevant for bioelectronics was set up by the signing of an international agreement between

Italy and Russia in December 1990. This agreement, signed by Italian Minister Ruberti and Russian Vice-Minister Bortnik, for cooperation in the fields of Biomolecular Engineering and Bioelectronics, resulted in the "Elba Project" between the Russian Academy of Sciences and Polo Nazionale Bioelettronica. As the result of this project a foundation has been established recently involving Polo Nazionale Bioelettronica, the Italian Ministry of Foreign Affairs, the Italian Ministry of University and Scientific and Technological Research, the Russian Ministry of Science and Technical Policy, leading research institutions from Italy, Russia, and the United States, with the participation of the Commission of European Union as an observer.

More recently, the same island and acronym (**El**ectronics and **B**iotechnology **A**dvanced), have given the name to a Scientific and Technological Park, which is becoming a key location of an expanding scientific and technological project of European dimensions focused on Bioelectronics.

I would like to express my gratitude particularly to Dr. Sergey Vakula. I would also like to acknowledge the precious assistance of Drs. Paolo Occhialini and Pietro Ragni of AF FORUM, and Dr. Isabella Zolfino of TECHNOBIOCHIP. Thanks are also due to the Ministry of University and Scientific and Technological Research of Italy, which financed this workshop.

Claudio Nicolini
Member of the National Science and Technology Council

March 1, 1994

INTRODUCTION

The EL.B.A. Forum Series is intended as a series of proceedings of periodic workshops, held on the Isle of Elba, jointly organized by Technobiochip, CIREF, and Polo Nazionale Bioelettronica, starting in 1992. The first forums were held during 1992-1993 at Marciana, with numerous leading scientists participating in these meetings.

This series of conferences is intended to give a unique retrospective of what is currently being defined as "bioelectronics." Bioelectronics emerged several years ago at the interface of life and physical sciences and is gaining more and more attention and respect worldwide. Strong evidence of this fact is the interest expressed recently by the European Community, which resulted in two workshops on Bioelectronics, sponsored and organized by the Commission of the European Communities. The first workshop was held at the end of 1991 in Brussels, the second one at the end of November 1993, in Frankfurt (Main).

The Italian leadership in the field of Bioelectronics is now acknowledged by the scientific community. This leadership role was confirmed by the mandate given by the Commission of the European Community to Polo Nazionale Bioelettronica to carry out a feasibility study for a new Research and Development program on bioelectronics to be undertaken by the Commission in 1994-1998. The study group was composed of leading experts in the fields relevant to bioelectronics, including Nobel prize winners.

The very first meeting on Bioelectronics in Italy was held in Novara on April 4-7, 1987 and was sponsored by CIREF, an Association for Industrial Strategic Research, upon the initiative of Professor C. Nicolini (at the time Science and Technology Advisor of the Italian Prime-Minister) and three top industrial Chief Executive Officers: Professor Amilcaro Collina (Montedison), Professor Umberto Rosa (SNIA-FIAT), and Dr. Pasquale Pistorio (SGS-Thompson).

Highly innovative and daring was the decision by Professor Antonio Ruberti in 1988 (who at that time was the Minister for Science and Technology and is now Commissary for Research in Brussels and Vice-President of the Commission of the European Community) to start a National Program of Research on "Technologies for Bioelectronics," chaired by Professor C. Nicolini.

The objectives of training and research activity of this program are constantly and soundly supported by some of the most noted experts worldwide in various fields of bioelectronics (Capasso, Faggin, Croce, Karplus, Barraud, Hopfield, Aizawa, Wüthrich, Skryabin, Varfolomeev, Bykov) and are confirmed by their participation in the EL.B.A. Forum.

More than twenty academic and industrial centers have been involved in the extensive training of 57 young researchers working in the field of bioelectronics, initially selected in the Isle of Elba among four hundred graduates and Ph.D.s, all of whom participated in the EL.B.A. Forum Workshops.

An analysis of the initial activity in the three different areas of the national program–neural and submicron electronics, protein engineering, and first generation bioelectronics–

confirmed the industrial importance of the results and prototypes obtained by the companies, research institutions and universities, the most competitive of which were present at the III CIREF Workshop at Bergamo on October 26-27, 1993.

An island and an acronym (**El**ectronics and **B**iotechnology **A**dvanced) have given the name to a Scientific and Technological Park– a natural location of an expanding scientific and technological project of European dimensions.

Technobiochip (a research consortium of five leading industries operational in the area of electronics and biotechnology) played the most significant role in organizing the EL.B.A. Forum Workshops on the Isle of Elba, where it's research laboratories are located (in Marciana). The level of the research activity performed there is undoubtedly of the highest European standards and importance in the field of bioelectronics.

The numerous participants who attended the EL.B.A. Forums held in 1992-93 represented the leading industrial research organizations and academic institutions, and presented a wide and deep outline of what is now being defined as Bioelectronics, at the crossing of advanced electronics and biotechnology.

CONTENTS

FROM PROTEIN ENGINEERING TO BIOELECTRONICS

Claudio Nicolini

Institute of Biophysics, University of Genova
and Polo Nazionale Bioelettronica, Portoferraio (LI), Italy

INTRODUCTION

In recent times, new scientific and technological areas have emerged at the frontier of biotechnology and electronics, namely biomolecular engineering and bioelectronics which can have a notable impact on the industrial and economic development of Europe. One major feature of modern biology relevant to this attracts considerable attention, namely the coded self-assembly of biopolymers leading to complex structures in the human body. Searching for the specific amino-acid sequence that will yield a protein with both the three-dimensional structure and self-assembly properties needed to obtain monolayers with the desired packing order, function and electronic properties, we aim to make the first step towards the development of bioelectronics.

The object of this presentation is to outline the main route to bioelectronics from protein engineering. Initial emphasis is placed on the most recent striking achievements in the development of thin film biotechnology, by means of mono- and multi-layers of bi- and/or tridimensionally oriented metalloproteins, using both auto-assembly and Langmuir-Blodgett technique (or its modified version by Langmuir and Shaeffer) with or without the aid of lipids and of suitable chemical modifications. One of the main drawbacks in producing protein thin films is the necessity to arrange globular proteins on the surface of water typically volatile and non-polar, without precipitating them and without denaturation. By appropriate procedures we have obtained polar sufficiently rigid monolayers of proteins, that are highly compressed and strikingly stable with the maintenance of globular native structure up to 200° C. We have also been able to overcome the second drawback due to the properties of naturally existing proteins unsuitable for bioelectronic devices and which needed to be properly engineered at molecular level through recombinant DNA technology, from *a priori* knowledge of their solution structure determined at atomic resolution via 2DFT NMR. The final objective is indeed the development of new materials with unique geometric structure and electronic properties optimal for the construction of molecular biodevices, namely biotransistors and biosensors. In this work we show how the formation and the function of protein monolayers can be first engineered by site-specific mutations, then tied to substrate (with covalent or other type of bonds) and autoassembled, with new exceptional properties. The relevant most recent papers resulting from the activities here reported and from others directly or indirectly related to bioelectronics are summarized in the attached list of references.

1. PROTEINS OPTIMAL FOR BIOELECTRONICS

Selection of biological molecules for construction of biochips and biosensors has to meet many requirements:

From Neural Networks and Biomolecular Engineering to Bioelectronics
Edited by C. Nicolini, Plenum Press, New York, 1995

- the detailed knowledge at atomic resolution of structure of the protein to be used;
- the gene relative to the protein must be cloned so that large quantity of protein may be produced in a cell organism used for expression;
- the best possible knowledge about biophysical, biochemical and enzymatic data of the utilized system;
- the possibility to engineer the protein so it can adapt itself better to required "bioelectronic" application;
- asymmetric charge distribution which is useful to be able to organize the proteins, by applying an external electric field.

Few metalloproteins containing the heme group, have all of the above mentioned characteristics, and which can, moreover, have other properties with high bioelectronic potential:

- enormous "strength" of heme-associated oscillation which permits to measure the orientation on a very small surface (both in area and in thickness);
- the high resolution with which the structure of these proteins is known that permits to obtain, starting from the measurements of the heme orientation, the position of other sites in the protein.

To realize the development of these new materials we have therefore decided to investigate in a systematic way their capacity to organize on ordered thin film by several alternative procedures or to establish an optimal self-organizing system strictly utilizing protein engineering. Alternatively we have pursued the route to utilize redox proteins like thioredoxin, which, whenever naturally engineered in thermophilic microorganisms, appear to be also a good candidate for bioelectronic devices. Particularly intriguing for possible bioelectronic application of protein engineering is the observation that in *R.viridis* photosynthetic reaction centers the electron transfer across the molecule (about 30 Å path) occurs in the picoseconds range.

1.1 Metalloproteins

All known cytochromes are hemoproteins which take part in the process of electron transfer. Their classification is based on their function and the number refers to wavelength of absorption maximum of the reduced form. The metalloprotein system chosen, cytochrome P450scc with its respective proteins, adrenodoxin and adrenodoxin reductase, constitute a good "model system" to reveal the basic rules correlating the primary structure (i.e. amino acid sequence) with the tertiary structure, and thereby function, to engineer proteins for the desired functional application. For the chosen proteins data have been already obtained at the atomic resolution, though insufficient at present for the proper reconstruction of the 3D-structure. An understanding of the relationship between protein three-dimensional structure and function is critical in the development of the "new medicine" where the above cytochrome is tailor-made in the laboratory to carry out the desired function. The other protein that might be studied is azurin, with the aim of designing redox protein regeneration systems, since azurin is believed to mediate the electron transfer between cytochrome c551 and cytochrome oxidase/nitrite reductase. However, apart from the final goals, the approaches and methodology used will be similar to those used in the studies of the cytochrome P450 system. The aim of the azurin studies (azurin coupled to a dehydrogenase) would be the design of novel enantioselective biocatalysts for specific transformation of "environmental chemicals" and/or fine chemical synthesis. The supplementary goals are to define the structure/function relationship of the above protein products of interest for the European industry by combining molecular biology, including protein chemistry and nucleic acid biochemistry, with crystallographic X-ray diffraction analysis of the protein components and Nuclear Magnetic Resonance correlation spectroscopy of proteins in solution.

Cytochrome p450. The name "cytochrome p450" is assigned to proteins that have an iron atom bound to sulphur, and that form complexes with carbon monoxide, which have a band of absorption around 450 nm. This enzyme is present in nature in many organisms. In cells, the protein is located in mitochondria of the cortex and microsomes of hepatocytes and its function is associated with the reaction of detoxification. This protein uses high energy electrons from NADPH, which are transferred by a special reductase present in endoplasmic reticulum, to catalyze the reaction of addition of hydroxyl groups to numerous hydrocarbons that are water-insoluble and potentially dangerous dispersed in the lipid double layer.

The typical reaction which involves the cytochrome p450 is as follows:

$$RH + O_2 + NAD(P)H + H^+ \rightarrow ROH + H_2O + NAD(P)^+$$

Other enzymes add other negatively charged groups to the hydroxyl group thus rendering soluble (and therefore transferable) a medicine or harmful metabolic, that otherwise would remain confined in the plasmatic membrane. Among different known cytochromes, P450cam has the advantage of being soluble and being available in large quantity. This protein catalyzes the stereospecific hydroxylation of camphor carbon group, which leads to formation of 5-exo-hydroxycamphor,

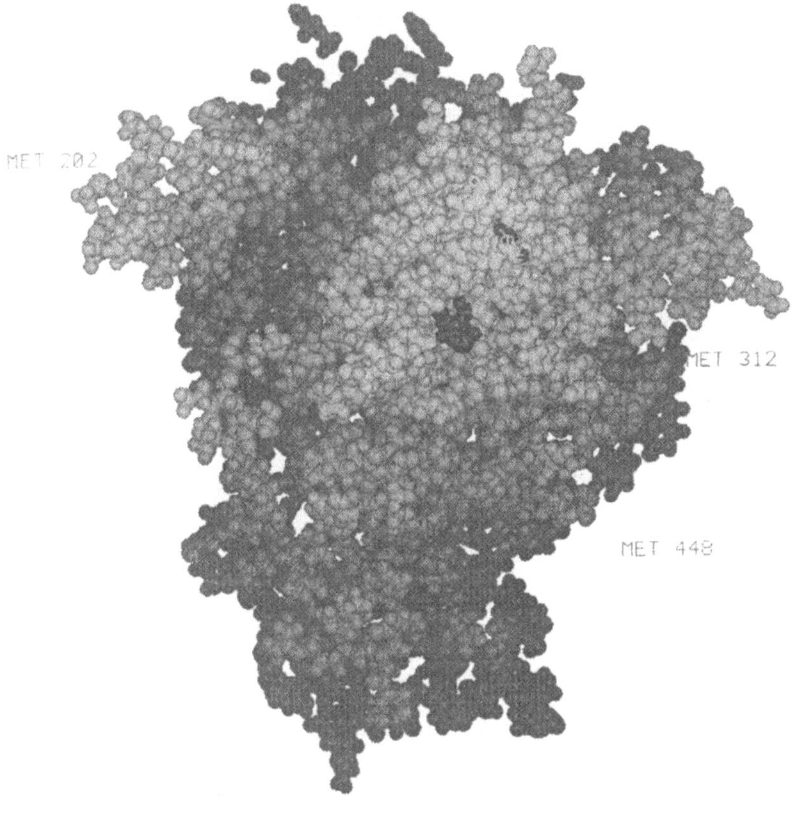

Figure 1 Atomic resolution structure of P450 scc.

A cytochrome p450 of particular interest is indeed the cytochrome p450 scc ("scc" means "side chain cleavage"), the component of the redox chain responsible for cholesterol oxidation. The components of this chain are cytochrome P450scc, adrenodoxin (a metalloprotein of ferrodoxin type) and adrenodoxin reductase (similar to flavoprotein), which should be used *together* to create the chain of electron transfer necessary for hydroxylation of cholesterol. Cytochrome P450scc extracted from bovine adrenocortical mitochondria, the primary sequence of which is known, is specific in the reaction of side chain cleavage of cholesterol. One of the reasons why this metalloprotein is of particular interest for bioelectronics (in particular, for the development of biosensors) is based on its oxidizing action and not mere electron transfer. Their form of trigonal prism makes it particularly interesting for the formation of highly packed monolayers in which the space for the water molecules (which are absolutely not favoured by modern electronics) can be conveniently reduced to zero. In contrast to other typical cytochromes, the heme of p450 is not directly accessible form the surface of the protein. For this reason it is quite feasible that one or more lateral amino acid chains could be directly implicated in the electron transfer from heme to its redox partner, the putidaredoxin, bound electrostatically to the surface by some basic amino acids. The molecular weight, determined from knowledge of the primary sequence is about 56.4 kDa and its 3D structure has been recently determined by X-ray crystallography. Cytochrome P450scc is present in the human genome in a single copy in the chromosome 11. Cytochrome P450scc is a globular metalloprotein which is formed by a heme group and a long polypeptide chain, 70% of which is constituted by α -helix [Waterman M.R., 1992]. The molecular weight is about 56.4 kDa. The cytochrome P450scc in intact human cells is localized in the mitochondria on the internal membrane. The transcription of the gene of cytochrome P450scc proceeds with higher activity in mitochondria of cells of surrenal glands, but takes place also in liver. The main function of cytochrome P450scc in surrenal glands is biosynthesis of steroids, whereas in liver it is the metabolism of xenobiotics. Despite a relatively low rate of electron transfer, our original idea of using cytochrome P450 scc in bioelectronics has been confirmed: the regular triangular form (quadrangular prism shown in Fig. 1 has proven itself to be ideal for the formation of highly packed and stable monolayers as revealed by the autoassembly and Langmuir-Blodgett technology [see later].

In order for the cytochrome P450scc to function as a monooxygenase, the presence of 2 binding sites is necessary. The first binding site is the "oxygen pocket" that binds the O_2 molecule, whereas the second site is the substrate pocket. Under normal conditions, the necessary requirement for the monoxygenation process to proceed is the cyclic electron transfer according to the a scheme which utilizes Adx (adrenodoxin (Fe-S protein)) and AdxR (adrenodoxin reductase (flavoprotein)). The reduced and oxidized states of the iron atom in the cytochrome P450scc can be determined spectrophotometrically in the model system with sodium dithionate and CO by the appearance of the peak at 450 nm in the oxidized state.

In addition to the construction of thin films with optimal stability and electron transfer, as bioelectronic application of the cytochrome P450scc in health care we may envisage an assay system for determination of the steroid level in blood and the precise determination of blood cholesterol with the use of a biosensor based on cyt P450scc. The change in the steroid level of blood is an important parameter for the diagnosis of endocrinic diseases (widely spread and hard to diagnose). The determination of the ratio cholesterol -cholesterin* is the key point in the diagnostics and efficient control of the therapy of atherosclerosis. Cyt P450scc has been cloned in strains of yeast that express and secrete the protein. Some of the clones are the property of Technobiochip, others of the Institute of Biophysics. One clone expresses P450scc in large quantity and the protein has the same electrophoretic mobility and the absorption ratio OD_{393}/OD_{280} of the wild-type. The gene of cytochrome P450scc has been cloned also in *E.coli* to produce, by site-directed mutagenesis, new proteins with the desired properties.

Cytochrome c. Cytochrome c is the most studied of all cytochromes. It is present in the cells of all aerobic organisms, and in the animal cells in the mitochondrial protein-lipid complex. Since a transition between the ferric and ferrous states takes place in all cytochromes, they are efficient electron transporters and are generally considered as catalysts of the respiration essential for the oxidation of substrates with O_2. The electron transfer chain in which the cytochrome is involved is as following:

NADH --->	NADH-Q --->	Q --->	cytochrome --->	cytochrome c-
	reductase		reductase	
	I complex		II complex	

---

* cholesterol = 5-cholestene-3--ol
  cholesterin = 5-cholestene

-->cytochrome c ---> O_2
oxidase (Q = ubiquinone)
III complex

The transfer of the heme from the oxidized form to the reduced one seems to depend on some changes in the bending of the backbone of the cytochrome in the proximity of the heme. The function of the iron atom in the cytochrome is to be oxidized and reduced by specific molecules and related complexes. The structure consists of about 104 amino acids or more, with the heme group attached by residues of cystein in the 14 and 17 positions. The primary sequence is known for different kinds of mammals and their molecular weight is about 13 kDa. Cytochrome c is a very interesting model since it has been found almost unaltered during billion years of evolution. Cytochrome c from grain or from rice, for example, if extracted and purified, can react with cytochrome oxidase from horse, and the cytochrome from horse reacts with the cytochrome oxidase from baker's yeast. Even if other parts of the molecule can undergo immediate changes, the part essential to the activity as electron transporter has remained unaltered. The primary sequence of various cytochromes c is known in much more detail that of any other protein (Atlas of Protein Sequences), hence it seems that the changes effected regard the substitution of amino acids with similar ones. In particular, both the bonds of the heme with Cys 14, Cys 17, and Hys 18 remain unaltered and not subject to mutations, as well as the long sequence between the residues 70 and 80.

Cytochrome c551. The cytochrome c551 is a small protein which is present in periplasmatic space of *Pseudomonas aeruginosa*, and its function is to yield the electrons to nitrite reductase either directly or via azurin. The protein is constituted by 82 amino acids and has a molecular weight of about 8.6 kDa. The tridimensional structure was determined by means of X-ray crystallography. The refolding of cytochrome c551 is very similar to refolding of other more heavy cytochromes and also the pocket which contains the heme is the same. The more evident differences between the larger cytochromes and cytochrome c551 consist in a large deletion at the bottom of the fixation of the heme, since one string of 15 amino acids (from 41 to 55) is absent, which is, however, present in cytochrome c of tuna fish. This gap in the bottom part of the molecule makes the heme group more inclined laterally in respect to cytochrome c of tuna fish and exactly in this zone the heme is more disclosed, thus favouring the ionization of propionate group. In the cytochrome c551 the heme is bound to the protein via Cys 12, Cys 15, while the Hys 15 together with Met 61 bind axially the Fe atom. The cytochrome c551 looks like a sphere with a defect where the heme is enclosed; the perimeter of this cavity, like for other cytochromes c is surrounded by positive charges (Lys). The cloning and expression of cytochrome c551 has been recently reported.

Azurin. The three-dimensional structure, electrostatic characteristics and hydrophobicity of azurin from *Pseudomonas aeruginosa* bacteria has been already analyzed in more detail. The molecule of azurin has a certain asymmetry of charge between the regions of the protein, including the copper site, dominated by positive values of the electrostatic potential, and the rest of the molecule in both oxidation states of the metal. From the structural viewpoint the region characterized by the positive electrostatic potential is found in the proximity to the hydrophobic surface area, although not being completely superimposed on it. This molecules of azurin attached to the substrate which would permit to modify the specificity of the homo and hetero protein-protein interaction. An additional approach is to introduce mutations that would reduce the redox potential of the protein system. A strategy has been adopted to design the mutations thereby modifying amino acids exposed into the solvent. These would not intervene directly in the process of electron transfer with the aim of not altering the stability and, thereby, retaining of functional activity of the protein. The substituted residues could be chosen also for their position in the inversions or loops of the polypeptide chain thus minimizing the damage to the structural topology of the molecule. Attention will be turned towards removal of strongly polar residues around the hydrophobic zone that surrounds the copper atom. This mutation would have the double aim of increasing the hydrophobic interaction surface and simultaneously extending the region of negative potential located in the part opposite to the copper site, thus favouring the intermolecular interactions. A second possibility of mutagenesis foresees the accentuation of the intrinsic dipole of the azurin molecule, increasing the positive charge density in the region of the copper site and the negative one, by neutralizing or inverting the charge of one amino acid residue located in the area opposite to the copper site (Asp 11 is situated in an inversion and Lys 74 in al loop of polypeptide chain). The prosthetic heme group is a crucial component of all six metalloproteins described above for its capacity to interact with the electromagnetic radiation which renders it particularly suitable for photosensitive devices, for commutation and for electronic calculation.

1.2 Thioredoxin and Alcohol Dehydrogenase

Thioredoxin and alcohol dehydrogenase for the present studies were either purchased in the oxidized mesophilic form or isolated from thermophilic microorganisms. For the 2DFT NMR studies the protein was dissolved in 0.1 M potassium phosphate at pH=5.7 (90% H_2O 10% D_2O) to a final concentration of 3.6 mM. The solution was equilibrated under N_2.

In fact, archaebacteria are important to bioelectronics since they have evolved molecular mechanisms, metabolic pathways and compounds that are unique and contain macromolecules as lipids and proteins with increased stability to heat and to all common denaturants and unique properties such as resistance to proteolytic enzymes, organic solvent and high salt concentration. They have lipids, the chemistry of which is based on the occurrence of ether linkages between glycerol or more complex polyol with isoprenoid alcohols of 20, 25 or 40 carbon atoms. Even if little is still known about the biosynthesis of such lipids, the lipids themselves and the enzymes involved in ether bond formation may have important applications.

Future goals in this field are the study of the primary and tertiary structure of archaebacterial proteins. While the thermophilic Eubacteria have presumably evolved from mesophilic lines, by inventing different ways to turn mesophilic macromolecules into more thermophilic ones, the exclusively thermophilic Thermoproteales may be considered as the common ancestor of the lineage, that "may have invented and further transmitted a unique, most successful principle to construct thermostable proteins allowing growth even in boiling water". Thioredoxin isolated from Archaebacteria and in general from thermophilic bacteria are examples of thermostable and thermophilic proteins, which, after determining the primary and tertiary structure, can be used as models for designing enzymes that could then be produced by genetic engineering. The resolution by NMR of the structure of thioredoxins isolated from thermophiles and extremophiles (archaebacteria) is under way. Should we understand from the above studies the principles that give stability to proteins, we could modify only few amino acid residues in known proteins and render them stable.

1.3 Photosynthetic Reaction Centers

Photosynthetic reaction centres from *Rhodobacter sphaeroides* is a membrane protein with a molecular weight of 100 kDa, consisting of 3 subunits L, M and H. The crystal structure of the protein was obtained [J.P.Allen et al, 1984]. The function of the protein is light-induced electron transport with 100% quantum efficiency. The electron is transferred by tunnelling mechanism through cofactors, attached to it, namely, dimer of bacteriochlorophill, bacteriopheophetin and two quinones. The first stage of the charge separation process is very fast (less than 4 ps).

2. PROTEIN SOLUTION STRUCTURE AT ATOMIC RESOLUTION

2.1 Automation of protein structure determination by NMR correlation spectroscopy and by neural network

Two-dimensional Nuclear Magnetic Resonance (2DFTNMR) is presently the most powerful tool to determine protein structure in solution. Peak assignment (an interpretation of the two-dimensional spectra that leads to the individuation of pairs of hydrogen atoms that are involved in a NOE peak) is a corner-stone of such use of 2DFTNMR. Manual peak assignment of a protein often requires months of work by a specialized equipe. An automation of this task could speed up the protein study process, or alternatively allow to

study previously unmanageable proteins. This article describes PEPTO, an expert system for the interpretation of sets of 2DFTNMR spectra on proteins. The present version of the program deals with spectra obtained from NOESY and COSY experiments. Tests of PEPTO on simulated spectra of five proteins with known assignments were described and discussed [Catasti et al., 1990]. Peak-picking is the lowest-level task of the interpretation of two-dimensional, and multidimensional Nuclear Magnetic Resonance (NMR) spectra in general for protein structure determination. It consists of individuating peaks on two-dimensional frequency spectra, destined for further elaboration. The performances of several feedforward artificial neural networks trained with back propagation with temperature on the task of peak-picking are compared [Carrara et al., 1993a]. The best one averages less than an approximate 5% error on well-defined spectral regions. The performances of the network are comparable with those of a human expert: the consequences of this fact on the possibility of improving the performance of the network have been further explored in later work [Carrara et al., 1993b]. In Carrara et al., 1993b, the training set and the testing set came from the same protein, namely BPTI; now we are studying the performance of FF4 upon other proteins. The preliminary results shown in Fig. 2 are the following:
- the network trained on BPTI and tested on NOESY and TOCSY spectra of cytochrome and thioredoxin gives the 25% of error;
- the network is able to learn the patterns on the new proteins without any modification in the network or in the training methods.

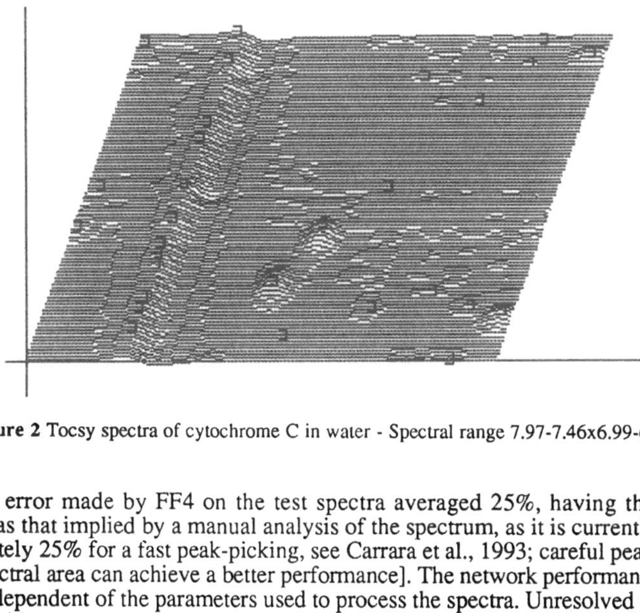

Figure 2 Tocsy spectra of cytochrome C in water - Spectral range 7.97-7.46x6.99-6.50

The error made by FF4 on the test spectra averaged 25%, having thus the same magnitude as that implied by a manual analysis of the spectrum, as it is currently performed [approximately 25% for a fast peak-picking, see Carrara et al., 1993; careful peak-picking on a small spectral area can achieve a better performance]. The network performances appear to be fairly independent of the parameters used to process the spectra. Unresolved problems are concentrated in ambiguity-rich regions, where even to a manual peak-picker would result difficult to achieve better results than the network. To improve the network performances in these regions, it is necessary to achieve a better, certain way of discriminating between real peaks and artefacts. The second part of the task concerns the automation of tertiary-quaternary structure determination from NMR data. The Double-Iterated Kalman Filter (DIKF) is a new method recently introduced to address the problem of the three-dimensional structure determination of polypeptides from NMR experimental distance constraints. It is a non linear bayesian estimator (that is, probability condition-based) which compares the experimental geometrical constraints with the corresponding values measured on the structure. The structure is updated weighting the uncertainty in the experimental constraints and in the atom positions; the algorithm also gives a direct estimate of the uncertainty in the result (Fig. 3).

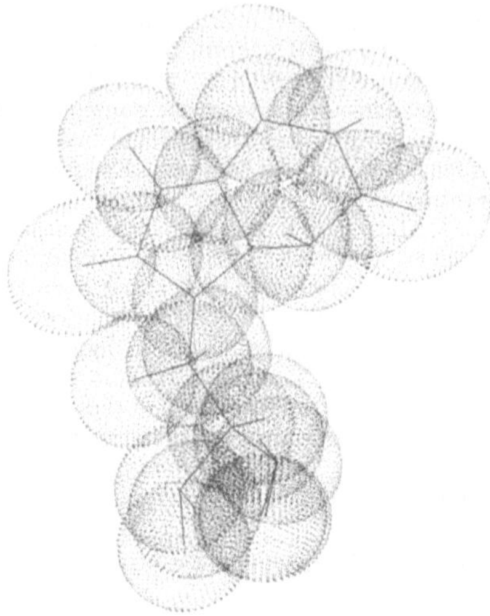

Figure 3 Kalman Filter reconstruction of a tryptophan, starting form a randomized map: each ellipsoid represents the uncertainty on the atom position. The constraints used by the algorithm are the distances among all the atoms involved in covalent bonds or in angle bonds (75 distances), and the dihedral angle H_{α}-C-C_{β}-N, which determines the chirality of the aminoacid.

The problem is mathematically set in the following terms. Given a set of N points and a list of M geometrical constraints on positions, distance angles and dihedral angles between some of the points, it is iteratively updated with all the constraints according to the following formula:

$$x_{k,i+1} = x_{k,0} + K_{k,i} .[Z - (h(x_{k,i}) + H_{k,i} . (K_{k,0} - x_{k,i}))] \qquad (1)$$

$$C_{k,i+1} = C_{k,0} - K_{k,i} . H_{k,i} . C_{k,0}$$

where $K_{k,i}$ is a 3NxM matrix called gain matrix, defined as:

$$K_{k,i} = C_{k,0} . H^{T}_{k,i} .(H_{k,i} . C_{k,0} . H^{T}_{k,i} + v)^{-1} \qquad (2)$$

$K_{k,i}$ is the Jacobean 3NxM matrix of the function h, defined by:

$$H_{k,i} = \frac{\delta h(x)}{\delta x} \Big|_{x_{k,i}}$$

h: R^{3N} x R^{M} is the function that models the so called 'sensor' of the system: for each constraint, that is for each $l=l,..., M$, h_l is a one-value function of the state vector, which gives the measure of the geometrical property to be compared with the constraint.

The gain matrix $K_{k,i}$ is the heart of the DIKF as bayesan estimator, allowing a full handling of the uncertainty in the problem at hand. In fact, as you can see from the equation (2), the gain matrix may be interpreted as a comparing term between the uncertainty on the positions (matrix C) and the uncertainty on the experimental constraints (matrix v). For example, in the figure 3 you can see the Kalman Filter reconstruction of a tryptophan, starting from a randomized map (that is a map in which the atoms coordinates have been shaked with a fixed level of noise): every atom is surrounded by an ellipsoid which represents the uncertainty on its position. The power of the DIKF lies in the possibility of obtaining such an estimation of the uncertainty on atoms positions directly from the

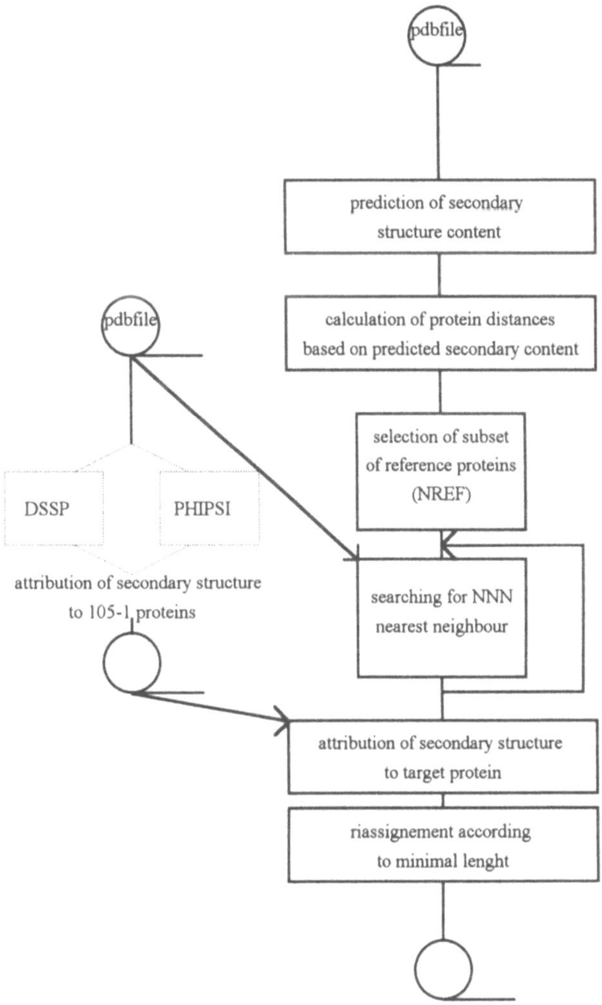

Figure 4 Secondary protein structure prediction procedure

algorithm for every generated structure during the computation, while other methods need a set of trials to build a statistic on the atomic coordinates.

2.2 Ab initio structural prediction

We have devised and implemented a new method for secondary structure prediction based on nearest neighbour classification (Fig. 4). This employs preliminary selection of a reference protein subset for each protein to predict, that allows to increase the speed and the accuracy of predictions. Performance was optimized by varying the window length and the number of nearest neighbours used for the assignment of secondary structure. The absence of a pre-processing learning step makes the algorithm faster and the results more objective. Two different algorithms for secondary structure assignment in 3D structures have been used, one based on torsion angles and the other on hydrogen bond pattern, in order to understand their influence on the performance of the algorithm. It was found that the accuracy of predictions is dependent on the method of secondary structure assignment, being always worse when the torsion angles are used. The maximum predictive ability we obtained (70.5% right predictions, with correlation coefficients 0.545; 0.503; 0.459 for helix, strand and coil) is better world-wide than for recent investigations using the same dataset of 105 proteins (Table I).

Table I Structure prediction methods

METHOD	YIELD	DATASET
Carrara et al. [1992]		our dataset
GOR METHOD	55.4 %	[unpublished]
REFINEMENT ACCORDING TO CD DATA	60.0 %	
Qian, Sejnowski [1988]		our dataset (slightly
NEURAL NET	64.3 %	different)
Kneller, Cohen, Langridge [1990]		our dataset
NEURAL NET ENHANCED BY ADDING SPECIFIC		
INPUT (hydrophobic moment)	65 %	
Stolorz, Lapedes, Xia [1992]	64.4 %	our dataset
NEURAL NET AND STATISTICAL METHOD		
Zhang, Mesirov, Waltz [1992]		our dataset (slightly
STATISTICAL METHOD	63.5 %	different)
NEURAL NET	63.5 %	
MEMORY BASED REASONING [nearest neighbour)		
EXPERT COMBINED	64.5 %	
	66.4 %	
Slazberg, Cost [1992]		our dataset
NEAREST NEIGHBOUR WITH WEIGHTED		
COEFFICIENTS	65.1 %	
Gabrielian, Anselmino, Catasti, Carrara, Nicolini [1993]		our dataset
ASPANNE		[submitted to J. Mol.
	70.5 %	Biol.]

2.3 Solution structure of recombinant versus wild-type c551

X-ray crystallographic structure of c551 cytochrome has been recently resolved at the Å resolution and the corresponding solution structure is being presently determined by NMR correlation spectroscopy utilizing a Varian 600 MHz spectrometer.

2.4 Solution structure of thioredoxin oxidized versus reduced and mesophilic versus thermophilic

Spectra were obtained on a Varian Unity 600MHz spectrometer at 298 and 308 K. Standards methods were used to obtain 2QF-COSY, TOCSY, DQ-2D and NOESY for the oxidized states. The TOCSY spectra were acquired with a spin-lock period of 30, 70 and 80 ms. The DQ spectra with the excitation period of 8 and 15 ms; NOESY spectra acquired with the mixing times of 30, 60, 120 and 150 ms. Spectra were acquired with 4096 complex data points and the spectra width was generally 7600 Hz (digital resolution 3.7 Hz/point). Quadrature detection in ω_1 was achieved by using States et al. method; 512 t_1 points were acquired for two dimensional spectra. In some cases spectra were folded in ω_1 for better digital resolution. Two-dimensional spectra were processed by using a Sun 1 workstation, with software FELIX 1.1 and 2.0. Phase shifted sine-bell window function were used in both dimensions. For NOESY, TOCSY and DQ spectra, considerable improvement was obtained by base-line correction. Proton resonance assignments were made by using a strategy which has been successfully used with several proteins and analyzes DQF-COSY, TOCSY and DQ-2D [Wüthrich, Chazin & Wright]. In this strategy spin systems are first obtained as completely as possible by utilizing the favourable dispersion of the amide proton resonance and relaying magnetisation from the backbone NH proton to the side-chain resonances. Not all NH-αCH cross-peaks appear in the DQF-COSY and TOCSY spectra as a result of overlap of some αCH resonances with that of H_2O. The complete set of NH-αCH cross-peaks was obtained by using a DQ-2D spectra and all the spin systems were identified by using the TOCSY spectra (30, 70 and 80 ms). Assignment of individual amino-acid spin systems to their correct position in the sequence was achieved by using the sequential assignment procedure of Wüthrich which analyzes the NOESY spectra (30, 60, 120 ms). The interresidue sequential NOE connectivities δαN(i,i+1) (for helical conformations) were observed. Strong, medium and weak connectivities are shown in figure.

Table II Secondary structure of reduced versus oxidized thioredoxin from mesophilic microorganism

	Location of β-Strands, Helices and β-Turns, residues		
secondary structure	X-ray oxidized	NMR, oxidized	NMR reduced
β-strands			
β1	2-8	4-8	4-8
β2	22-29	22-29	22-29
β3	53-58	53-60	53-60
β4	77-81	77-82	77-82
β5	88-91	88-92	88-92
helix			
α1	11-18	11-19	11/12-19
α2	35-49	33-49	33-49
α3	59-63		
α4	95-107	96-107	96-108
β10	66-70		
reverse turns			
1	8-11		
2	32-35		
3	49-52	49-52	49-52
4		74-77	74-77

The constrain obtained from NOE provides ample bases necessary for obtaining the three-dimensional solution structure of oxidized thioredoxin, using a distance geometry (DSPACE) followed by simulated annealing with the program DISCOVER, with significant differences with respect to the known reduced form (Fig. 5 and table II). The solution structure of reduced thioredoxin from mesophilic versus thermophilic microorganism is summarized in Table III.

Table III Secondary structure of thioredoxin fom mesophilic versus thermophilic microorganism, as obtained by circular dichroism data.

	α–helix	β–sheet	β-turn	random coil
mesophilic	0.30	0.29	0.06	0.34
thermophilic	0.13	0.40	0.06	0.41

3. ENGINEERING OF AD HOC MUTANTS

Our attention is concentrated on cytochromes P450 and on azurin. As described in the above paragraphs, at this initial stage of research, two ways can be identified for the optimization of the properties of the biocatalysts:

1) mutations to increase the stability and electron transfer at the level of a single cytochrome molecule, also using thermophilic/extremophilic microorganisms and intervening at the level

of sequence of amino acids responsible for electron transfer; these are the side chains and the basic amino acids at the surface of P450;
2) mutations to enhance the catalytic activity of the protein, its affinity towards specific substrates and changes of substrate specificity of the system towards environmental pollutants or hazardous metabolites of toxic substances;
3) mutations to reduce to zero the interprotein spacing and to favour the maximum packing of the film, possible in proteins that have the shape of a trigonal prism, like P450; also to enhance the protein-protein interactions by putting at the surface the amino acids of the required electrical charge, hydrophobicity and/or geometrical complementarity.
4) mutations to 2D order the protein in the thin film (see also section on chemical modification).

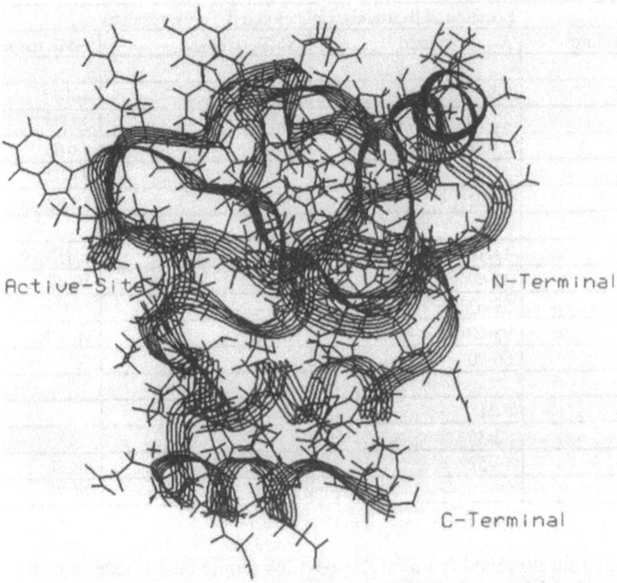

Figure 5 Solution structure at atomic resolution of oxidized thioredoxin, as obtained by 2DFT NMR [Nicastro et al., 1994].

A special attention must be, finally, paid to the modulation of the accessibility of the heme via "ad hoc" site-specific mutations. The accessibility of the electron transfer of the redox group is very important for the retention of the functional activity of the protein and for the conductivity of thin films of redox proteins. For what concerns cytochrome P450, the initial specific activity in molecular biology has involved optimization of expression/secretion of S.cerevisiae strain YPH857 (pYepsec1P450) by selecting conditions of induction of GALUAS promoter and gene-engineering versions of various host strains; site-directed mutagenesis on the basis of E. coli cloning collection with the use of synthetic oligonucleotides; creation of fusion proteins on the basis of already available recombinant vectors (pYepsec1P450, pYAdx, pYAdxR); study of compartmentalization and processing of recombinant metalloproteins in the case of YPH857 strain using the Western - blotting analysis; study of processes of cholesterol degradation in a model system with recombinant components of natural degradation of steroids and toxins (cytochrome P450scc, adrenodoxin, adrenodoxin reductase).

3.1 Identification and construction of P450 mutants

Protein hydrophobicity and dense packing of protein residues in the molecule are generally considered important factors in the stabilization of the protein structure. The structure of the protein presents a delicate balance between packing and flexibility required for the functional activity of the protein. Thus, it has been suggested that cavities at the inside of the protein are necessary elements in the protein architecture to provide both for functional properties (binding of the ligand) and structural properties characteristic of real proteins, but the cavities are energetically unfavourable because the surrounding residues have a lesser degree of Van der Waals interaction compared with residues localized in a densely packed environment. It has been demonstrated that proteins may adjust the packing of their hydrophobic nucleus in order to accomodate numerous mutations, thus sustaining the hydrophobicity. However, these mutations frequently introduce negative effects, like, for

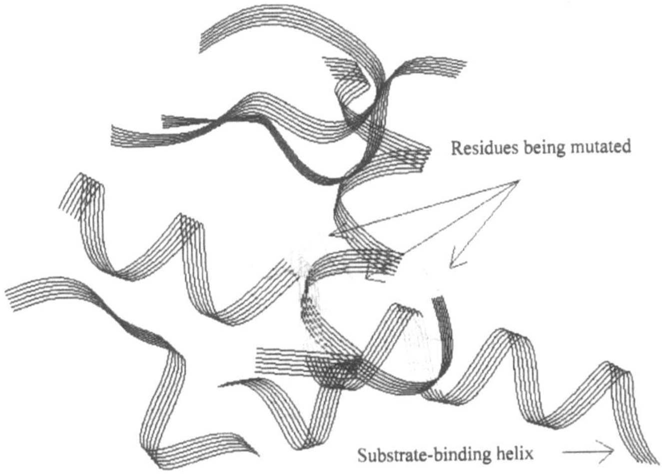

Figure 6 Secondary structure elements surrounding residues being mutated (cut-off distance 10 Å) in the substrate-binding helix.

example, the destabilizing introduction of the superposition of Van der Waals potentials. Hence, the conventional strategy to optimize the stability of proteins is to fill the cavities and to remove the internal water molecules. Unfortunately, quite a few molecules are not sensitive to mutations in the internal cavities but seem to be sensitive to those of the residues at the surface of the protein. In case of enzymes, the cavity which could be evidently "filled" is the substrate-binding pocket, but this results in the loss of enzymatic activity and in the changes in enzyme specificity. The enzymatic activity of cytochromes as components of bioelectronic device, with the only exception of their utilization in the construction of biosensors, need not be preserved. Therefore, we chose a strategy that would optimize the stability of the cytochrome P450scc and that consisted in the introduction of mutations into the region of the substrate-binding pocket (Fig. 6).

Up to the present, there are no data on the structure of the substrate-binding pocket of the cytochrome P450scc. Moreover, studies of site-specific mutagenesis have not been conducted and in the tridimensional crystallographic structure recently deposited in the Brookhaven Protein Data Bank it is not possible to identify the heme and the substrate site. Moreover, the water molecules have not been localized, for which it does not seem to be possible to optimize the cytochrome structure by removing water from the inside of the protein globule after having introduced voluminous hydrophobic residues. The only way to estimate the importance of every single amino acid in the functioning of P450scc is to compare them with isostructural residues of cytochrome P450cam, where the heme, the bound substrate and the water molecules are present in the 3D crystallographic structure. However, even for the cytochrome P450cam there is no precise information on how the substrate inserts itself into the binding pocket. The access channel postulated for the substrate leads to the distal face of the heme. However, in the case of cytochrome P450scc there are two possibilities to arrive at the distal face of the heme. In any case, the detailed information on the substrate-binding pocket is not required to be able to fill the cavities. The only requirement is to select the internal residues that possess sufficient conformational freedom and can be replaced with voluminous hydrophobic residues without Van der Waals overlaps with surrounding residues. The conservative element of cytochrome P450 is a small inner helix involved in the binding of the substrate. This helix is extremely flexible and has a significant conformational freedom to induce local deformations necessary for an efficient binding. The helix is composed of small residues that could be mutated into larger ones to increase the packing density and, hence, the stability of P450scc molecule. We used the alignment of the secondary structures of cytochrome P450scc and P450cam to estimate the correspondence of residues in the region of the substrate-binding helix:

P450scc	**Leu** ala **Gly Gly** val asn **Thr** thr ser met	323-332
P450cam	**Leu** val **Gly Gly** leu asp **Thr** val val asn	246-255

However, some precautions should be observed, since the electron transfer properties of the cytochrome should be retained. Thus, the following mutations were avoided:

1 in the vicinity of the heme-binding pocket
2 along the possible channel of heme incorporation
3 of aromatic residues that may be very important in the electron transfer.

In Fig. 6 the zone suitable for mutagenesis in the proximity of the substrate-binding helix is indicated. Moreover, several obvious considerations were taken into account for the mutagenesis:

1 - the local secondary structure should not be destroyed. Thus, mutations that introduce lower propensity in formation of the secondary structure (in this case, α-helix) should be avoided.

2 - the internal residues should give rise to minimal possibilities of forming hydrogen bonds and charge-charge interactions.

3 - simple geometrical considerations were used to control whether the mutated residue would fit properly into the structure, thus reducing the CPU time that otherwise would have been required for the minimization of the P450scc structure.

The adequacy of the volume increment and the bond length in each case were estimated using the bond length and the residue volume and calculating the distances to atoms nearest to each of the mutated residues. All the aforementioned criteria led to the suggestion to mutate three residues in the inner helix. These mutations are Ala324Leu, Asn328Leu, and Thr329Leu (the numbering is according to the original sequence; in the PDB file *1scc* the numbers are 286, 290 and 291, respectively), although it is possible to mutate each of these residues (not contemporarily) to Phe because the latter could be implicated as an electron transfer intermediate.The above mutations have been carried out both separately as well as jointly, guaranteeing thus a synergistic effect.

3.2 Identification and isolation of Azurin mutants

The system chosen permits to produce azurin in the periplasm (in *E.coli* K12) with a high efficiency (about 30 %) and, as a consequence, of a high purity (57 %, corresponding to 16 mg per liter of culture medium). The three-dimensional structure, electrostatic characteristics and hydrophobicity of azurin from Pseudomonas aeruginosa bacteria has been analyzed in detail. Given the low resolution of the crystallographic data available for this protein (2.7 Å), some particularly mobile residues are not clearly resolved on the diffraction map, for the reason of which it became necessary to reconstruct graphically the protein structure using the Sybyl software from Tripos Associates on the graphics system PS390 of Evans & Sutherland. The analysis of the graphically reconstructed protein has evidenced a significant concentration of hydrophobic residues at the surface near the copper site, residues that could render possible the interaction with a lipid support of the Langmuir-Blodgett type. The analysis confirmed the observations made earlier with azurin from P. aeruginosa. There is a certain asymmetry of charge between the regions of the protein, including the copper site, dominated by positive values of the electrostatic potential, and the rest of the molecule in both oxidation states of the metal. From the structural viewpoint the region characterized by the positive electrostatic potential is found in the proximity to the hydrophobic surface area, although not being completely superimposed on it. This consideration enables to suggest that interaction of electrostatic type takes place between diverse molecules of azurin attached to the substrate which would permit an ordered assembly of the molecules to form a protein layer. Basing on the previously described results, the possibility to introduce mutations that are able to improve the interaction of the protein with an eventual lipid layer and/or enhance the asymmetry of the intermolecular charge has been studied. A strategy has been adopted to design the mutations, which would see for the modification of amino acids exposed into the solvent that would not be involved in interaction that stabilize the β cylinder or the geometry of copper ligands, and that would not intervene directly in the process of electron transfer with the aim not to alter the stability and retaining of functional activity of the protein. The substituted residues were chosen also for their position in the inversions or loops of the polypeptide chain to minimize the damage to the structural topology of the molecule. For what regards the mutants with better characteristics of adhesion to the substrate, our attention has been turned towards removal of strongly polar residues around the hydrophobic zone that surrounds the copper atom; for this sake, the possibility has been studied to mutate the two charged residues Asp 69 and Lys 70 situated in an inversion of the polypeptide chain into two apolar amino acids. This mutation would have the double aim of increasing the hydrophobic interaction surface and simultaneously extending the region of negative potential located in the part opposite to the copper site, thus favouring the intermolecular interactions. A second possibility of mutagenesis foresees the accentuation of the intrinsic dipole of the azurin molecule, increasing the positive charge density in the region of the copper site and the negative one, by neutralizing or inverting the charge of one amino acid residue located in the area opposite to the copper site (Asp 11 is situated in an inversion and Lys 74 in a loop of polypeptide chain). The previous considerations are confirmed by calculations of electrostatic potential made for the mutated protein (the structure of which has been graphically reconstructed), which show an extension of the positive potential zone around the copper site and the opposite negative zone. Other possible mutations are studied that favour the adhesion and auto-assembly of azurin, even though the possibility of chemical modifications such as attachment of apolar chains and intermolecular cross-linking aiming at anchoring the protein to the support and favouring its ordered auto-assembly is not to be excluded.

4. ELECTRON TRANSFER

The mechanisms responsible for electron transfer become clear with quantum mechanics both for relatively simple organic molecules and for proteins like cytochromes. The electron transfer in non-proteic molecules, like for example organic molecules, is allowed for the π-electrons of the molecules separated by the sigma-bond. The interesting aspect of such organic compounds is that different parts of the same molecule may function as conductor and insulator (non-conductive zone). Moreover, these two regions may interchange their function during redox reactions. The model for functioning of the conducting and non-conducting zones has been proposed by Ari Aviram [1988]. In the non-conducting zone (insulator) the HOMO orbital of the molecule is completely occupied by the electrons (electron pairs) which are below the Fermi level of metal electrons; this prevents the passing of the electric current in case when the organic molecule is linking two metal conductors. In the conducting zone (conductor), the HOMO orbitals of the molecule is not completely full, which implies the possibility of the electron exchange between two levels of the molecule upon application of electric field. The difference between insulator and conductor lies in the fact that while the insulator behaves itself as such only up to a certain limit of applied voltage, the conductor does not have any limits of functioning. To prevent the tunneling effect (the insulator transforms itself into a conductor), the two zones of the molecule that have different conductivity are separated and placed perpendicularly to each other by means of a sigma-bond. Similar redox processes are typical in the mitochondrial respiration chain, where the energy needed for the reaction $H_2+1/2\ O_2 = H_2O$ is gradually freed. The various steps are catalyzed by enzymes or electron transport molecules. The H_2 molecule to be oxidized is carried by NADH (in the form of H^-) and H^+ is already dissolved in water. The electrons, prior to being caught by the oxygen, pass by diverse molecules of which 3 are of the cytochrome family. The first molecule capable of accepting the two electrons of NADH is ubiquinone. Ubiquinone transfers its electrons to the cytochrome reductase complex which comprises 9 different polypeptide chains (the cytochromes cyt c1, cyt b562 and cyt 566 are included) and exists as a subform of dimer of 500 kDa. This complex can transport two electrons to two molecules of cytochrome c (ca. 10 kDa). Cytochrome c gives, in turn, the electrons to the complex cytochrome-oxidase (cytochrome aa3) constituted by 8 different polypeptide chains and isolated in the form of a dimer of 160 kDa. This complex then gives the 2 electrons taken from the cytochrome c and enables oxygen to form one molecule of water. The influence of the oxidation state of the redox center on the three-dimensional structure of the analyzed protein has been studied comparing, wherever available, (like in the case of cytochrome c, plastocyanin and mioglobin), the crystallographic structures at various oxidation states, or using experimental data known from literature. Such an analysis permitted to determine that the protein conformation and the geometry of the ligands of the active redox center remain practically unaltered in the case of cytochromes, plastocyanin and azurin, whereas small variations can be observed in the case of superoxide dismutase and mioglobin. Both theoretical and experimental studies indicated that variations in long range electron interactions might play the key role in the control of the rate of electron transfer in proteins. The most interesting theoretical predictions are those that explicitly include the 3D structure of intervening lipids; the electron interaction maps introduced by Beratan, Betts, and Onuchic (BBO) revealed themselves to be extremely useful for the design of experimental "ab initio" systems of electron transfer, in particular in the determination of the points where the heme is attached to the surface of the protein. This constitutes a good reference point for engineering of cytochromes optimal for bioelectronics, as recently illustrated for cytochrome c.

4.1 Theoretical ab initio estimate

Four steps are involved on our calculation of the interaction energy between two proteins in a thin film, starting form 3D atomic co-ordinates of proteins.

Finding the electron transfer pathways through proteins. (procedure by Beratan, Onuchic, Regan. available in C-source code). This procedure consists in two steps.

The first one starting from 3D coordinates and covalent bonds among atoms creates a connection file adding hydrogen bonds and through space jumps to molecules.
INPUT: file of Brookhaven Protein Data Bank
OUTPUT: connection file: network of coupling factors. The second one finds electron transfer pathways and coupling reports between any single atom and all other atoms.
INPUT: connection file output of previous step.
OUTPUT: Ordered pairs of distances and interactions (couplings) between each atom and all the others in the molecule. This output file can be visualised by InsightII.

Calculate the electrostatic fields around proteins. (procedure by Fedoseyev et al. available only in the executable code for 386-486 PC).

This program calculates electrostatic field around a molecule solving a Poisson-Boltsmann equation by mean of a multigrid method. The region of the solution is limited by an artificial rectangular boundary and is discretized using method of finite elements.
INPUT: file of Brookhaven Protein Data Bank. List of parameters like region of solution, discretization, dielectric constant inside and outside the molecule, etc.
OUTPUT: values of electric field in every node of 3D network around the protein. This output file can be visualised by InsightII.

Calculate the interaction energy between proteins in solution. Amber package by U.C. Singh et al., Universtity of California, San Francisco. Available in FORTRAN-source code for Silicon Graphics).

This procedure is concerning with use of several modules of standard Amber package (in a version slightly modified) to individuate the coordinates of the donor-acceptor system and a total description of the force fields. From these one can get interaction energies between different parts of the system, i.e. interaction energy between any two given proteins.
INPUT: file of Brokhaven protein Data Bank.
OUTPUT: every module has different output used as input for following modules. Intermediate results are files of coordinates and force fields. Final output is file of interaction energy between proteins.

Calculate the interaction energy between proteins in the thin film. (Work on progress by I. Kurnikov. No code is available at this time.)

The idea of this procedure is, starting form 3D coordinates of proteins in the thin film (3 spatial coordinates and 3 rotational angles) and utilising results of previous procedures, to search for set of configurations of possible position of proteins and to find among these the one which optimizes packing, using a Montecarlo analysis. It can be built an artificial energy according to the model:

$$\varepsilon = c_1 \int_{Hda} + c_2 \int_E$$

where $\int_{Hda}$ is a function concerning coupling between donor and acceptor (provided by procedure 1) and $\int_E$ concerns the energy of interaction between proteins (provided by procedures 2 and 3). The goal is to minimize this expression and the minimum of energy corresponds to a maximum of electron transfer.

4.2 Experimental determination by NMR spectroscopy

Cytochrome C551 (called C551 thereafter) is an electron carrier extracted from *Pseudomonas aeruginosa*. The protein consists in one polypeptide chain with 82 residues. Prosthetic group is a heme C group. Bimolecular electron exchange (BEE) results from the transfer of one electron form reduced C551 to oxidized C551. In this way a dynamic equilibrium in the solution is present and the fraction of reduced and oxidized C551 is time independent. Autoxidation (i.e. oxidation by the solvent) can modify the C551 red/ox ratio, but at rate much slower than the electron transfer rate. The method of R. M. Keller and K. Wüthrich has been adopted for the BEE measurement by proton NMR. The high value of the rate constant respect to 1/T1 and 1/T2 makes difficult the use of methods from the rate constant determination based on such parameters. In addition, spectra frequencies of oxidised (paramagnetic) and reduced C551 are chemically shifted of less than the value (in Hz) of the rate constant. This results in changes on chemically shifts of the observed resonance of ferro-C551. Cytochrome C551 has been extracted from *Pseudomonas aeroginusa*. The sample has been dissolved in D20 with phosphate buffer 67 nM, 0.9% NaCl (physiologic solution), pH 7.4. C551 concentration was 2 mM, measured spectrophotometrically. The temperature was maintained at 210 K.

Spectra have been obtained on a Bruker AMX 500Mhz (Fig. 7), equipped with a variable temperature unit. Data are loaded on a digital size of 16 Kbyte. Scans for noise reduction and phase cycling have been 256 (about 8 min of experimental time). Data have been Fourier transformed with the program WinNMR (Bruker) on a 486 PC computer . No use of window function has been done, for not biasing subsequent elaboration. Spectra simulation have been obtained by a PC program built for this scope using C language. For the study of electron exchange in C551 we have used resonances belonging to Met-61 alfa, beta and gamma protons, known to have significant differences in the chemical shifts of oxidized versus reduced species. It must be noted that these differences for the Met-61 protons can be expected from the contribution of pseudo-contact shift term due to the unpaired electron of Fe(III) heme. These differences in paramagnetic shifts permit to improve the accuracy on the determination of the rate constant on a broader interval of BEE values. It can be noted that resonances differ not only in the absolute value of frequencies difference red-ox, but also in sign. This fact influences the shift of resonance at the varied BEE and red-ox fraction, and improves the sensibility of the method. Three different lines of Met-61 are simulated. The fraction of reduced C551 and the BEE are changed interactively. This resulted in a value of 0.983 for the fraction of reduced form, and of $6.3 \times 10^6\ M^{-1}\ s^{-1}$ for the BEE value (fig.7). This value differs from that obtained by R. M. Keller and K. Wüthrich ($12 \times 10^6\ M^{-1}\ s^{-1}$) by a factor 2, due to the different ionic strength of the solution. For error estimation of the BEE measure, we have used a bidimensional numerical fitting. Spectra are simulated in a two dimensional array of BEE and reduced fraction values. Quadratic difference point by point is taken from real and simulated spectra. Values are arranged in a two dimensional array and the total minimum value of quadratic differences is found. Method of manual comparison and automated search has been proved to be equivalent, with the advantage of the last method to give an estimation of the error of both the parameters. This fact is of considerable importance for further experiments which require a measure of BEE. Moreover, automatic fitting prove the unimodal form of the BEE values, because of the presence of only one minimum (fig.8).

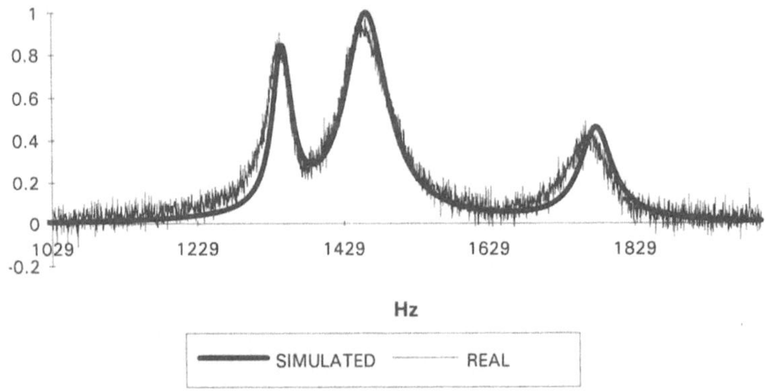

Figure 7 Real and simulated spectra overlaid. Simulation is done with BEE=$6.3 \times 10^6\ M^{-1}s^{-1}$ and red/ox=0.983. C551 concentration 2mM, pH=7.4, phosphate buffer 67 mM, T=210 K. Sample in D_2O. Both spectra are normalised between 0 and 1. Zero reference of real spectra is shifted by 10 percent for accounting of noise.

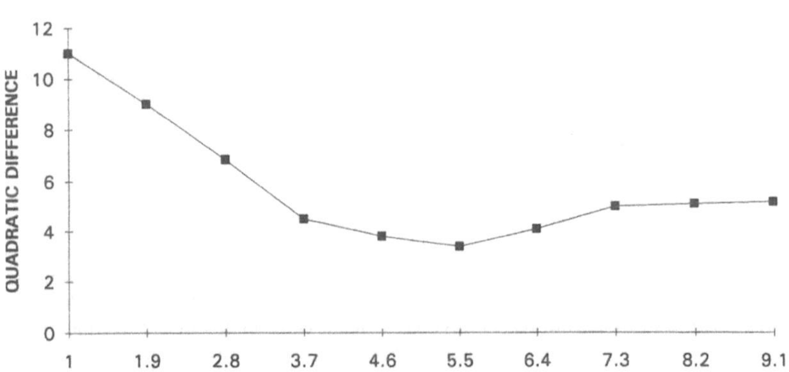

Figure 8 Quadratic difference between real and simulated spectra. A) red/ox fixed at 0.98 and BEE varying between 1 and 9.1 steps of 0.9. B) BEE fixed at $5.5 \times 10^6\ M^{-1}s^{-1}$ and red/ox comprised between 0.97 and 0.988.

The method has proven to be reliable for subsequent studies of BEE constant, like measurement for mutated C551. In this case, pseudo-contact shift interaction with paramagnetic electron of porphirin ring can also be studied. Pseudo-contact shift give information on the protons distance from heme. In this way, the BEE value can be correlated with structure changes induced by specific mutation of aminoacid composition.

5. ALTERNATIVE STRATEGIES FOR THIN BIOFILM FORMATION

Formation of thin films of metalloproteins by Langmuir-Blodgett technique and/or self-assembly has been successfully achieved. The resulting films of engineered metalloproteins have been properly oriented with LB (or LS) technique by anchoring to lipids, by chemical modifications and/or by reversed micelles.

5.1 Langmuir-Schaefer technique

The films were formed and studied in a LB trough (Asse-Z/MM-MDT Corp.) by a modification of well-known LB technique [Langmuir-Blodgett Films, Ed. G. Roberts, Plenum Press, 1991, New York] the trough size is 240x100 mm^2 and its volume is 300ml; there are two barriers which compress the monolayer from two sides resulting in improvement of the homogeneity of the monolayer in the central part where the deposition and the measurements of the surface pressure take place. This instruments is equipped with a Wilhelmy balance with a surface pressure sensitivity of 0.05 mN/m. The water used for the subphase was purified by Milli-Q system till the resistivity was 18 MΩxcm. Films were transferred onto quarts substrates by Langmuir-Schaefer method (horizontal lift) at 356 mN/m surface pressure. As RC is a membrane protein with two hydrophilic regions in opposite positions it is rational to suppose that, being spread on the water surface, molecules form a monolayer in which one hydrophilic region is in the water and the opposite one in the air. After monolayer formation a hydrophilic substrate horizontally touches the film. The best deposition is observe using substrates just treated with weak oxygen-containing plasma (very bad deposition takes place when hydrophobic substrates are used). Molecules attached to the substrate are closely packed and interact by hydrophobic bonds with neighbours. Touching the film with the substrate results in a slight immersion into subphase of the part of RC monolayer under the substrate results in a consequent separation of this area from initial layer. During the upward motion, some amount of water is also transferred onto the substrate and some amount of RC molecules, not connected with the substrate, is also transferred and can face to the upper meniscus of water droplets on the substrate. To obtain a good deposited monolayer is, therefore, necessary to remove the excess water and protein molecules not attached directly to the substrate surface (they are not regularly distributed and due to the drying process they will be absorb irregularly on the substrate); we use for this aim a nitrogen flow which remove the water droplets without let them drying on the substrate. This technique was shown to provide a good order of the deposited RC films. The quality of the deposited films was controlled looking to the homogeneity of the interference colours on Silicon substrate after deposition of RC multilayer film and by ellipsometry. The amount of the deposited mass was measured by nanogravimetric technique using quartz resonators and found to be reliable in the successive layer depositions. Initially, hydrophobic regions of RC are surrounded by the detergent molecules. Being placed onto the air/water interface, the detergent molecules leave the proteins and transfer themselves to the water surface. Thus, just after spreading, a mixed monolayer of RC and detergent is formed. The presence of detergent molecules resulted in irreproducibility of π-A isotherms during compression and expansion for the first two cycles. Initially the increase of the surface pressure is due not only to the presence of RC but also to detergent molecules. After the compression to rather high pressures, the surface concentration of the detergent is decreased due to the formation of microcollapses (the length of hydrocarbon chains of the detergent is too short to form stable monolayers at high surface pressure) which can be solved in the volume of the subphase in shape of micelles. Performing two compression-expansion cycles makes the isotherm reproducible; this behaviour is in good agreement with data from literature, indicating the improvement of order and homogeneity of the monolayer. It is possible to assume that, after such action, mainly RC molecules enter the process of the monolayer formation.

5.2 Lipid monolayers

Various compounds have been deposited in monolayers and their properties studied; these compounds are derivatives of fatty acids (e.g. octadecyltriethoxysilane), of aromatic compounds (e.g. 4-octadecylamine), of chinone (e.g. 3,5,6-trichlor-2-octadecyl-1,4-benzochinone), of heterocyclic compounds (e.g. porphirine), of diacetylene (e.g. octadecyl-2,4-diionoic acid), of ethylene (e.g. omega-tricosenoic acid), of aromatic polycyclic compounds (e.g. 5-carboxy-1-butylantracene), of polymers (e.g. polyester). The use of these complex molecules requires various techniques (IR spectroscopy, H-NMR, mass spectrometry) and purification methods (gas chromatography and HPLC). With the help of the LB film technique it was made possible to obtain the monolayer disposition of newly synthesized molecules, such as biotinyloctadecylamine (BODA), in monolayers.

This lipid, synthesized at the Institute of Biophysics [Antolini et al., 1993] has as a head group a molecule of biotin (Fig. 9). This group has the ability to bind strongly ($K_d=10^{-15}$) the streptavidin (a protein with the molecular weight of 60 kDa). This lipid, synthesized at the Institute of Biophysics [Antolini et al., maunscript in preparation], has as a head group a molecule of biotin. This group has the ability to bind strongly ($Kd=10^{-15}$) the streptavidin (a protein with the molecular weight of 60 kDa). We have obtained lipid monolayers of the BODA and ODA (octadecylamine) mixture, as confirmed by measurements of isotherms surface pressure/area and surface potential and by CCD camera. LB thin films of nine different lipids from archaebacteria were recently obtained displaying highly uniform multilayers for few of them [Troitsky et al. manuscript in preparation]. Ad hoc chemical modifications prove able to make these films stable in water solutions, while electron micrographs and electron diffraction patterns of corresponding monolayers reveal an U-shape for these lipids with close packing of isoprenyl chains. Two of these thermophilic matricxes were able by a vertical lift to yield films with uniform ordered distribution of valinomycin (tipically quite unstable) stable at high temperatures and in water solution to ensure the desired functions, namely to absorb ions selectively; the small-angle X-ray diffraction of a mixture of alternative monolayers of and valinomycin displays Bragg reflections which do not disappear up to 373 K [Troitsky et al., in preparation].

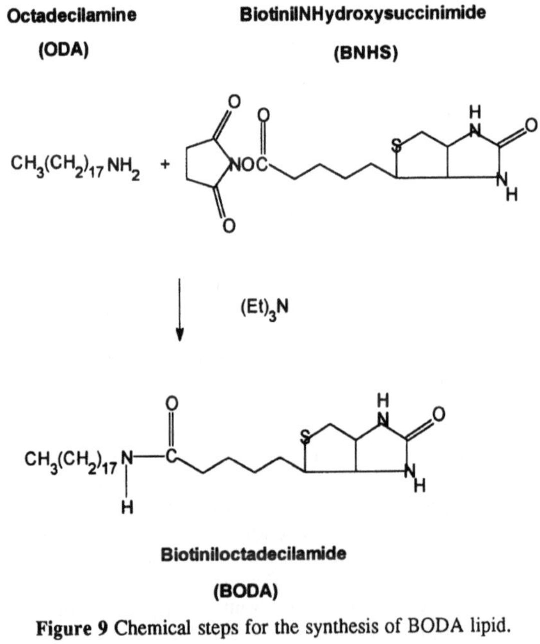

Figure 9 Chemical steps for the synthesis of BODA lipid.

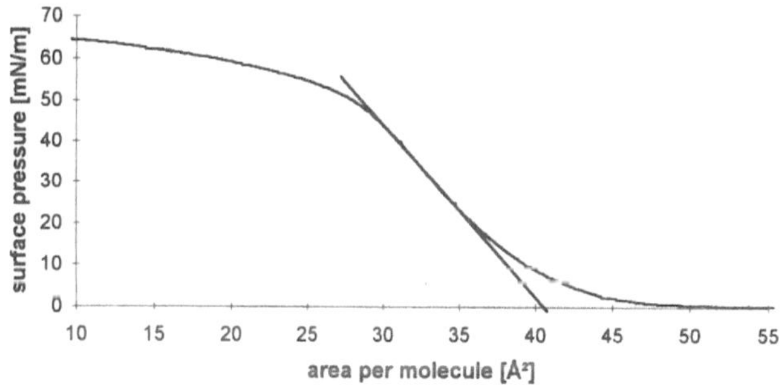

Figure 10 Isotherm of BODA

5.3 Chemical modifications

Another route pursued is the functionalization of metalloproteins in order to obtain thin films oriented at the air-water interface. At the Institute of Biophysics of Genoa, an arginine residue located at the surface of the cytochrome c (Fig. 11) is being derivatized chemically by an ad hoc synthesized lipid [functionalized).

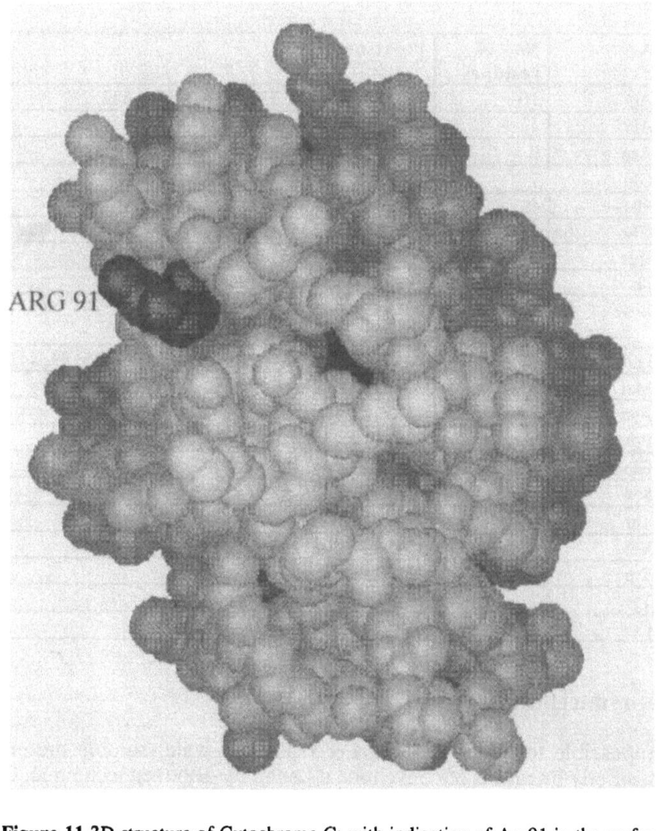

Figure 11 3D structure of Cytochrome C; with indication of Arg91 in the surface.

Similarly in cytochrome C551 and azurin the unique arginine residue being present at the surface will be derivatized, respectively Arg79 and Arg47. This procedure could be valid for any other amino acid for which it might be possible to synthesize the relevant modified lipid: some interesting examples are methionine, cysteine or, alternatively, tryptophan. In case of cytochrome P450, considering its large dimensions, there is more that one residue of each amino acid at the surface (see Table IV).

Indeed the strategy of functionalization requires the presence at the surface of a single amino acid and in the required position, both of which can be attained only by means of a single or multiple point mutations. For instance, the optimal orientation of P450 at the air-water interface is being attained through following mutations:

Met312 ==> Arg312
Met448 ==> Arg448

derivatizing methionine 202 with the synthesized lipid; alternatively, given the fact that no cystein residues are present at the external surface of cyt P450scc, the desired result can be achieved through a single-point ideally mutation of any amino acid at the surface into a cystein residue, ideally the one localized in one of the angles of the pyramidal structure, in particular through the mutation:

Met202 ==> Cys202

The choice of mutations between those indicated above is determined solely by the simplicity of synthesis of a corresponding lipid; indeed, we have to stress the significant complexity of various procedures to be implemented in every single case for the modification of the amino acids in question [C. De Nitti, F. Antolini and C. Nicolini, in preparation].

Table IV External surface amino acids of bovine cytochrome P450scc.

N°	AA	No. of residues	Position
1	Arg	10	473, 82, 305, 29, 252, 35, 93, 321, 322, 257
2	Ala	4	324, 231, 388, 372
3	Asp	9	327, 362, 405, 367, 84, 394, 340, 233, 197
4	Glu	7	186, 255, 124, 81, 311, 270, 185
5	Phe	4	256, 475, 203, 469
6	Gly	7	452, 32, 17, 326, 158, 20, 107
7	Hys	6	204, 26, 228, 92, 145, 450
8	Ile	9	328, 409, 85, 13, 451, 99, 36, 2, 196
9	Lys	18	404, 406, 253, 146, 373, 110, 330, 31, 395, 445, 161, 201, 241, 271, 194, 268, 111, 54
10	Leu	12	273, 143, 318, 184, 266, 164, 307, 22, 55, 89, 441, 332
11	Met	3	202, 312, 448
12	Asn	7	117, 211, 413, 308, 237, 134, 19
13	Pro	9	208, 131, 135, 87, 479, 480, 14, 9, 474
14	Gln	6	138, 193, 248, 150, 481, 34
15	Ser	5	3, 392, 391, 269, 329
16	Thr	6	254, 6, 118, 399, 4, 234
17	Val	6	114, 454, 207, 212, 120, 364
18	Trp	2	401, 21
19	Tyr	7	10, 95, 51, 200, 264, 95, 79
20	Cys	0	only 2 internal residues

5.4 Reversed micelles

It is quite impossible to obtain ordered L&B films of water-soluble proteins, such as cytochromes, strictly by means of conventional Langmuir-Blodgett technique. One way out to obtain it is through the formation of reversed micelles with cytochromes (like cytochrome c from horse heart which we used in the reported experiment), in which the molecules of cytochrome are surrounded by lipids or surfactant molecules, soluble in organic solvent but insoluble in water (Fig. 12). When such complexes are placed at the water/air interface, they form a monolayer of lipids with single metalloprotein molecules attached to their head groups [Erokhin et al., 1993].

By applying numerous cycles of compression and expansion it is possible to obtain a lipid monolayer with high packing in air, with closely packed protein monolayer in water. These films, once transferred onto a solid substrate, have evidenced a high degree of orientation by X-ray low-angle scattering (Fig. 13), superior to any other monolayer of proteins so far being produced.

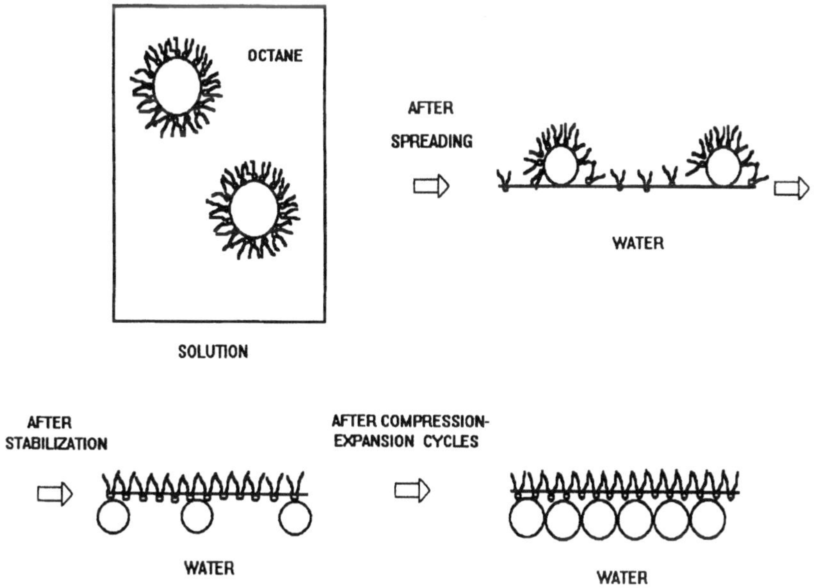

Figure 12 Model of the formation of LB films using reversed micellar solutions. Initially, cytochrome c molecules are solubilized in the solution of AOT reversed micelles in octane. Spreading of the solution reorganizes the micelles in such a way that AOT molecules form a monolayer at the air/water interface and cytochrome molecules are attached to their head-groups. Application of compression-expansion cycle several times results in the removal of excess molecules of AOT from the water surface and in the formation of closely packed monolayers both of AOT at the interface and of proteins in the water in the vicinity of the surface [Erokhin et al., 1993].

5.5 Self-assembly engineering

The production of protein monolayers can also be effected by other means such as protein mutagenesis and/or its chemical functionalization in order to optimize the protein-protein and protein-substrate interactions directly on a self-assembled monolayer. This type of approach requires a strong effort towards determination of structure of wild-type and mutant proteins (2DNMR) and large scale protein production. The packing of the protein layer, the electron transfer and orientation are simultaneously optimized by applying the recombinant DNA techniques which enable to obtain mutants with required properties at the surface and in terms of electron transfer characteristics. Using the mutants of cytochromes discussed in the earlier paragraph alternative ways are pursued towards the formation of metalloproteins layers through 2D and 3D ordered auto-assembly. The striking effect obtained with wild-type P450scc makes evident the unique properties of self-assembled films.

6. NEW MATERIALS OF UNIQUE PROPERTIES

New materials of uniquely exceptional electrical, structural and functional properties are obtained by the formation of ordered thin protein film.

6.1 Structural

The structural characterization of films, whether mono- or multilayer, is made by a combination of the following biophysical methods. In particular: - **small angle X-ray diffraction.** The use of small angle technique is due to a higher resolution in respect to classical X-ray analysis - 20-100 Å for a film, 1-2 Å for conventional crystals. This technique can be divided into two methods: diffractometry and reflexometry. The samples for X-ray diffractometry contain usually 6 or more monolayers (Fig. 13), for reflexometry - one layer or more.

The diffractometry enables to obtain several informations, namely:

- determination of the spacing of the film (thickness of mono- or bilayer);
- calculation of electron density profiles;
- determination of long range order;
- characterization of complex structures formed at molecular level;
- estimation of dimension and form of protein molecules in solution.

The reflexometry permits to obtain the same data, even though not directly from experimental data. In this case, a theoretical model of the film structure should be made, the theoretical X-ray curve resulting from the model should be calculated and compared with experimental data. The best experimental approach is that to combine the two techniques, obtaining on the one hand information on the film structure from diffractometry experiments on a relatively thick film (10 monolayers or more) and, on the other hand, refining the structure for a single monolayer by means of reflexometry experiments.

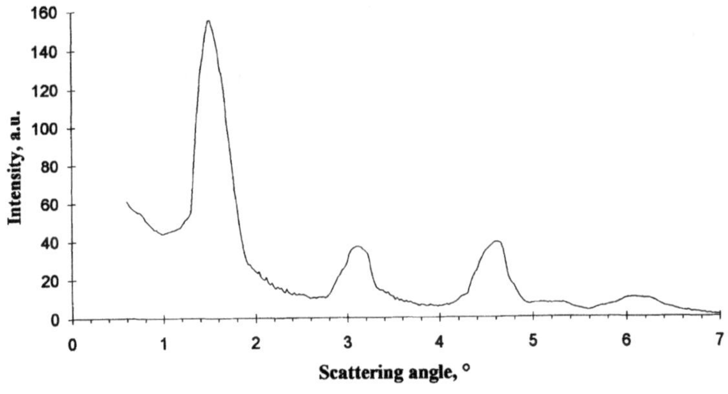

Figure 13 X-ray pattern of the sample, containing 10 periods (AOT-cytochrome c-AOT), deposited using reversed micelles as the spreading solution.

Infrared spectroscopy. Permits to determine:

- mutual orientation of molecules (and parts of molecules);
- preferential orientation of active groups;
- investigation of the presence of various bonds in the molecules that form the film (prior and after the treatment).

Fluorescence microscopy. A CCD camera with an extremely high signal-to-noise ratio has allowed to visualize the structure of protein domains anchored to BODA lipids and to estimate the concentration of protein molecules labelled with fluorescent marker [Antolini et al., 1993].

Scanning tunneling microscopy and atomic force microscopy. Permit to determine the degree of order and bidimensional orientation of molecules in the monolayer. Data obtained with an STM device from MDT Corporation on films of lipids and a film of reaction centers are discussed in Facci et al., 1994, and shown in Fig. 14. Langmuir-Blodgett monolayers of photosynthetic Reaction Centres from Rhodobacter Sphaeroides have been studied by scanning tunneling microscopy. Freshly deposited films were studied both in the dark and in the light. In the dark, images revealed molecular structure with 64 Å and 30 Å periodicities which correspond to protein and subunit sizes known from X-ray crystallography, while no periodic structure appeared in the light duc to the tip action on excited proteins. STM voltage-current measurements showed the charge separation in single protein molecules in the film and their different behaviour in the dark and light. Together with surface potential measurements at macroscopic level, they pointed out the preservation of Reaction Centres activity in the monolayer.

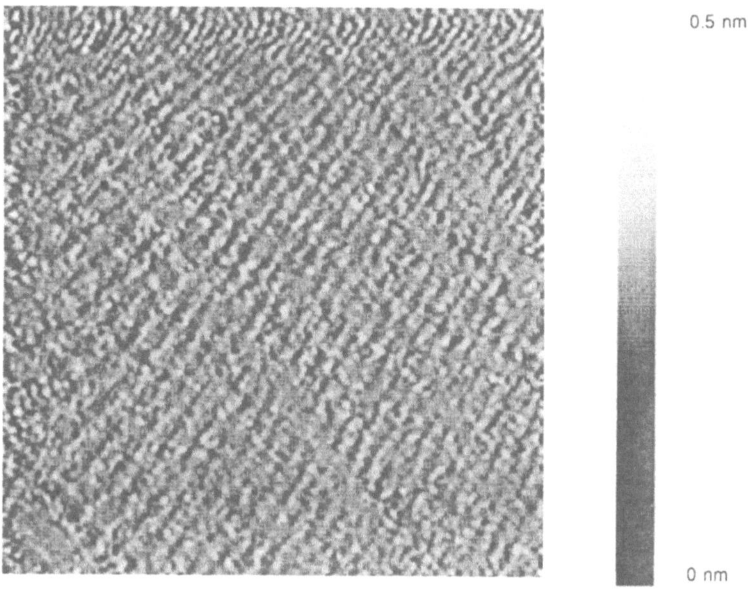

Figure 14 STM of a photosynthetic reaction center monolayer.

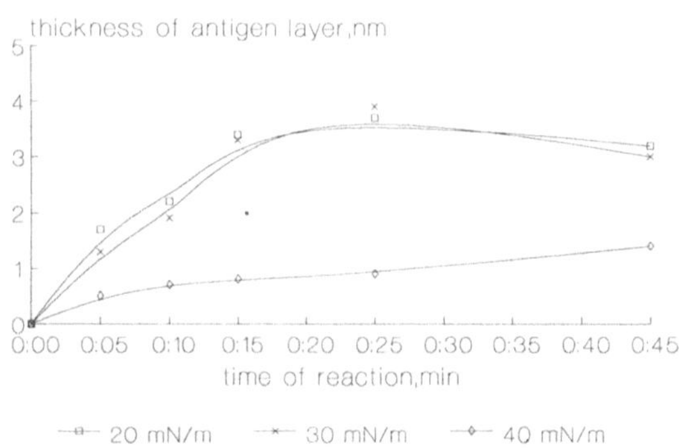

Figure 15 Thickness of the antigen layer (nm)

By fixing the protein layer with glutaralhdeyde, it was possible to prevent the perturbing tip action and obtain a periodic molecular structure with 30 Å spacing even in the light. After heating at 150° C, the unfixed film reorganized itself into a long-range ordered state with a hexagonal structure of 27 Å spacing but without any activity.

Ellipsometry. Enables the determination of the thickness and anisotropy of the film; the comparison of this thickness with the ideal one (thickness of a monolayer obtained from X-ray experiments, multiplied by the number of monolayers present) gives information on the cavities in the film [Tronin et al., 1993] (Fig. 15).

Circular dichroism. Permits the determination of the secondary structure of proteins in solution and directly in the layer, by measuring the molar ellipticity between 190 and 260 nm. Besides the information on the secondary and tertiary structure of metalloproteins in solution, the spectropolarimetry enabled to determine, for the film deposited on a quartz plate, the conservation of the percentage of different types of the secondary structure after the deposition process [Antolini et al., 1994; Nicolini et al., 1993]. LB films containing streptavidin were studied in the same way and the above phenomenon was confirmed, additionally demonstrating that the exposure of this protein to heat treatment (100° C) in solution causes its denaturation.

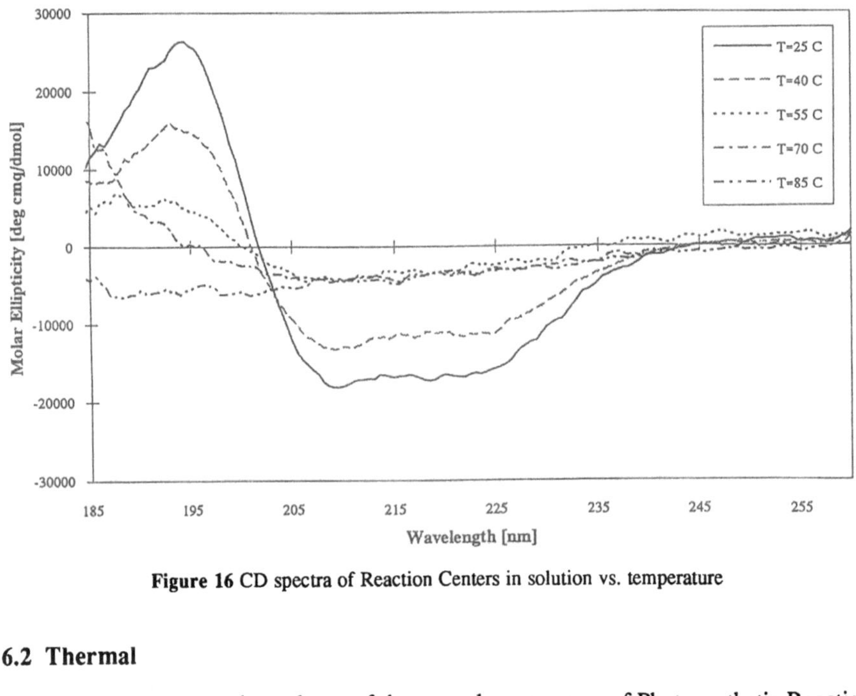

Figure 16 CD spectra of Reaction Centers in solution vs. temperature

6.2 Thermal

The temperature dependence of the secondary structure of Photosynthetic Reaction Centers from *Rhodobacter sphaeroides* in solution and in Langmuir-Blodgett film was studied by circular dichroism measurements. It was shown that the secondary structure of the protein was not affected in Langmuir-Blodgett films by heating up to 200° C, while it was completely lost in solution already at 55°C. The main factor responsible for the phenomenon was found to be molecular order rather than decreased hydration degree [Nicolini et al., 1993]. At the same time the spectrophotometric study of the proteins in LB films was carried out. The CD spectra of RC LB film are shown before and after heating, in solution (Fig. 16) and in ordered thin film (Fig. 17). The CD spectra of RC in LB films after boiling for different time in water display a decrease in the relative intensities of the peaks. Table V summarizes the above results in terms of secondary stucture.

Table V. RC secondary structure at room temperature in solution and in LB film. Percentage of α-helix, β–sheet, β-turn and random coil determined from X-ray measurements with a standard error of ±5 %. The coincidence of the data suggests that secondary structure in ordered layers is preserved.

Techniques and/or conditions	α-helix	β-sheet	coil*
X-ray	51	15.6	33.4
CD in solution Room Temp.	54	11	35
CD in film (18 Layers) Room Temp. (+)	46	14±1	40

(*) This value is comprehensive of random coil and β-turn.

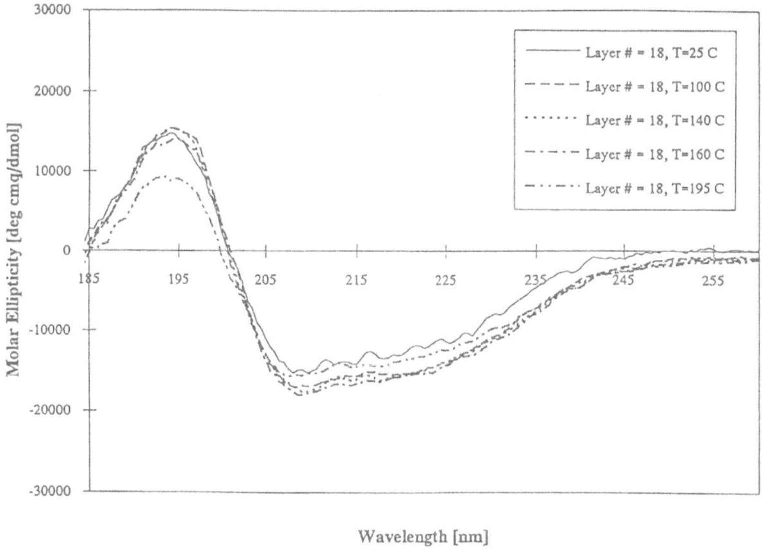

Figure 17 CD spectra of Reaction Centers on Films vs. Temperature

Thereby, it would appear that the stability of the RC secondary structure is strongly affected by the conditions under which the proteins are organized. Namely, the insertion of RC proteins into LB film confers unique stability to their secondary structure, as a result of both specific intermolecular interactions and decreased hydration. This is evident also from the significant enhancement of α-helix stability in the air-dried samples (up to 100° C). To sum up, LB appears to be an efficient and easy technique for stabilizing protein secondary structure, which is necessary but certainly not sufficient for the preservation of function. The main stabilazing factor of secondary structure was found to be the order induced by LB film formation. Dehydration, being of course a stabilizing factor too, in this case plays not a primary role, as it was shown by the experiment on the dried sample. This fact is in agreement with other experiments on antibodies where heating Lyophilized IgG molecules resulted in their denaturation, contrary to what happened in the case of LB film, where protein secondary structure was preserved together with antigen recognition properties.

6.3 Electrical

These techniques permit also to determine the orientation of electronic moments of transition and of polarizability of molecule up to the tenth of the thickens of the monolayer. The films are made by photosynthetic reaction centers (RC) isolated from *Rhodobacter sphaeroides*. These proteins can conduct an electron in only one direction in presence of a suitable light source (photoconduction) or of an external electric field (conduction). Both these conductive processes are due to the tunneling of a single electron into the structure of the protein. Our first measurements are obtained in dark in order to exclude all the photoconductive processes, by applying a modulated electrical field. Each film consisting of ten RC layers is deposited over a suitable interdigitated electrode, which is made by sapphire support on which aluminium conductive tracks are fabricated by a photolitography process. The output waves and all significative signals have been acquired by means an acquisition board, installed into a PC. All signals are visualized by means of an analogic oscilloscope and the input waves have been generated by a functions generator, with the desired shape (square or sigmoidal), frequency (500 Hz - 350 K Hz), amplitude (CO-IV) and rise-time. In absence of protein film, the frequency characteristic of the electrode displays cut-off frequencies in close correspondence with those simulated by a circuit simulator (SPICE) installed into a Personal Computer (PC) utilizing the known value of the R-L-C components. Depositing ten layers of RC proteins on the electrode, the measured cut-off frequencies and band widths hate quite different value depending on their orientation. In frequency the oriented layers of RC showed a behaviour that is similar to the one of the electrode. This is due to the particular property of the considered proteins that let the electron tunneling in only one direction. In relation with the equivalent electrical circuit of the electrode it is possible to verify that we have not a variation in the value of R, L and C components. When we subjected the electrode with non-oriented layers to the same frequency assay we saw a change in the final response. In particular, there is an increase of the amplitude of the output signal and a shift of the characteristic. It is possible to evaluate a variation in the value of R, L and C. The resistance R decreased and the reactive component increased (i.e. L and C component). This can be explained if we consider that proteins can conduct in only one direction. In fact the random orientation of these molecules gives place to a lateral conduction of the electrons. Moreover the change in the capacity C and in the inductance L probably is due to a variation respectively in permettivity and permeabilty of the film. These preliminary data, if confirmed, would be interesting because this underlines one of the electrical properties that can be used to make a biodevice.

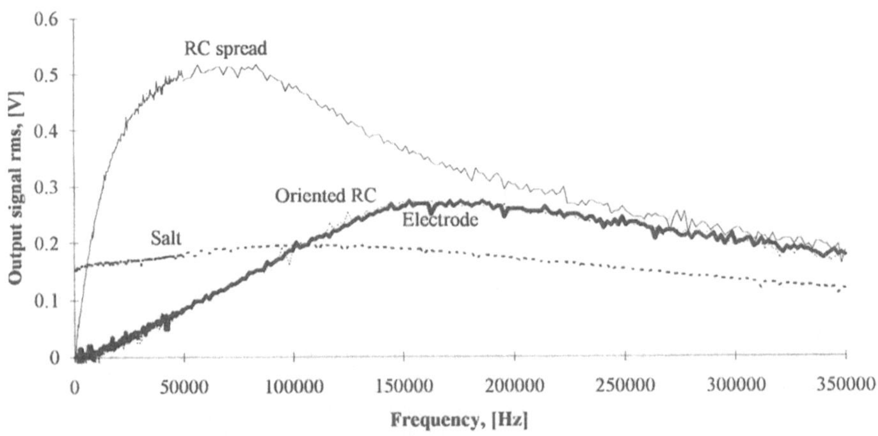

Figure 18 Frequency dependency of metalloprotein conductivity

To confirm the obtained results with the RC oriented layers, we built oriented layers with another type of proteins which have more or less the same characteristics. These proteins are the viridis obtained by *Rhodobacter viridis*. The experimental data showed the correctness of our hypothesis related to the lateral conduction induced by the electrical field applied to non oriented layers. This fact is very important because it puts in evidence the property of the proteins that now can be easily confirmed by measurements of vertical conduction on oriented layers. To obtain this vertical conduction of the ordered films it is necessary to use other types of structure. One of these methods uses a single-track electrode on which are deposited the ordered films, secondarily on these films we evaporated a metal to have the second electric contact. This can be obtained also using a mercury drop (work in progress). Figure 18 shows the output signals obtained imposing respectively sinusoidal and square waves in the two cases (oriented versus non oriented RC). While the first one shows the increase in the (lateral) conductivity and the variation in the phase the second confirms the presence of the double reactive component (L and C) in a better way.

6.4 Geometric at the interface air/water

The studies of the properties of monolayers at the air/water interface are made by using pressure-area isotherms. The measurement of the surface pressure is made with a Langmuir (or Wilhelmy) balance. The acquisition of such isotherms (e.g. Fig. 10) permits to have information regarding:

- formation of the monolayer;
- phase transitions of bidimensional system;
- time stability of the monolayer;
- interaction with substances (such as DNA, proteins, etc.) dissolved in the liquid lower phase (effecting simultaneously the measurement of the surface potential);
- the search for optimal deposition conditions;
- the compressibility of the monolayer.

Surface potential measurements have been made using a vibrating electrode (according to Kelvin) in order to obtain data on:

- charge redistribution during the compression of the monolayer;
- difference in the charge distribution after an attachment of protein molecules (or other substances) dissolved in the liquid subphase, to the monolayer;
- optimization of the pH and ionic strength of the liquid subphase during the deposition.

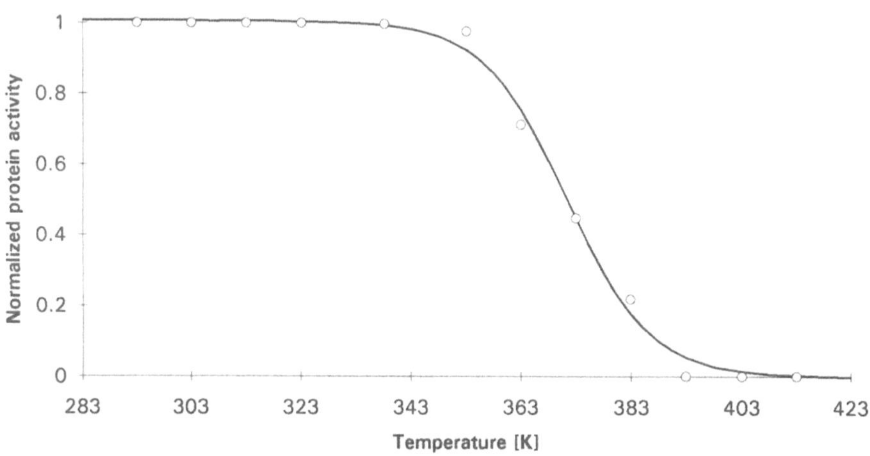

Figure 19 Functional activity of reaction center in LB film and in solution, as function of temperature

Usually, two types of microscopy are used for the characterization of monolayers at the water surface: the fluorescent microscopy and Brustver microscopy. In the case of fluorescent microscopy a small quantity of a fluorescent dye is added to the monolayer; the molecules of the dye, given that they cannot penetrate into the regions where the hydrocarbon chains are densely packed, are localized in the liquid or liquid-crystalline regions of the monolayer. The monolayer is illuminated with the light incident at the angle of Brustver, thus the reflected light is polarized. The polarization degree is different for the light

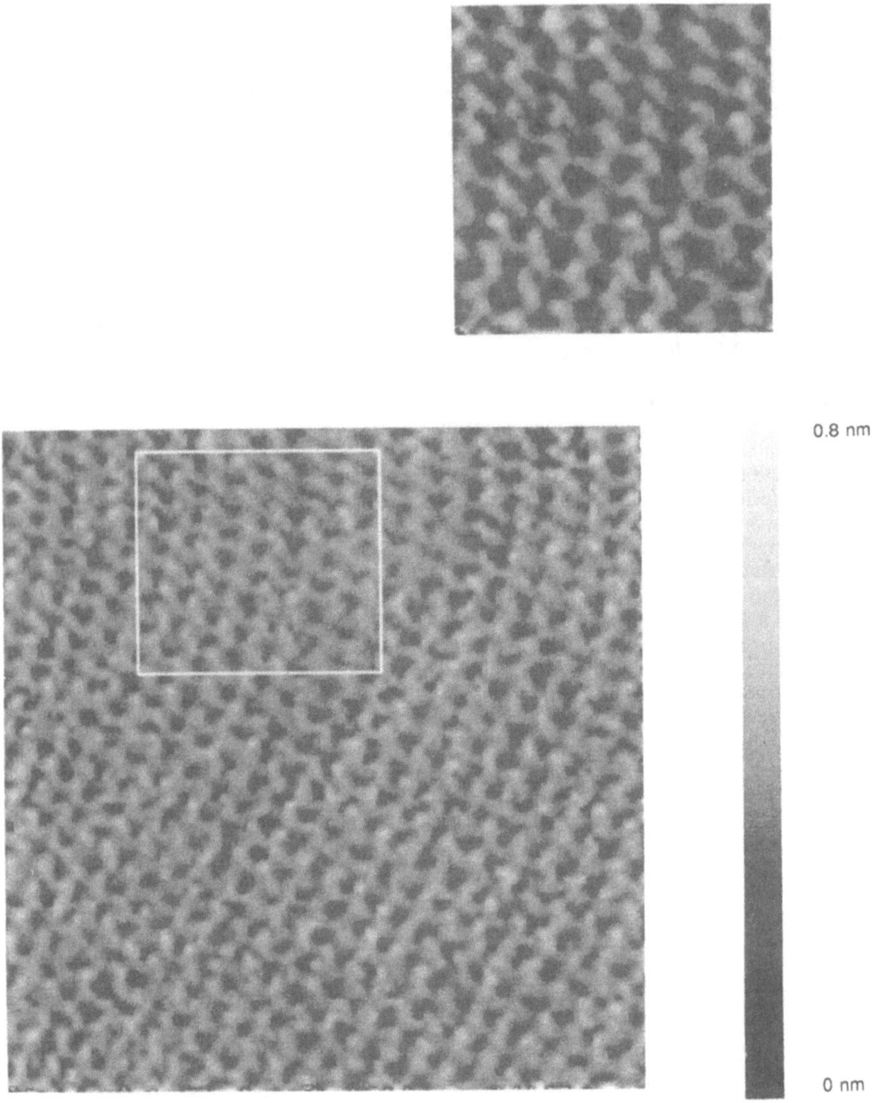

Figure 20 STM Structure of LB film of reaction centers at 160° C.

reflected from densely packed monolayers, from quasi-liquid region of the monolayer and from the water which makes it possible to visualize the organization of various domains of the monolayer.

6.5 Functional tests

Functional and kinetic tests are still in progress utilizing the EPR (Electron Paramagnetic Resonance) in the case of redox proteins with paramagnetic properties. At the moment the functional activity of photosynthetic Reaction Centers from *Rhodobacter sphaeroides* arranged in monolayer and deposited by the Langmuir-Blodgett technique onto a solid substrate has been monitored after exposing it to increasing temperatures. Surface potential measurements have been used for monitoring directly Reaction Centers activity in Langmuir-Blodgett films, while spectrophotometric assay was applied for estimating indirectly the activity of not oriented films. The protein functional activity in Langmuir-Blodgett films turned out to be unaffected up to 75° C, whereas in non-oriented film it began to decrease already at 35° C.

Moreover, by means of the nonogravimetric technique, it was possible to estimate the amount of water desorbed by the Langmuir-Blodgett film while exposing it to the various temperatures and comparing this result with similar data obtained on not oriented films. The analysis of these data allowed to correlate the rate of water desorption from the film with the loss of protein functional activity. The obtained results correlate well with earlier evidences on the thermal behaviour of this protein when arranged in Langmuir-Blodgett film, namely, increased thermal stability of protein secondary structure and loss of the quaternary structure [Facci P. et al., 1994] (Fig. 20).

7. BIOELECTRONICS DEVICES PROTEIN-BASED

7.1 Optoelectronic Devices

Membrane-proteins such as bacteriorhodopsin can be seen as main components of bioelectronics devices because of their specific properties as a kind of proton pump through the membrane with optical properties, which allows to use it for optical memory and optical signal treatment. The property is the different light absorbance with respect to the state in which the protein is: two main states are the most interesting for application B and M absorbing and yellow light respectively. The principle of the optical storage is based on the recording with the light of one wavelenght, which transfers the BR from one state to the other, and reading with the other wavelenght, making no effect to the state of BR. In principle, such memory must have molecular resolution, but technical limitation will be the sizes of the recording optical beams. The material can be also used for recording the image information. The system in this case must be organised as in the case of holography. The initial beam must be separated into two.

The field of nanotechnology, which is leading to true molecular electronic and photonic devices, offers immense promise for the near future. The results obtained up to now provide a clear demonstration of organic films, with highly specific programmable self-assembling properties, organizing into molecular structures with useful electronic properties. J-aggregates phenomenon can be used in order to obtain optical performances from LB films. This fact could lead to the following applications: optically based storage devices and optical filters. These two molecular devices present several advantages with respect to the conventional ones (i.e. high level of miniaturization, fabrication process). Furthermore they could be used for a more challenging task such as parallel optical computing. In many cases optical filters have been realized using inorganic materials, as these offer a wide range of refractive indices and useful mechanical properties Notwithstanding this it is often difficult to accurately control the thickness of such optical components at short wavelenght (UV/visible). Nowadays the LB technique allows to deposit monomolecular layers of organic materials producing very organized structures and mantaining a fine control of stack thickness. Simple and more complicated high performance optical filters can be constructed, namely quarter wave and rugate filters. In this case, by the repeated cycling of the substrate a highly organized multilayer stack can be realized of precise thickness. Optical filters operating at 530 nm have been realized by using ω-TA (a hydrocarbon-based carboxylic acid) and PSA 50 (a partially substituted side-chain polysiloxane) monolayers deposited on a BK7 glass substrate. Furthermore interesting properties could be added by modulating or changing the optical properties of such molecular devices by means of electrical applied fields or of specific deposited layers. Interesting applications could also be addressed by the utilization of BR (Bacteriorhodopsin) film. BR is the key protein in the halobacterial photosynthesis and its photochromic properties can be used in optical information processing techniques. Due to its interesting optical transitions in a wide timescaling it is possible to realize holographic memories and biodevices with application pattern recognition such those implementable in terms of associative memories noise-like structures.

7.2 Enzymatic Biosensors

Biosensors are used for a wide range of applications, from environmental control to optimisation of bioprocesses, from phytochemistry to biomedical analysis. For what concerns the ions, heavy metals, like copper, cadmium and mercury, are often monitored in industrial waste waters. Among the biologically active substances, we find glucose, fructose, ethanol, urea, aminoacids, lactate, piruvate, oxalate, purinic nucleotides, creatinine and cholesterol. The coumpounds used in phytochemistry, often at nano- or picomolar concentrations, represent another class of possible targets for a biosensor. They are usually monitored in drinkable waters: the maximum limits allowed, for instance of some chlorophenoxyacetic derivates, are of the order of ppb. Moreover, among the substances detectable in traces, hormones, neurotransmitters and some compound of pharmacological interest (metabolites of drugs) are monitored with the biosensors. In the last year, we carried out a theroretical and experimental study in the field of biosensors, which led us to realize a potentiometric sensor [Adami et al., 1991; Sartore et al., 1993] based on a silicon transducer identified with the acronym PAB (Potentiometric Alternating Biosensor). PAB has been built from a substrate of silicon ("p" or "n" type): as can be seen from Fig. 21, in the upper part a layer of silicon dioxide and a layer of siliscon nitride have been deposited while in the lower part, conductive material has been sputtered to form an ohmic contact which allows to acquire an outgoing signal; the remaining area of the substrate has been left unchanged to be lighted by an adequate source (IR-LED). The silicon nitride surface is exposed, directly or by means of a membrane, to the test solution: the interaction between silicon nitride and the ionic species present in solution lead to surface potential variations which are detected by the transducer. The simple "sandwich" structure of this device makes it extremely resistant to electrical and mechanical stress.

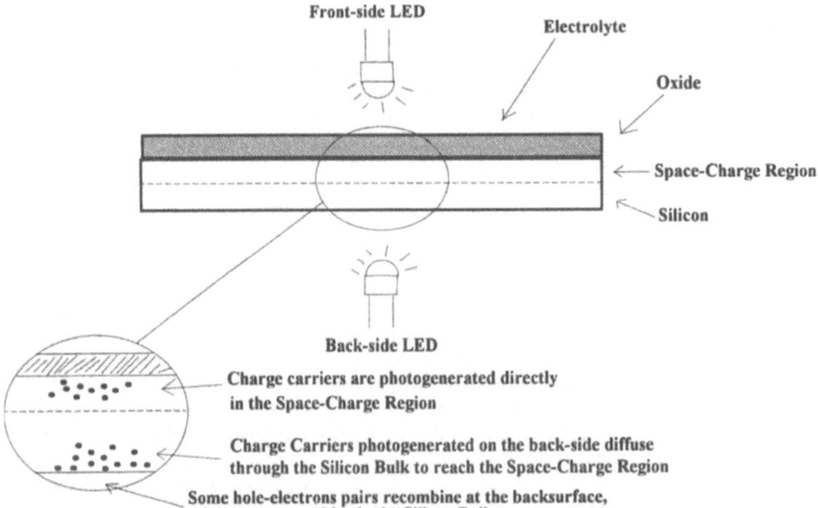

Figure 21 Schematic view of an electrolyte/oxide/silicon system, with either backside or frontside illumination. the magnified area shows the location of the minority carriers injected by the LEDs.

In the application as enzymatic sensor, the transducer is able to monitor chemical changes induced within a microvolume reaction chamber by an enzymatic activity. By means of the production of protons (ions H^+) it is possible to go back to the amount of enzyme which catalyze the reaction or to the concentration of the substrate on which catalyze the reaction or to the concentration of the substrate on which the enzyme acts.

The enzyme has been immobilized with four different techniques:
a) In a gel of 0.6 % Agarose;
b) Using LB films method with reverse micelles;
c) On preactivated membrane of nylon;
d) Silanizing the surface of PAT and then depositing the layer of the enzyme by LB method.

In all these cases, the enzymatic reaction produces protons which causes either an increase ("p" silicon) or a decrease ("n" silicon) in the output signal which allows to evaluate the pH value and, thus, the substrate or the enzyme concentration [Adami et al., 1993]. This has allowed us to obtain a new enzymatic sensor able to monitor, in real time, the enzymatic reaction, namely alcohol concentration by utilizing alcohol dehydrogenase (Fig. 22). These experiments have been carried out at a concentration of YADH of $7x10^{-8}$ M, constant for every experiment; this rather high value was chosen in order to reach quickly the final equilibrium. The ethanol (EtOH) concentration was varied between 0.8 and 352 mM, and the same kind of acquisition (output signal vs. time) has been performed as previously described.

In these experiments the total acquisition time was 22 min; an example of such acquisitions is shown the upper corner of Fig. 22, relative to an EtOH concentration of 352 mM. This curve is used to calculate the net variation in the output signal, after reaching the equilibrium; the measured value is essentially the difference between the baseline signal level (at $t_o \approx 5$ min) and the steady level (at $t_f \approx 22$ min). The same figure 22 represents data values at different EtOH concentrations; each point in the plot is measured as described above. The biosensor here described has been tested to monitor enzyme reactions in order to detect both enzyme and substrate concentrations. Other enzymatic biosensors for pesticides and for cholesterol determination (utilizing thin film of P450) are being implemented in our laboratories.

8. FINAL CONSIDERATIONS

Many studies tend to undervalue the quality and quantity of the requirements necessary until a physical and/or biological system could reasonably be considered eligible to become the inspiratory principle of a new technology of the future for the information processing. If we look at the computational systems as we know them to be today, we recognize immediately that an "elaborator" comprises at least 4 functional units with diverse tasks: the processor, which elaborates the information by executing instructions contained in the program, the central memory, where the information is momentarily imaged during the execution of the program; the input/output units to interact with the operator, and rigid memory units (discs, tapes etc.) where the data contained in the processing system are memorized in a permanent form. It is evident that a new technology of a calculating system, like the bioelectronics discussed here, must meet all the demands described above, if its ambition is to substitute in an integral form the already existing systems. Alternatively, the task could be to make only one part of the ones described above, like, for example, the central memory or the processor for the elaboration of the information. In this case it is necessary that the new subsystem should be interfaced with the conventional parts to establish the communication between them. Thus, the need to generate signals compatible with the present electronic standards. From the above considerations, it is clear that is not sufficient to determine a physical effect or a biological structure (e.g. a suitable molecule) that could memorize information: it is necessary that this information can be reliably written and read in the molecule, and this requires an addressing technique and an interaction with the outer world which today is overwhelmed by electronic systems.

An ultimate consideration is the following: with the increase in the complexity of electronic circuits, the space occupied by active components or transistors becomes a small fraction of the space occupied by interconnections. For example, it can be observed how a static RAM cell with 6 transistors with the 0.5 μm size, with 4 interconnection levels, occupies an area of 60 μm^2. Of these, only 2 μm^2 are effectively occupied by 6 transistors, which is 3 %. All the rest is occupied by interconnections. Similarly, in a transputer with 2.2 million of transistors with a total area of 250 mm^2, the total area occupied by transistors is 5 mm^2, or 2%. These considerations show that it is not enough to devise a fast and compact device if a technology for transmission of the processed information is not found. In addition, it is understood that a new technology aimed at the realization of new transistor-

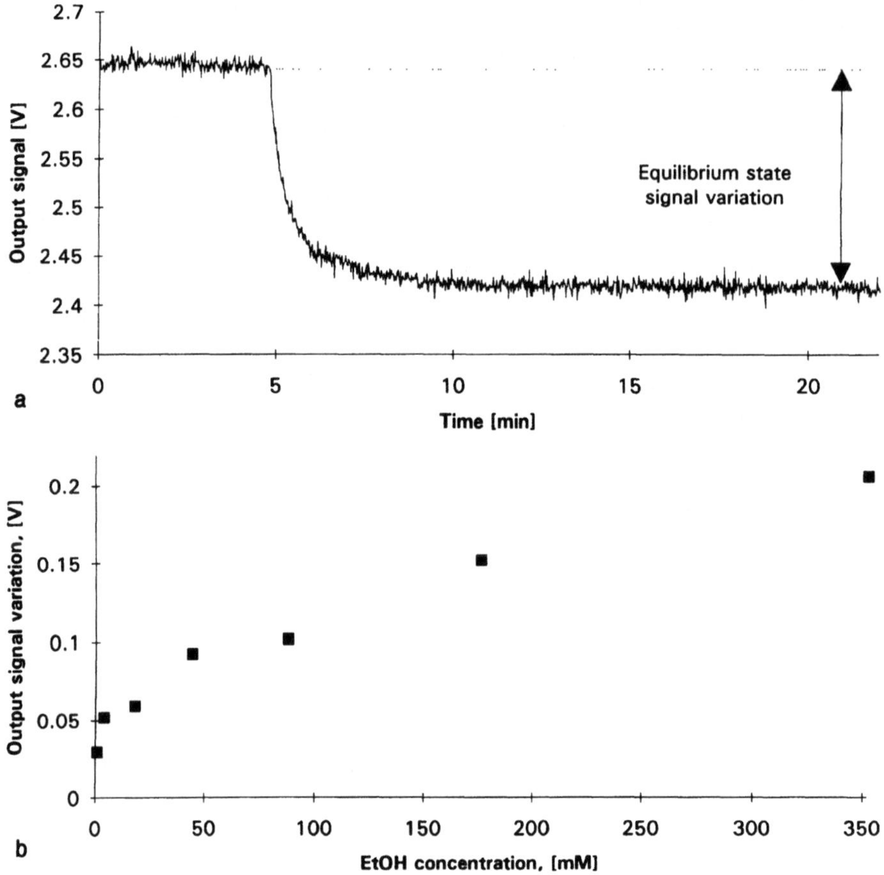

Figure 22 Enzymatic biosensor response.

based elementary analog devices will not be able to similarly regain the disadvantages accumulated in respect to the silicon technology which will continue to develop rapidly due to enormous investments existing in this field.

How To Overcome The Actual "Bottle Necks"

From what has been said previously, if the memory is inserted into a nano-hardware, but the access to the nano-hardware is to be gained via microelectronics, a bottle neck of significant size could develop itself in the transition nano-micro. Thus, the access could be really slow. A slow random-access to the memory must not give trouble. The reason for which the biological memory , although slow, is extremely powerful is that the biological memory is not and addressable memory, but is an intelligent memory in which the access is attained via similarity. Thus, in the brain what we call "memory" is really a combination of memory and calculation, and every access to the memory implies a significant calculation. Our "memory" is powerful, despite the slowness, exactly for this reason. So, going back to extensive nano-calculations, the obstacle of the nano-micro communication becomes tolerable in the areas such as memory if the memory is an "intelligent memory", like associative one. A similar argument can be presented for any computational use of the nano-hardware which communicates with macro-hardware; operationally, to overcome the bottle neck, one should make a notable number of calculations at the nano level for every passage through the macro-micro interface. In other words, the bioelectronics can lay a significant role only to the extent in which it will succeed in make proper use of functional architectural organization of biological systems, or where we will be able to introduce alternative modality of signal transmission (solitons rather than electrons) thus reducing in all cases the space occupied by interconnections. This can take place in various ways: i) developing systems that would emulate such an functional organization in a technology that would be diverse and more subject to control; ii) solving the problem of interfacing of biological systems with conventional electronic ones making use of the nanocomputation; iii) implementing the signal transmission via solution waves. For an "advanced biosensor", which most probably will be the first intermediate point for the bioelectronics to arrive at, it is interesting, in addition to molecular recognition, to study the possibility to communicate electrically with physical transducers with have the suitable sensitivity and hence are sensitive to electron transfer or to the release of protons associated with phenomena of molecular association. For computational circuits, the properties of commutation between the states, the associated electron transfer, the possibility to address a change in a state electrically of via an electromagnetic excitation are of interest. Regarding the latter aspect, proteins that have a prosthetic heme group are particularly interesting. Relative to the electron transfer, is seems interesting to explore also the possibility to develop devices that would utilize solution waves.

Naturally, the difficulty in developing a new generation of general purpose information processing systems does not imply that the intermediate goals of bioelectronics should not be pursued. For example, a biosensor that could combine the process of detection with a process of intelligent information elaboration would represent a good intermediate result, although not being a biochip in a general sense described above.

ACKNOWLEDGMENTS

Part of this work was supported by research contract from TECHNOBIOCHIP (Marciana, Italy) and FARMITALIA CARLO ERBA (Milano, Italy) within the framework of the National Program "Technology for Bioelectronics" sponsored by the Italian Ministry of Universities, Scientific and Technologic Research. The author would like to thank Mr. Fabrizio Nozza for the technical assistance in preparing this manuscript.

REFERENCES

Adami M., Piras L., Lanzi M., Faniugliulo A., Vakula S. and Nicolini C., 1993, *Sens. and Act. B*, in press.

Adami M., Sartore M. and Nicolini C., 1994, *Biosens. and Bioel.*, , in press

Allen J.P., Feher G., Yeates T.O., Komiya H. and Rees D.C., 1987, *Proc.Nat.Acad.Sci.USA* 84:5730-5734, 6162-6166, 6438-6442

Almassy R.J. et al., 1978, *Proc.Natl.A.cad Ssci. USA* , 75(6):2674-2678; 2Å resolution

Ambler R.P., 1963, *Biochem. J.*, 89:341-349

Antolini F., Erokhin V., Mascetti G. and Nicolini C., 1993, *B.B.A.*, submitted

Ari Aviram, 1988, *J. Amer. Chem. Soc.*, 110:5687

Bein T., 1993, *Nature*, 361:207-208

Beratan D. N., J. N. Betts, J. N. Onuchic, 1991, *Science* 252:1285

Beratan D. N., J. N. Onuchic, J. N. Betts, B. E. Bowler, H. B. Gray, 1990, *J. Amer Chem. Soc.* 112:7915; J. N. Onuchic, P. C. P. de Andrade, D. N. Beratan, 1991, *J. Chem. Phys.* 95:1131; J. N. Onuchic, D. N. Beratan, J. R. Winkler, H. B. Gray, 1992, *Annu. Rev. Biophys. Biomol. Struct.* 21:349; J. N. Betts, D. N. Beratan, J. N. Onuchic, 1994, *J. Am. Chem. Soc.*, in press; J. R. Winkler and H. B. Gray, Chem. Rev., in press

Bertrand P., in "Structure and Bonding", 1991, G. A. Palmer, Ed. (*Springer-Verlag*, Berlin,), 75:1-47; A.Kuki, ibid., pp. 49-83

Carr N., Goodwin J., Harrison K.J., Lewis K.L., 1993, *Thin Solid Films*, 230:59-64

Carrara E.A. and Nicolini C., 1992, *Biomedical Modeling And Simulation*, 183-188

Carrara E.A., Gavotti C., Catasti P., Nozza F., Berutti Bergotto L., Nicolini C., 1992, *Arch. Biochem. Biophys.* 294:107-114

Carrara E.A., Pagliari F. and Nicolini C., 1993, *IJCNN'93-Nagoya Japan*, 1:983

Carrara E.A., Pagliari F.and Nicolini C., 1993, *Neural Networks*, in press
Dubrovsky T., Tronin A., Vakula S. and Nicolini C., 1993, *Sensors and Actuators*, submitted
Erokhin V., Vakula S., and Nicolini C., 1994, *Thin Solid Films*, in press
Facci P., Erokhin V. and Nicolini C., 1993, *Thin Solid Film*, 230:86-89
Facci P., Erokhin V. and Nicolini C., 1994, *Thin Solid Films*, in press
Farazdel A., 1990, *J.Amer.Chem.Soc.*, 112:4206-4214
Farver O. and I. Pecht, 1989, *FEBS Lett.* 244:376; ibid., pp. 379; 1990, *FEBS Lett.*, 1-2:33-35
Frontiers in Biotransformations, Vol. 3, pp 1-59, *Ed. Ruckpaul R.*, Berlin-London
Gabrielian A., Anselmino A., Catasti P., Carrara E.A. and Nicolini C., 1993, *J.Mol. Biol.*
Gourdon A., 1992, *New Journal of Chemistry*, 16(10):953-957
Groeneveld C.M., Canters G.W., 1988, *J. Biol. Chem.* 263:167-173
Hampp N., 1993, *Proc. XV IEEE-EMBS, IEEE Press*, 3/3:1525
Heller M.J., Tullis R.H., 1991, *Nanotechnology*, 2:165-171
Jacobs B. A. et al., 1991, *J. Am. Chem. Soc.*, 113:4390
Janzen A.F. and M. Seibert, 1980, *Nature*, 286:584-585
Keller R.M., Wüthrich K., 1976, *FEBS. Lett.*, 70:180
Kemp J.C., 1969, *J. Amer. Chem. Soc.*, 59(8):950-954
Lukashev E.P. , A.A. Kononenko, P.P. Noks, V.I, Gaiduk, B.M. Tseitlin, A.B. Rubin and O.V. Betskii, 1989, *Proc. Acad. Sci. USSR* , Vols. 304-306
Marcus R.A. and Sutin N., 1985, *Biochim. Biophys. Acta*, 811:265
Marcus R.A., 1963, *J. Phys. Chem.*, 67:853
Matsuura Y., 1982, *J. Mol. Biol.*, 156:389-409, 1,6Å resolution
Matthew J.B. et al., 1983, *Nature*, 301:169
Nicolini C, Erokhin V., Antolini F., Catasti P. and Facci P., 1993, *Biochim. Biophis. Acta*, 1158:273-278
Nicolini C., Adami M., Antolini F., Beltram F., Sartore M., Vakula S., 1992, *Physics World*, 5(5):30-37
Nicolini C., Towards the Biochip, 1989, *World Published Co.*, Singapore-London
Nicolini C., Adami M. and Sarote M., 1993, patent pending
Nordling M., 1990, *FEBS letters*, 259(2):230-232
Pashkevitch P., Di Silvestro I, Vakula S., Gabrielian A. and Nicolini C., 1994, in preparation
Porter and Coon, 1991, *J.Biol.Chem.*, 266:1369-72
Poulos T.L. et al., 1985, *J. Mol. Biol.*, 260:161-72; 1987, *Biochemistry*, 26:8165-8174
PoulosT.L., 1983, *J. Biol. Chem.*, 258(12):7369
Ratner M. A., 1990, *J. Phys. Chem.*, 94:4877
Rodgers K., 1988, *Science*, 240:1675
Sabo Y., Kononenko A.A., Zakharova N.I., Chamorovskii S.K. and Rubin A.B., 1991, *Proc. Acad.Sci. USSR*, Vols. 316-318
Sandstrom J., 1982, "Dynamic NMR Spectroscopy", *Academic Press*
Sartore M., Adami M., Nicolini C., Bousse L., Mostarshed S. and Hafeman D., 1992, *Sens. and Act.*, A32:431-436
Sartore M., Adami M., Bousse L., Hafeman D., Mostarshed S. and Nicolini C., 1994, *J. Appl. Phys.*, in press
Sigel H. and Sigel A., Eds., Metal Ions in Biological Systems, 1991, *Dekker*, New York, vol. 27
Tachibana H., Azumi R., Nakamura T., Matsumoto M. and Kawabata Y., 1992, *Chemistry Letters*, pp.173-176
Taylor and Francis, 1989, *Biochim. Biophis. Acta* , 998(1):189-195
Therien M. J., Chang J., Raphael A.L., Bowler B.E., Gray H.B., in "Structure and Bonding", 1991, G. A. Palmer, Ed., *Springer-Verlag*, Berlin, 75:109-129; Sykes A. G., ibid., pp. 175-224
Tieke B., 1990, *Advanced Material*, 2(5):222-231
Tronin A., Lvov Y. and Nicolini C., 1994, *J. Polym. Colloid. Sci.*, submitted
Wuttke D. S., Bjerrum M. J., Winkler J. R., Gray H. B., 1992, *Science*, 256:1007-1008

ON HAND-PRINTED CHARACTER RECOGNITION

Zs. M. Kovács-V, R. Guerrieri and G. Baccarani

Dipartimento di Elettronica, Informatica e Sistemistica (D.E.I.S.)
University of Bologna
Viale Risorgimento 2
40136 Bologna
ITALY

INTRODUCTION

Handwritten digit and alphabetic character recognition has been an active area of research for many years. Character recognition techniques associate a symbolic identity with the image of a character. The problem of replication of human functions by machines involves the recognition of both machine and hand-printed or cursive-written characters. Character recognition is better known as Optical Character Recognition (OCR) since it deals with recognition of optically processed characters rather than, for example, magnetically processed ones. The origin of character recognition can be found as early as in 1870, though the modern version of OCR appeared in the middle of the 1940s with the development of digital computers. Its main area of interest was data processing with application to the business world. The principal motivation for development of OCR system is the need to cope with the enormous flood of paper such as bank cheques, commercial forms, government records, mail sorting generated by the expanding society and international partnerships. Since the origin of OCR systems extensive research has been carried out and a large number of technical papers have been published by researchers in this area. Many techniques and experimental results have been published in the literature. In recent years, new classification approaches, such as neural networks, and design methodologies, such as classifier combination and parallel feature networks extraction, have renewed interests in this fields. As OCR technology seeks applications in more complicated, real time environments, the criterion for successful algorithm becomes more rigorous. New feature extraction and classification techniques are developed with parallel hardware implementation in mind. Several recognition methodologies are combined in order to improve the recognition quality. State of the art reports on character recognition research have been presented by Nagy [1968], Harmon [1972], Suen et al. [1980], Mori et al. [1984], Mantas [1986], Davis and Yall [1986], Govindan and Shivaprasad [1990], Lee and Srihari [1993].

Optical character recognition technology has many practical applications. For example, in Govindan and Shivaprasad [1990] several practical applications are introduced. The following are some of the most interesting applications for which OCRs have been used or suggested.

- Use by blind people as reading aid.
- Use in postal department for address reading, and mail delivery speeding up.
- Use in direct processing of documents as a multi-purpose document reader for large-scale data processing, in order to change text into a computer-readable form.
- For business applications, financial business applications like cheque sorting strategy optimization.

From Neural Networks and Biomolecular Engineering to Bioelectronics
Edited by C. Nicolini, Plenum Press, New York, 1995

- For shorthand transcription, and in electronic package industries, reading characters stamped on metallic parts.

On the basis of the application nature character recognition can be grouped in two main schemes, namely, off-line character recognition and on-line character recognition. In off-line systems the recognition is not done at the time of preparing the documents. In on-line recognition, the recognition is done as and when the characters are hand-drawn and hence the timing information of each strokes are also available along with the characters images. In this work only off-line recognition is considered. In order to realize a system to face one of the above applications, several building blocks are needed, among which the isolated character recognition module is an essential one.

This paper only deals with the isolated character recognition task. The paper is organized as follows. The general classification scheme of a character recognizer is presented. Some considerations on the evaluation of classification quality follow. Finally, a hybrid system is introduced, which is capable of high quality character recognition.

GENERAL CLASSIFICATION SCHEME

An OCR classifier usually comprises the blocks of Fig. 1. Here circles represent inputs and outputs; rectangles represent processing modules and arrows indicate the logical process flow. The block boundaries in a real system are not always sharp, but their operation can often be identified easily. A bidimensional binary raster contains the input image. Its size may or may not be fixed. The output of the system is an ordered list of character classes the input image is associated with and the corresponding class-confidence values.

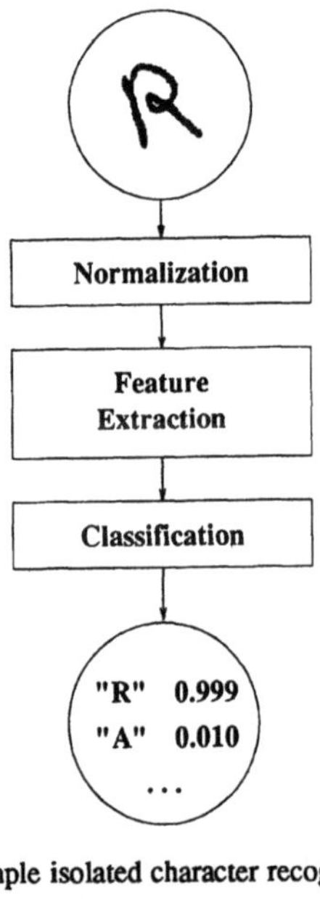

Figure 1. Simple isolated character recognition system

The first processing module normalizes the image in some geometrical sense. It is used for noise filtering and to reduce the variance of each class of characters the image may belong to. Operations like stroke width normalization, deslanting, size normalization are carried out in this module.

The second block is the feature extraction module which transforms the bidimensional human-recognizable image into a vector containing some relevant information on the image geometry. The vector is called feature vector and it represents a point in the feature space.

The goal is to map character images of the same class into clusters in the feature space as separated as possible from other character classes. Transformations employed in this module vary from a raw pixel subsampling to more sophisticated mappings as the Distance Transformation or the Karhunen-Loeve Transformation. This block has to be carefully designed because it has to separate the clusters in the feature space in order to help the further classification module. If the feature extraction is able to generate a complete separation of character classes in the feature space, a very simple classifier is able to produce a zero recognition error. Unfortunately, the optical handwritten character-recognition problem is intrinsically ambiguous, which means that a complete separation among clusters is not possible. The minimum recognition error is bounded by the bayesian error as usual for overlapping-class probability distributions, as shown by Duda and Hart [1973].

The last module is the classifier which reads the feature vector and identifies the class the vector belongs to. The feature space usually contains many clusters of the same character class in different regions. The classifier, therefore, has to find separation boundaries between clusters of different classes and it has to associate the most likely class with the incoming feature vector. An OCR system designed as part of a real document analysis system is required to generate a confidence value associated with the output class, in order to keep the quality of the answer under control. For example, handwritten "U" and "V" images without a context may be very similar and their classification as "U" and not "V" (and vice versa) can not be done safely. In this case, when the classifier generates multiple hypotheses, both classes will associate a similar confidence value, while if the classifier is allowed to create just one answer, it will create a low confidence output.

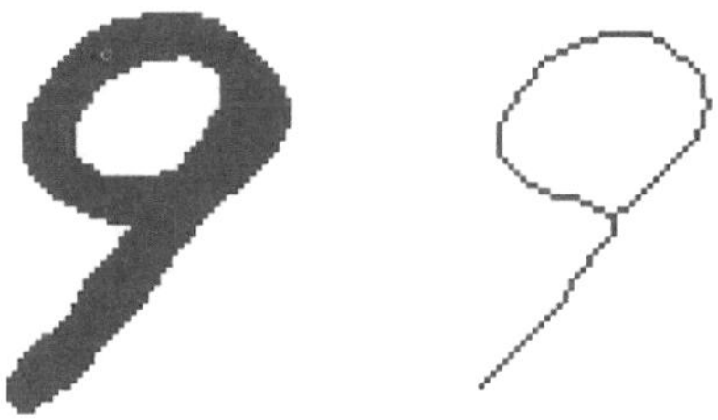

a b

Figure 2. Original image (a) and thinned image (b)

NORMALIZATION MODULE

The input image can be obtained using a camera, a scanner or any other image acquisition device. The raster representing the image itself can be in a grey scale or binary form. Usually OCR systems use binary images, needing thresholding of the grey scale image after some image enhancement if the quality is poor, as shown by Ghaderi and Liff [1992]. Image filtering using median or gaussian filters, see f.i. Brownrigg [1984], helps to remove the noise present on the paper or collected during

The acquisition phase

Writers may use any writing device creating thick or thin strokes. In some systems stroke normalization as shown by Guo and Hall [1989] is used to reduce the class variance and to help the subsequent classification stages. In Fig. 2 the original "9" and the stroke-normalized character is shown. The skeletonization process is not unique and may severely deform the image: as an example, occluding for some reason the center hole of a "O" leads to an "I".

The slant of handwriting is sometimes useful to infer about the writer's character, but it may degrade automatic reading. Usually a deskewing algorithm, see f.i. Casey [1970], is applied in order to normalize the character slant. Also, this type of operation may produce distortion, but the latter is not worrying so long as preprocessing of characters of different classes does not lead to very similar images. In Fig. 3 a slant normalized "9" is shown using a share transformation. The vertical stroke of the normalized image is slanted, but it does not create any classification problem, being the same for all characters of this class and being different for all other classes.

a b

Figure 3. Original image (a) and deskewed image (b)

When the raster input is presented to the system, its size is known but it could not be correct for further processing. An image size normalization may be carried out in order to produce a fixed image size independent of the input dimensions. There are three size normalization methods used in OCR. The first one simply encloses the image by a white frame of suitable size and discards the image, i.e. does not classify it, if the original size is too big. the second method computes a scaling factor which increases or decrees the image with a linear stretching until the horizontal or vertical dimension equals the final one. In this way the original aspect ratio is maintained and no distortion occurs. The third method operates an horizontal stretching and then a vertical one, making the image touch all four final boundaries. The latter operation leads to the best class-variance reduction, but also to a distortion which can be serious for one dimensional characters such as "1", "l" or "I" amplifying their irregularities. It has to be found experimentally if, in a given application, this procedure helps or harms instead. In Fig. 4 an example of the three size normalization techniques is shown applying them to a handwritten "9".

Feature Extraction Module

Once the input image has been geometrically normalized, its feature vector has to be computed in order to perform the classification. The selection of the best features depends on the previous normalization stage and on the subsequent classification method. The features have to be defined in order to cluster as much as possible the character classes in the feature space and to create decision boundaries among clusters compatible with the further classification methodology. In pattern recognition the use of the raw normalized images as feature vector produces extremely complicated decision boundaries and a coarse separation among classes, leading usually to a low quality-recognition performance. In Fig. 5a a hypothetical raw-image clustering is shown while in Fig. 5b the correctly defined feature space clustering is reported. Theoretically, it is possible to define a multilayer perceptron architecture which is able to find very complicated decision surfaces but the network size may be too big if compared with the practically manageable training database size and the strength of the learning algorithm. All the successfully working OCR systems described in the literature use some sort of feature instead of the row normalized input image. The description of some features follows.

A possible feature extraction can be based on a distance transform of the normalized image. The distance transform is described for example by Rosenfeld and Pfaltz [1966]. Let Ω be the two-dimensional region where the pattern ζ lies. The distance transform associates with each point $(x,y) \in \Omega$ the distance to the nearest point $(\alpha,\beta) \in \zeta$, according to a suitable metric. The advantage of the grey scale image obtained by distance transform compared with a normal grey-scale image is that the information is propagated in Ω instead of the neighboring area of ζ only as shown in Fig.6. The image a) shows the original grey scale pattern and b) the related grey scale curve. In c) the distance transform applied to the thresholded image is reported. Thus, this procedure maps a binary pattern into a point in an N-dimensional space, where N is the number of pixels of the normalized image. Thanks to the distributed information, it is possible to subsample the distance transformed image, obtaining a lower dimensional feature space. The feature vector obtained in this way contains the position of white and black pixels as relevant information. Due to the final subsampling, resolution problems may occur. For example, if a handwritten "5" has the lower half nearly closed, with this type of feature the cluster of the "6" class is very close and may give rise to classification errors. A similar feature vector can be obtained convolving the normalized input image with some gaussian template.

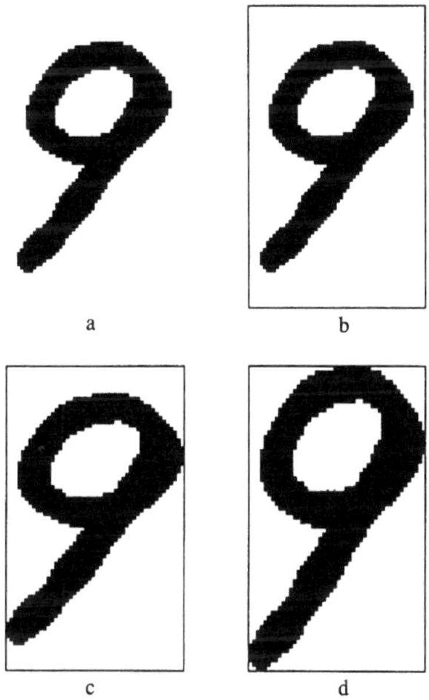

Figure 4. Original and size normalized images: a) Original image;b) Size normalization adding a white frame; c) Size normalization with constant aspect ratio; d) Size normalization expanding both vertical and horizontal dimensions.

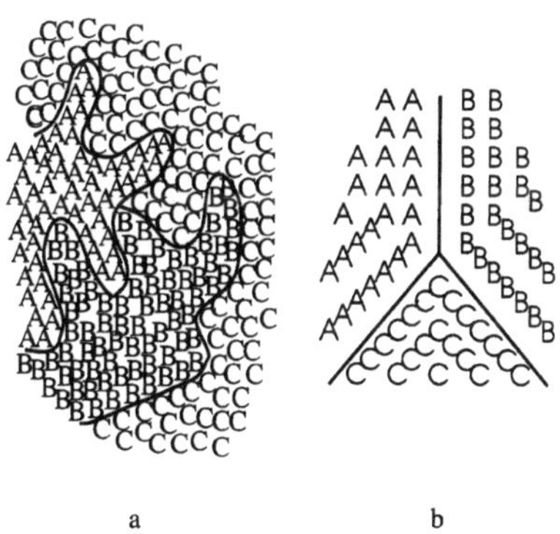

Figure 5. Raw-image clustering (a) and ideal clustering (b)

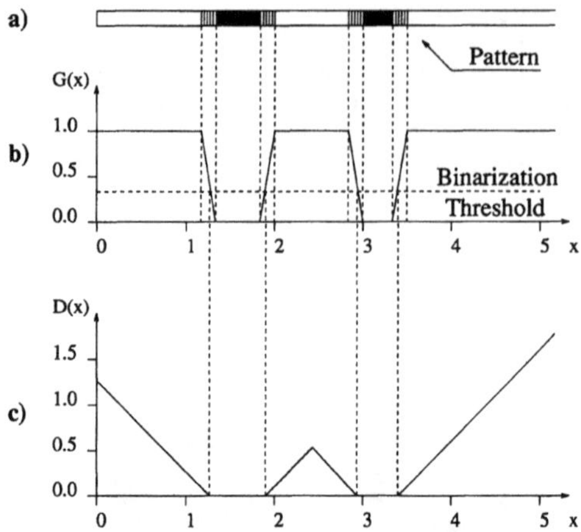

Figure 6. Distance transform versus standard grey scale: a) Original one-dimensional grey scale image; b) Grey scale curve of image a); c) Distance transform of image a)

The chain code histogram; shown by Cao et al.[1992] or Takahashi [1991], is another useful feature. The relevant information it extracts is the orientation of the character's border. In this process the rectangular frame enclosing the normalized contours is divided into the rectangular zones. In each zone, a local histogram of the chain codes is calculated by supplying 2x2 templates. The feature vector is composed of these local histograms. Since the contour direction is quantized to one of the 4 possible values (0, 45, 90, 135 degrees) a histogram in each zone has four components.

Once the character border has been extracted, its bending points form a known feature as shown by Takahashi [1991]. The contour in a character image is traced and strong curvatures are detected, i.e., bending points with certain attributes such as position, orientation of the curvature, convex or concave, and acuteness. This feature represents a geometrical feature such as stroke terminals (usually convex with strong acuteness). It is sensitive to local curvatures that are useful to distinguish similar characters. A similar feature is described by Miyamoto [1988].

The Karhunen-Loeve (K-L) transformation of the image is used to extract a low dimensional feature vector as shown by Grother [1992]. The K-L transformation is an orthonormal function and corresponds to a projection of the images onto the eigenvectors of the covariance matrix of the image data. The production of this transform is also known as "principal factors" or "principal component analysis". The covariance matrix is diagonalized producing the largest eigenvalues and corresponding eigenvectors. Feature extraction is thus the application of an affine function to the image. The mean image vector is subtracted and the result is premultiplied by the matrix whose rows are the eigenvectors. Thus each element of the K-L transform is the projection of the image onto a basis vector. The sample variance of that element is the respective eigenvalue, which is therefore a measure of its "size". The K-L transform is, among the unitary transforms, an optimally compact signal representation of the original data.

Several other features can be used, such as position of holes and other geometric items, connected components in several directions, foreground and background bar-features, as described by Gader et al. [1992], and so on.

Classification Module

The task of the classification module is that of finding the class boundaries in the feature space, by separating clusters belonging to different classes. According to the classification algorithm, each classifier is characterized by the type of discriminant function it implements. The classifiers may be separated into several categories, as detailed by Grother and Candela [1993]. It is, however, notable that the category names are arbitrary and that some classifiers have attributes of more than one category. In the statistical pattern recognition literature the Parametric classifiers use variables such as the expected means and covariances to express the class density functions. In assuming, for example, linear and quadratic forms for the discriminant functions, one obtains a simple Euclidean Minimum Distance and the Normal classifier as parametric classifiers, or some of their more sophisticated versions. The Non-Parametric classifiers do not adopt a structured expression of the density functions. Nearest-Neighbors classifiers, Mutli-Layer Perceptrons and Radial Basis Functions classifiers are the most popular in this category.

The Euclidean Minimum Distance classifier is perhaps one of the most simplest classifiers that can be designed. Its discriminant function is the squared euclidean distance between an unknown point and the closest cluster center. This is equivalent to using the class label of the estimated cluster-mean that is closest in the euclidean distance sense to the unknown element. In the one cluster per class case the hypothetical class regions are convex polygons.

The Normal classifier is based on parametric density estimation that assumes a multivariate normal distribution for each class. Given a particular loss function, the optimal or bayesian classifier is the one that minimizes the expected loss. Defining a symmetric loss function, the correct classifications produce no losses and all kind of incorrect classifications produce equal loss values. In this case the bayesian classifier is the one that classifies each unknown to the cluster for which the a posteriori probability is higher.

There are several Nearest Neighbors based classifiers. The simple l-Nearest Neighbor classifier is an elaboration of the previously outlined Euclidean Minimum Distance classifier, where instead of using just the cluster centers as single prototypes for each class, all elements of the training examples are used. The class of the nearest neighbor defines the classification result. The rule is intuitively appealing and Cover and Hart [1973] have shown it to have good asymptotic behavior. Each class region is the union of many convex polygons each containing a single prototype of the class. Hence a class region is a very complicated polygon, not necessarily convex or even connected. In the more general case voting between the k nearest neighbors is used. The voting rule further characterizes the classifiers. The majority voting is one of the most popular voting rules. A more elaborate form of the nearest neighbor method is to allow k to be a random variable such that the number of voting neighbors is different for each unknown. This classifier finds the closest prototype to the unknown, the defines the neighboring prototypes to be those whose squared euclidean distance from the unknown is less than a function of the distance of the nearest prototype. Further the number of votes received by a class is divided by the square root of the sum of squared distances of the same class near neighbors from the unknown, so as to diminish the importance of neighbors that are relatively far away compared to other neighbors.

The Multi-Layer Perceptron classifier is also known as feed-forward neural net. It is composed of a number of layers, each formed by the union of a number of perceptrons. Each perceptron has a number of inputs, whose numerical values are multiplied by weights and these products are summed. The sum is passed into a saturating function, called activation function, defining the perceptron's output. Each perceptron of a layer can receive input from the immediately previous layer only and its output can be used as input to the immediately subsequent layer only. For the training of the weights of this network, a reasonable procedure is the use of an optimization algorithm to minimize the error, i.e. the mean squared error over the training set between the discriminant values actually produced and the target values, as shown by Rumelhart et al. [1986].

Neural nets of the Radial Basis Functions type get their name from the fact that they are built form radially symmetric gaussian functions of the inputs. If the radial basis functions use gaussian functions that are more general than radially symmetric functions, their constant potential surfaces are ellipsoids whose axes are parallel to the coordinate axes, whereas radially symmetric gaussian functions have spherical constant potential surfaces. However, Radial Basis Functions have become customary for any neural net that uses gaussian functions in its first layer. Training of the network is done by optimization similar to that of the Multi-Layer Perceptron, as shown by Musavi et al. [1992].

EVALUATION OF OCR SYSTEM PERFORMANCE

Classification, rejection, confidence and error are general ideas of importance in OCR, as outlined in Nistir 4912 [1992]. A classification process assigns an ASCII character, i.e. the class, to an image of a character. The classification may be correct or incorrect.

A rejection process divides a set of classifications into rejected classifications and accepted classifications. Only the accepted classifications are considered useful. Usually the rejection mechanism is applied after the classification process. However, some rejection processes work in parallel with the classification process and no classification is assigned when a character is rejected. Most systems that carry out the rejection process after the classification process produce what is called a confidence for each classification. This is a number, usually between 0 and 1, that orders the classification according to the expected reliability.

The rejection rate for a set of character classifications is defined as the ration of the number of characters rejected by the rejection process to the total number of characters presented for the classification.

If a confidence is associated with each classification, any desired rejection rate can be obtained by choosing the correct value for the confidence threshold and rejecting any classifications having confidences less than or equal to the threshold and accepting any classification having confidences greater than the threshold.

The error rate for a set of classifications can be defined in two ways: it can be defined as the fraction of the unrejected (accepted) characters that are classified incorrectly, or it can be defined as the ration of the characters that are classified incorrectly to the total number of characters presented for the classification. The first method always gives rise to numerical values greater than or equal to the second one, because the number of accepted character classifications is less than or equal to the total number of characters presented to the system. The error versus rejection curve is always decreasing if the error is calculated according to the second formulation, while this behavior is not necessarily true for the first method. In fact, a system which rejects correctly classified elements instead of misclassified elements, associates an increasing error vs. rejection curve.

HYBRID SYSTEMS

Unfortunately, the previously outlined simple recognition technique does not lead to high quality classification capabilities, because a simple feature is not able to cover all the subtle aspects of human hand-writing. For example, in some cases the proximity of the strokes contains the information, while in some others the global shape is the important issue. As an example of the former situation , let us consider the classification of two curves, representing a "5" and a "6", when the two curves overlap everywhere but on the pixel which closes the loop of the "6". In this case, a feature based on the global distance between the two curves, such as the distance transform, is not able to discriminate the two classes because the absence of one pixel produces a difference smaller than the difference among symbols belonging to the same class. In the same situation the chain code histogram works correctly, because the derivative of the strokes' border is extracted regardless of their proximity. Similarly there are situations where the chain code technique fails while a global measure succeeds. This situation occurs when a symbol is split into several parts due to poor digitizing or binariazation.

These examples show that if a high quality classification is required, the combination of features and classification techniques is the key point to address. Thus, a classification system becomes the combination of several classifiers through a supervisor coupling features with a different validity range. The combination of different features could be simply obtained by using a very large classifier, whose input space includes all the possible features. However, this approach is difficult to use since the training is hampered by the network size. Several researchers have recognized the importance of the cooperation for quality improvement. Nowlan and Hinton [1991] define a set of classifiers acting in parallel over the same data and the final result is chosen by another "gating" network, deciding which expert is more qualified to answer. An ensemble of networks is outlined by Hansen and Salomon [1990], performing the required classification when the majority of the networks makes the same choice. In Ho et al. [1992] several linear combination techniques are shown in order to extract the final information from several classifiers acting in parallel on the same input.

In the next section a system is presented, see Kovàcs et al. [1993] for details, which is composed of four single classifiers and a supervisor, designed for the recognition of isolated hand-written upper-case characters. The classifiers produce different output types, such as nearest neighbour classes generated by a *k*-NN based classifier and activation function values obtained by a MLP. The network, acting as a supervisor, combines the available information and provides the final results, rather than choosing the "best" expert as described by Nowlan and Hinton [1991]. This procedure is more flexible than the one proposed by Hansen and Salomon [1990] since the "rules" used by the supervisor are tuned to different output situations produced by the lower-level classifiers.

The Classification System

In Fig. 7 the structure of the isolated handprinted upper-case character classification system is reported. Four single-feature based classifiers are used, operating in parallel and presenting their results to supervisor. The four classifiers are based on two different classifiers. Several implementations of the supervisor are possible, based on rules, weighed, sums and multilayer perceptron.

The used normalization techniques are standard. The first step is the application of an area filter to the original binarized image, in order to remove noise, some small lines and spots due to the previous segmentation process. The second phase for all the classifiers, but the third, is slant normalization and, finally, the size is normalized to 32x32 pixels, with the character touching all four boundaries.

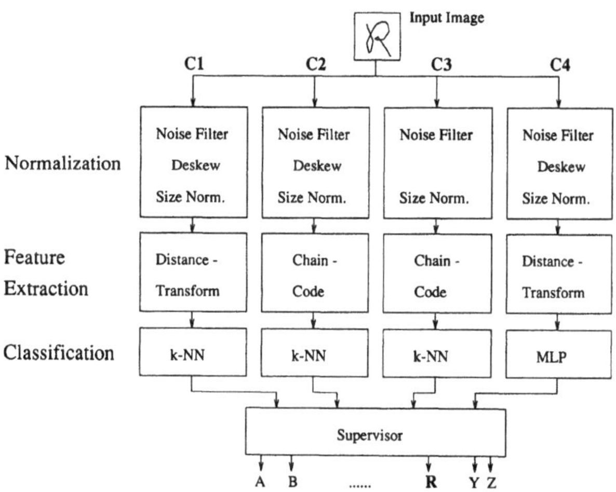

Figure 7. The complete hybrid recognition system.

The feature used in the first (C1) and fourth (C4) classifier is based on the distance transform as shown by Kovàcs and Guerrieri [1992]. The distance transform associates with each pixel the distance to the nearest pixel of the character, transforming the black and white image into a grey-scale one. The final 32x32 grey-scale image is subsampled on a regular basis obtaining an 8x8 map. The 8x8 = 64 numbers are considered as coordinates of a point in the 64 dimensional feature space used as input to the following classifier stage. The feature used in the second (C2) and third (C3) classifier is based on the chain code histogram, as described by Takahashi [1991]. The image is split into 16 square regions, and the four basic directions (horizontal, vertical, diagonal left, diagonal right) are extracted applying 2x2 templates to each pixel. For each region the count of the pixels where the templates match gives a value associated with each of the 4 directions, leading to the 4x16 = 64 dimensional feature space.

The first three classifiers (C1, C2, C3) are based on the *k*-NN rule, defined in detail by Duda and Hart [1973]. A specifically designed metrics, see Kovács et al. [1992] for details, is used to compute the distance between the two points in the feature space. A loop on the training elements allows to obtain the first seven ($k = 7$) neighbors, which are passed to the final supervisor stage. The number of nearest neighbors was chosen in order to allow a sufficiently accurate information on the neighboring elements, i.e. to allow a reasonable rejection criterion. The last classifier (C4) is based on a two-layer multilayer perceptron. The input layer contains 64 nodes, one for each dimension of the feature space, the hidden layer has 325 nodes with sigmoidal activation function and the output layer has 26 sigmoidal nodes, one for each class a character can belong to.

The four classifiers present different output types, according to the classification stage they implement. In fact, the first three (C1, C2, C3) are based on a *k*-NN classifier with $k = 7$, where the outputs are the 7 classes of the ordered nearest neighbors, while the fourth classifier (C4) is based on a MLP, presenting as output the activation function value for each class in the range $0 \div 1$. Several combination methodologies can be tried: based on rules, weighed sums, multilayer perceptron and Parzen windows as shown by Lee and Srihari [1993], or by Kovács et al. [1993].

To apply the rule set, the outputs of the different classifiers have to be homogenized. For this task a transformation is defined which maps the first *k* neighbors obtained from the *k*-NN classifier into the range $0 \div 1$ related to each of the 26 classes, so they add up to 1. The transformation applied to the output of the *k*-NN classifier creates a set of weights, which are computed as the inverse of the position in the list multiplied by a normalization factor. The likelihood associated with each class becomes the sum of the weights in the *k* neighborhood. The inverse function closely interpolates the results of an adaptive optimization process used to find the optimal set of weights. Finally, the combination rule is a sequence of conditions with heuristically defined thresholds, containing the results of highest rank, majority voting and Borda count as defined by Ho et al. [1992].

The classifier combination by weighed sums can be obtained using a perceptron layer, where a delta rule helps to define the weights adaptively. There are 7x26 (nearest neighbor x number of character classes) binary input elements for each k-NN based classifier and 26 floating point input elements for the MLP one. The multilayer perceptron has the same input elements of the previous perceptron layer and one hidden layer of 26 sigmoidal nodes.

Results

The evaluation of a pattern classifier, using supervised learning, requires the selection of a training set, which is made of known data, and a test set, which is classified using the first set. The isolated upper-case binary characters of NIST Special Database 3 and those of NIST Test Data 1, made available by Garris and Wilkinson [1992], were used for the upper case letter recognition, as training and testing data respectively. In Fig. 8 the error rate is reported as function of the rejection rate for the hybrid system, using the multilayer perceptron as supervisor. The error rate is computed normalizing the misclassified characters to the total number of elements in the test database. The NIST Test Data 1 was recognized by a human (P.Z.) and its hypotheses checked against the original classification, leading to an error rate of 2.1% when no rejection is allowed.

CONCLUSIONS

The recognition of isolated characters can be automatically done with a recognition rate comparable with human performance. A 1.5% difference in the error rate still exists, justifying ongoing research in this area. However, further lowering of the error rate requires very high costs in terms of computational power and algorithmic complexity. These higher costs have to be considered from the system's point of view, rather than from that of the simple recognizer alone. The isolated character recognizer is a module in an application solving system, where the connection of different modules defines the possible bottle-neck of the system and therefore where research has to be carried out. The focus in academia and industry on OCR research is now migrating toward the more difficult problem of recognition of structured documents. At its simplest this involves segmentation and recognition of test fields. The design of the building modules has to take into account all the necessary interaction among the blocks rather that just addressing the best quality of single modules alone. Hence, a system perspective is necessary if useful cognitive functions are to be pursued.

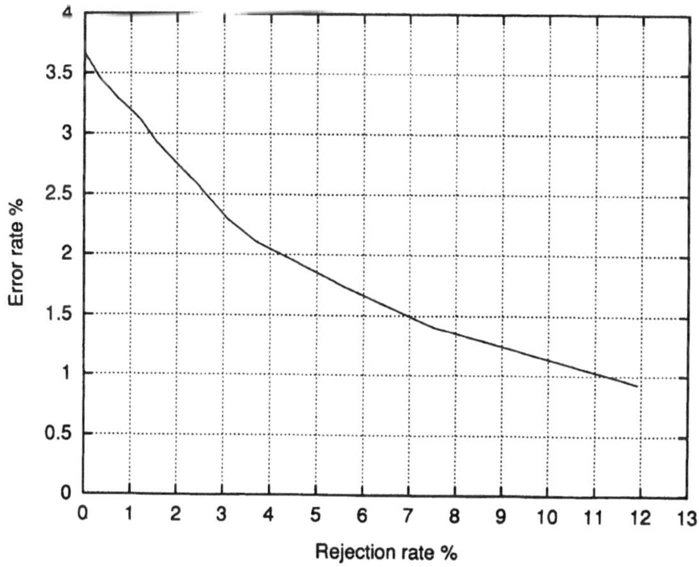

Figure 8. Error vs. rejection rate.

ACKNOWLEDGEMENTS

This work has been partially supported by Progetto Nazionale Bioelettronica. The first author acknowledges the support provided by a grant from SGS-Thomson. The authors are grateful to P. Zabberoni for her patient and valuable help.

REFERENCES

Brownrigg D.R.K., 1984, *Image Processing and Computer Vision*, 27(8):807-818.

Cao J., Shridhar M., Kimura F. and Ahmadi M., 1992, *Proc. 11th IAPR*, The Hague-The Netherlands, 2:643-646.

Casey R.G., 1970, *IBM J. Res. Dev.*, 548.

Davis R.H., Yall J.L., 1986, *Image Vision Comput.*, 4:208-218.

Duda R.O., Hart P.E., 1973, "*Pattern Classification and Scene Analysis*", John Wiley and Sons, New York.

Gader P., Mohamed M., Chiang J.H., 1992, *Proc. 5th USPS Advanced Technology Conference*, Washington DC, 1:215-225.

Garris M.D., Wilkinson R.A., 1992, *NIST Special Database 3 and Test Data 1*, NIST Advanced Systems Division, Image Recognition Group.

Ghaderi M., Liff H., 1992, *Proc. 5th USPS Advanced Technology Conference*, Washington DC, 2:1139-1152.

Govindan V.K., Shivaprasad A.P., 1990, *Pattern Recognition*, 23(7):671-683.

Grother P.J., 1992, *Proc. SPIE Applications of Artificial Neural Networks III*, Orlando.

Grother P.J., Candela G.T., 1993, Comparison of Handprinted Digit Clasifiers, *NIS-TIR* 5209.
Guo Z., Hall R.W., 1989, *Image Processing and Computer Vision*, 32:359-373.
Hansen L.K., Salomon P., 1990, *IEEE Trans. on PAMI*, 12:993-1001.
Harmon L.D., 1972, *Proc. IEEE*, 60:1165-1176.
Ho T.K., Hull J.J., Srihari S.N., 1992, *Proc. 11th ICPR*, The Hague - The Netherlands, 2:84-87.
Kovács Zs.M., Guerrieri R., 1992, *Electronics Letters*, 28(19):1825-1827.
Kovács Zs.M., Guerrieri R., Baccarani G., 1992, *Proc. 11th ICPR*, The Hague - The Netherlands, 2:96-100.
Kovács Zs.M., Guerrieri R., Baccarani G., 1993, *Proc. WCNN'93*, Portland-Oregon, 1:186-189.
Kovács Zs.M., Ragazzoni R., Rovatti R., Guerrieri R., 1993, *IEE Electronics Letters*, 29(14):1308-1310.
Le Cun Y., Boser B., Denker J.S., Henderson D., Howard R.E., Hubbard W., Jackel L.D., 1990, in "*Advances in Neural Information Processing Systems*", D. Touretzky ed., Morgan Kaufmann, 2:396-404.
Lee D.S., Srihari S.N., 1993, *Proc. IWFHR-III*, 153-162.
Miyamoto N., Tsutsumida T., Nakajima N., Kawatami T., 1988, *NTT Review*, 37:175-181.
Mori S., Yamamoto K., Yasuda M., 1984, *IEEE Trans. on Pattern Anal. Mach. Intell.*, 6:386-405.
Musavi M.T., Ahmed W., Chan K.H., Faris K.B., Hummels K.M., 1992, *Neural Networks*, 5:595-603.
Nagy G., 1968, *Proc. IEEE*, 56:836-860.
Nistir 4912, 1992, *The First Census Optical Character Recognition Systems Conference.*
Nowlan S.J., Hinton G.E., 1991, in "*Advances in Neural Information Processing Systems*", R.P.Lippmann, J.E.Moody, D.S.Touretzky Eds., Morgan Kaufmann, 3:774-780.
Rosenfeld A., Pfaltz J.L., 1966, *J. Assoc. Comput. Mach.*, 13:471-494.
Rumelhart D.E., Hinton G.E., Williams R.J., 1986, *Nature*, 332:533-536.
Suen C.Y., Berthod M., Mori S., 1980, *Proc. IEEE*, 68:469-485.
Takahashi H., 1991, *Proc. Int. Conf. on Document Analysis and Recognition*, France, 821-828.

SMART VISION CHIPS IN CCD AND CCD/CMOS TECHNOLOGIES

G. Soncini*, P.L. Bellutti, M. Boscardin, F. Giacomozzi,
M. Gottardi, and M. Zen

*Dipartimento Ingegneria dei Materiali - Università di Trento, Italy
IRST Microelectronics Division - Povo- Trento, Italy

ABSTRACT

IRST-VLSI technologies, i.e. the baseline CCD and the innovative CCD/CMOS fabrication processes to be used for the development of smart optical sensors and vision chips are outlined and examples of chips to be implemented by these technologies are given.

1. INTRODUCTION

IRST Microelectronics Division has been designed, equipped and staffed to be a state of the art research facility devoted to the development of smart sensors for optical and other "ionizing" radiation with integrated, analog or mixed full-custom signal preprocessing capabilities, and fabrication of special devices which require non-standard processing technologies. Core of the Microelectronic Division is the I.C./Special Devices Fabrication Facility. Designed to process 100 mm diameter Silicon wafers with minimum features of 2 micron over a 1 x 1 cm^2 chip size, it has been fitted out with highly versatile processing equipments in order to allow a continuous adjustment of the baseline technologies and the simultaneous development of innovative "custom-tailored" I.C./Special Devices processing on the same pilot-line. The facility is fully operational since January 1992. Non standard (i.e. non usually available from commercial Silicon Foundries) processing technologies have been developed and are now routinely used to produce devices at prototype level for a number of National/European sponsored research programs. Examples are:

- **CCD (Charge Coupled Devices):** specially tailored technology to develop optical imagers and ionizing radiation sensors, with analog storage and preprocessing capability.
- **CCD/CMOS:** innovative technology specially tailored for the implementation of "smart" optical sensors, capable of adding CMOS analog/mixed signal video-signal preprocessing to CCD image to charge conversion.

In this work the basic characteristics of these processes are outlined and examples of "smart optical sensors" to be implemented by IRST technologies are briefly described.

2. IRST-CCD BASELINE TECHNOLOGY

The IRST-CCD process uses three levels of n^+ doped polysilicon (the first to form the transfer gate, the second and third to form the CCD shift registers, the storage gates and the input/output circuitry) and two level of aluminium metallization (the first forms

From Neural Networks and Biomolecular Engineering to Bioelectronics
Edited by C. Nicolini, Plenum Press, New York, 1995

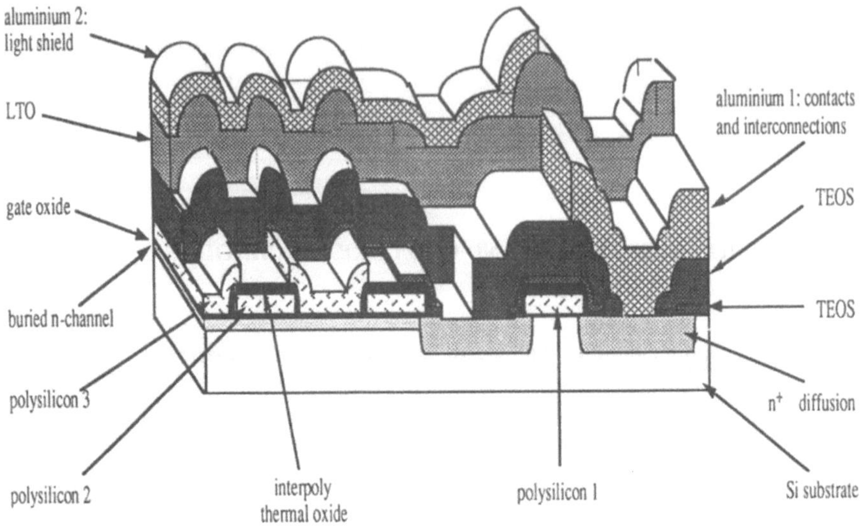

Figure 1 Schematic drawing of the IRST-CCD baseline technology. A brief outline of the main processing steps follows in table 1.

Table 1. IRST-CCD Process Outline

A) Starting material:
(100) oriented, 20 ohm-cm,
p type silicon.

B) Definition of the activ area:
pad oxide (dry oxidation);
nitride deposition;
1st photostep (active area);
nitride etching (dry etch);
local oxidation.

C) Buried channel formation:
2nd photostep (buried channel);
buried channel phosp. implant.

D) 1st Gate oxide growth and 1st polysilicon deposition:
gate oxide growth (dry oxidation);
phosphorous doped polySi dep.;
3rd photostep (first poly layer);
polySi gate oxide etch.(wet etch).

E) 2nd Gate oxide growth and 2nd polysilicon deposition:
gate oxide growth (dry oxidation);
phosphorous doped polySi dep.:
4th photostep (second poly layer);
polySi gate oxide etch.(wet etch).

F) 3rd Gate oxide growth and 3rd polysilicon deposition:
gate oxide growth (dry oxidation);
phosphorous doped polySi dep.;
5th photostep (third poly layer);
polySi gate oxide etch.(wet etch).

G) MOS source-drain and diodes formation:
oxide depostion;
6th photostep (MOS source-drain and diodes);
oxide etching (wet etch);
phosphorous deposition (liquid source);
phosphorous diffusion.

H) Contact holes:
oxide deposition;
7th photostep (contact holes, preliminar);
oxide etching (wet etch);
8th photostep (contact holes, final);
oxide etching (wet etch).

I) 1st Metallization (interconnections):
aluminium deposition (1.2 μm);
9th photostep (contacts, interconn.bond.pads);
aluminium etching (dry etch).

J) 2nd Metallization (light shield):
oxide deposition;
aluminium deposition (0.5 μm);
10th photostep (light shield);
aluminium etching (wet etch).

K) Vias:
11th photostep (vias);
passivation oxide etch. (wet etch).

L) **Back contact:**
backside aluminium metallization;
sintering.

electrical interconnects while the second one is used as a light-shield for the non optically-active areas of the circuit). A n-type buried channel is formed in the p-type silicon substrate to move away from the silicon surface the region of minimum potential where the photogenerated charge packets are collected and transferred, thus avoiding charge trapping by surface states. In addition the carrier mobility is higher since the transport is in the silicon bulk rather than at the surface. Finally these devices can be constructed to operate with larger fringing fields thus allowing operation with higher clock frequency. On the other hand, bulk channel devices involve a more complex processing and a somewhat more critical design. They also have reduced signal handling capability because of the reduced capacitance [Beynon & Lamb; Soncini et al.].

3. IRST-CCD/CMOS TECHNOLOGY

Fig. 2 explains how the CCD/CMOS process architecture has been achieved by merging the baseline n-buried channel IRST-CCD baseline technology with a conventional n-well CMOS technology. From this figure, where the main CCD and CMOS processing steps are presented, it is immediately observed that:

- active area, i.e. the chip regions where active devices are to be implemented;
- n^+ diodes, i.e. the n^+ doped regions required by diodes and by source/drain of n-channel transistors;
- metallization, for electrical interconnections and light-shield formation;
- passivation, for complete device protection;

are common to both CCD and CMOS technologies, while others, like:

- n-buried channel and gate oxide for CCD;
- n-well formation, MOS threshold/punch-through adjustment implants, gate oxide and p^+ diodes for CMOS;

are peculiar to each technology.

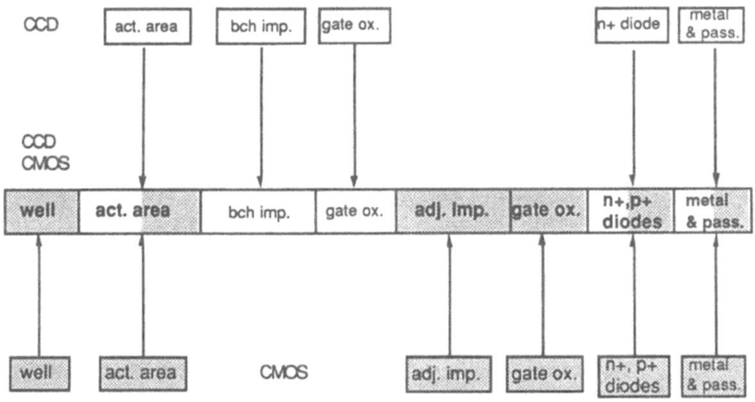

Figure 2 CCD/CMOS process architecture as a result of the merging the baseline n-buried channel CCD process and a conventional CMOS process.

The merging of CCD with CMOS to achieve a combined CCD/CMOS technology has been pursued with great care in order to ensure:

- a good electro-optical performances of the CCD section;
- an optimized analog signal processing capability of the CMOS section.

Both above mentioned requirements point out an incompatibility in terms of gate oxide thickness. Actually, in order to assure an acceptable production yield, the baseline 4 micron CCD process is characterized by a 85 nm thick gate oxide, while an optimized analog oriented 4 micron CMOS process requires a 40-50 nm thick MOS transistor gate oxide. Since the overlapped CCD electrodes configuration is obtainable with two polysilicon layers, it has been possible to attribute exclusively to the CMOS section the third polysilicon layer. The choice of the third layer has been suggested by technological constrains like "spacers" formation (these are unwanted residues of the polysilicon dry etching) and the need to proceed with threshold/punch-through adjustment implant for MOS transistors optimization. The final cross-sectional view of the CCD/CMOS combined process which results from the merging of the two distinct CCD and CMOS processes, as outlined in Fig 2, is shown in Fig 3. Table 2 lists the required photosteps while table 3 summarizes the technological parameters mean value.

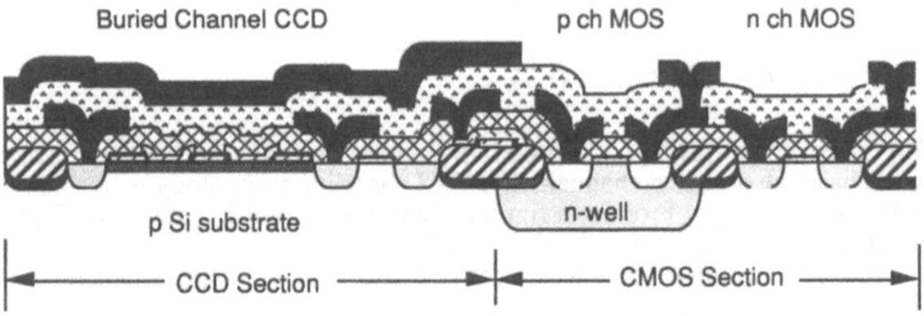

Figure 3 Cross sectional view of the basic CCD/CMOS microstructure.

Table 2 - IRST 4 μm CCD/CMOS process - Photosteps list

	CCD Section	CMOS Section
1		n-Well
2	active area	active area
3	p-channel stop	p-channel stop
4		n-channel stop
5	buried channel	
6	poly 1st CCD	
7	poly 2nd CCD	
8		n pun. thr. & thr. adj.
9		p pun. thr. & thr. adj.
10		poly 3rd CMOS
11	n^+Source/Drain	n^+Source/Drain
12		p^+Source/Drain
13	contacts: preliminary	contacts: preliminary
14	contacts: final	contacts: final
15	metal 1	metal 1
16	Via's (bonding pads)	P/I Via's
17	light shield	metal 2
18		overglass

4. SMART OPTICAL SENSORS AND VISION CHIPS: PRELIMINARY EXAMPLES

Three designs have been selected and will be developed as internal training and external demonstrators. The first design concerns a 200x100 pixel frame-transfer optical imager specifically tailored for automatic vision applications. The image and memory zone will be implemented in CCD technology, while the input/output registers will be implemented in CMOS technology. Square shaped pixels are used instead of the conventional rectangular configuration, and the images are acquired in a non-interlaced fashion. A preliminary scheme of the CCD/CMOS frame transfer optical imager is shown in Fig. 4.

Table 3 - IRST 4 μm CCD/CMOS process - Main technological parameters

p-substrate doping	$8\ 10^{14}$ cm^{-3}
n-well surface doping	$5\ 10^{15}$ cm^{-3}
field oxide thickness	1.1 μm
CCD gate oxide thickness	85 nm
CMOS gate oxide thickness	45 nm
n^+ junction depth	1.2 μm
p^+ junction depth	1 μm
n-well junction depth	4.3 μm
buried channel depth	0.9 μm
n^+ diode sheet resistance	20 Ω/sq
p^+ diode sheet resistance	76 Ω/sq
n^+ poly sheet resistance	23 Ω/sq
n-channel threshold voltage	+1 V
p-channel threshold voltage	-1 V
oxide fixed charge	$\leq 8\text{x}10^9$ cm^{-2}
surface states density	$\leq 1\text{x}10^{10}$ eV^{-1}cm^{-2}
minority carriers lifetime (bulk)	≥ 1 ms
minority carriers lifetime (n-well)	≥ 0.2 ms
CCD transfer inefficiency	$\leq 1\text{x}10^{-4}$

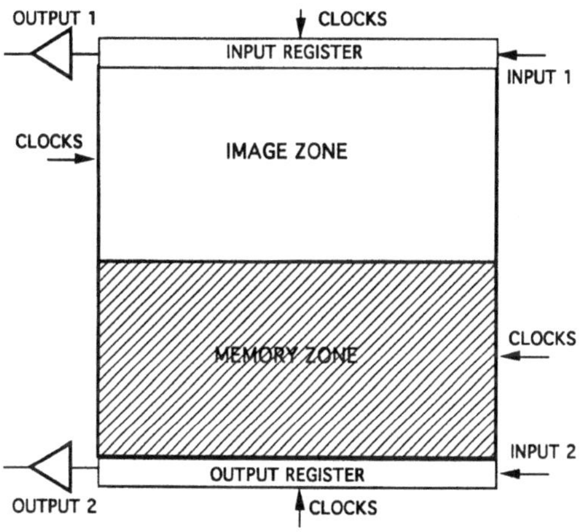

Figure 4 Schematic of the CCD/CMOS frame-transfer imager

The second design refers to a special linear optical motion sensor, which relies on a simple motion detection algorithm based on correlation between two successive images. Specifically, the motion sensor should be able to acquire and store two subsequent one-dimensional images, to shift and correlate them, and to determine the relative shift with the best correlation. A preliminary schematic of the CCD/CMOS motion sensor is shown in Fig. 5.

As third design, a CCD/CMOS camera with integrated A/D converters and digital interface has been proposed, whose layout is schematically shown in fig. 6. A preliminary version with 32 x 32 pixels image sensor array will be designed. The sensor will be a CCD interline with one charge amplifier and one A/D converter for each row. The digital value of each column of the sensor will be stored into 8 parallel CCD shift registers and transferred to the output by simply clocking them.

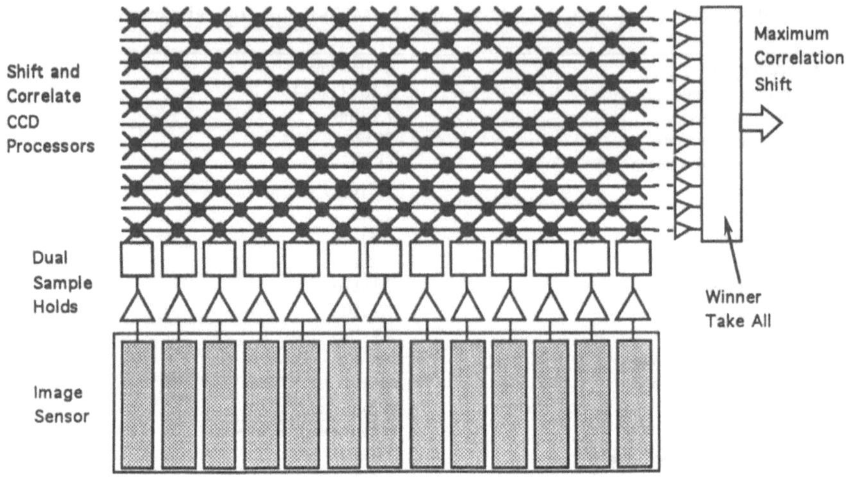

Figure 5 Schematic of the CCD/CMOS motion sensor

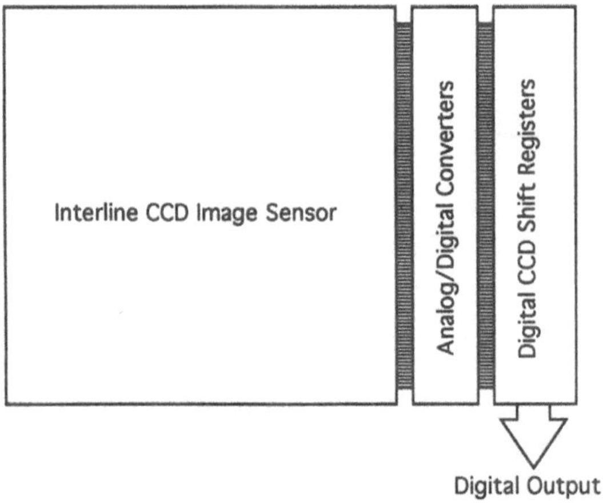

Figure 6 Schematic of the CCD/CMOS Camera with digital interface

5. CONCLUSION

An overview of the IRST baseline CCD technology and of the innovative CCD/CMOS technology, which combines Charge Coupled Devices with conventional, polysilicon-gate n- and p-channel MOS transistors, has been presented. Examples of special optical sensors to be implemented in silicon by using the preliminary 4 μm version of this innovative CCD/CMOS technology are also described.

REFERENCES

J.D.E. Beynon and D.R. Lamb "Charge Coupled Devices and their applications" Mc Graw-Hill, New York (1978)

G. Soncini et al. in "Process and Device Modelling for Microelectronics", 1993, G. Baccarani ed., Elsevier

ION SENSITIVE FIELD EFFECT TRANSISTORS: BASIC PRINCIPLES AND FABRICATION TECHNOLOGY

Giovanni Soncini*, Antonella Lavarian, Alberto Lui, Benno Margesin, Vittorio Zanini, and Mario Zen

*Dipartimento Ingegneria dei Materiali, Università di Trento, Italy
IRST Microelectronics Division, Povo, Trento, Italy

ABSTRACT: An overview of the Ion Sensitive Field Effect Transistor (ISFET) basic principles and fabrication technology, mainly with reference to devices currently under development at IRST Microelectronics Division is presented.

ISFET BASIC PRINCIPLES

Field Effect

Field effect is a well known phenomenon, largely used in microelectronics MOS type of integrated circuitry, which allows the control of surface electrical conductivity in silicon semiconductor crystals. In the most common case (a p-type thermally oxidised silicon substrate) the charge at the silicon surface is controlled by applying a suitable potential to a metal gate deposited on top of the oxide. When the applied gate to substrate potential V_{GS} is negative, positive holes are accumulated at the silicon surface thus increasing its electrical conductivity. When the applied V_{GS} potential becomes positive, holes are pushed away from the semiconductor surface, and a depleted region sets in. If the positive potential is large enough, (i.e. above a threshold voltage V_{TO} characteristic of the Metal/Oxide/Silicon Capacitor) electrons accumulate at the semiconductor surface, forming the so called "inversion layer". When inversion sets in, the depletion layer width remains constant with a further increase in the V_{GS} potential since the increased electric field decays over the insulator only, while the potential drop over the semiconductor remains at about the threshold $2\phi_F$ value, ϕ_F being the Fermi potential of the bulk semiconductor substrate. All the above statements are summarized in fig. 1, where the accumulation, depletion, inversion status of the silicon surface are schematically shown, together with the corresponding energy band diagrams. The physics of MOS Capacitors and Transistors is described in detail in a number of books: we refer the interested readers to ref. [S.M.Sze, 1981; R.S.Muller et al., 1977].

ISFET vs MOSFET

The functioning of an Ion Sensitive Field Effect Transistor (ISFET) can best be explained by comparing it to a conventional Metal Oxide Semiconductor silicon Field Effect Transistor (MOSFET). As shown in fig. 2 the ISFET is in fact a MOSFET in which the conventional metal gate is replaced by a reference electrode in an electrolytic solution with a given pH.

From Neural Networks and Biomolecular Engineering to Bioelectronics
Edited by C. Nicolini, Plenum Press, New York, 1995

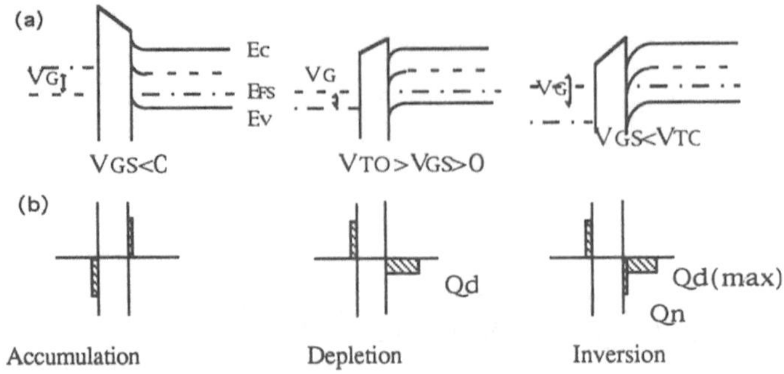

Figure 1 Schematic view of the semiconductor energy bands diagrams (a) and surface charge density (b). Q_d indicates the depletion region charge density and Q_n the inversion layer charge density.

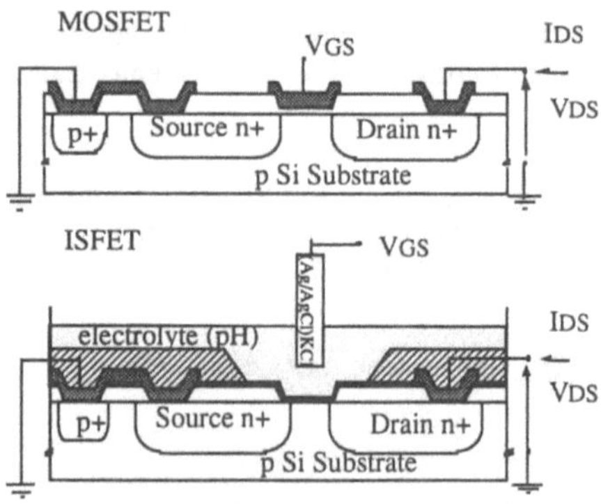

Figure 2 Schematic view of the MOSFET and ISFET devices.

Being the thermodynamic of the structure independent either on the geometry or size of any of its component, the ISFET basic functioning can be related to the electro-chemical processes taking place in each phase and at each interface. The sequence of contacting phases are as follows:

MOSFET: Metal gate/thermal oxide insulator/p-Silicon/substrate Metal
ISFET: electrode contacting Metal/reference electrode Metals/saturated KC1/electrolitic solution/pH sensitive insulator/thermal oxide insulator/p-Silicon/substrate Metal

When transforming the MOSFET into the ISFET, the interface Metal gate/insulator of the reference MOSFET has to be replaced with the ISFET corresponding interfaces. By following this statement, the conventional MOSFET threshold voltage

$$V_{TO} = \frac{\sqrt{2\varepsilon \, eN_A(2\phi_F)}}{Ci} + 2\phi_F + V_{FB} \qquad (1)$$

where:

$\sqrt{2\varepsilon \, eN_A(2\phi_F)}$	is the dopant ions charge within the depleted semiconductor region;
Ci	is the insulator capacitance per unit area;
f_F	is the Fermi potential of the semiconductor bulk;
V_{FB}	is the Flat-Band voltage characteristic of the MOS capacitor.

has to be replaced for ISFET by:

$$V^*_{TO} = E_{ref} - E_{lj} + X_e - \Psi_o(pH) + V_{TO} \tag{2}$$

where:

E_{ref}	is the potential of reference electrode, usually formed by Ag/AgCl in a KCl solution;
E_{lj}	is the liquid junction potential;
X_e	is the surface dipole potential at the insulator/electrolite interface
$\Psi_o(pH)$	is the potential difference between the insulator surface and the bulk of the electrolytic solution;
V_{TO}	is the potential contribution originating from the solid-state part of the ISFET, i.e. the same as eq. (1).

The potential difference function $\Psi_o(pH)$, which is actually the pH-dependent term in the ISFET threshold voltage eq. (2), is related to the surface proton dissociation reactions, as discussed in ref. [J.A.Davis et al., 1978; W. Smit et al., 1980; G. Massobrio et al., 1993], and can be assumed about linearly dependent on the pH of the solution. The description above makes it possible to understand the principles of chemical sensors based on field effect devices, since any chemically induced change in the terms of eq. (2) shifts the threshold voltage of the device and can be detected and measured. Referring to the conventional MOSFET operating with common source configuration in "on" condition (i.e. with the applied gate voltage above the inversion channel threshold) the drain I_{DS} current vs. gate V_{GS} and drain V_{DS} voltages characteristics are described by the following equations [S.M. Sze, 1981; R.S.Muller et al., 1977]:

$$I_{DS} = K\,[(V_{GS}-V_{TO}) - (V_{DS}/2)]\,V_{DS} \qquad \text{triode region: } V_{DS} < V_{GS} - V_{TO} \tag{3a}$$

$$I_{DS} = \frac{K}{2}(V_{GS} - V_{TO})^2 \qquad \text{saturation region } V_{DS} \geq V_{GS} - V_{TO} \tag{4a}$$

where

$$K = \mu^* \, Ci\,(W/L) \tag{5a}$$

represents the MOSFET conduction factor. This transistor key parameter depends on the inversion channel carriers mobility μ^*, on the insulator capacitance per unit area Ci and on the aspect ratio (i.e. channel width W divided by channel length L) of the device. The above equations (3a, 4 a) with the threshold voltage V_{TO} replaced by V^*_{TO} given by eq. (2) apply for the ISFET device as well. They become therefore:

$$I_{DS} = K\,[(V_{GS}-V^*_{TO}) - (V_{DS}/2)]\,V_{DS} \qquad \text{triode region: } V_{DS} < V_{GS} - V_{TO} \tag{3b}$$

$$I_{DS} = \frac{K}{2}(V_{GS} - V^*_{TO})^2 \qquad \text{saturation region } V_{DS} \geq V_{GS} - V_{TO} \tag{4b}$$

where

$$K = \mu^* \, Ci\,(W/L) \tag{5b}$$

represents the ISFET conduction factor, Ci being the "double layer insulator" capacitance per unit area.

ISFET as pH sensor

ISFET devices operate normally in the constant I_{DS} current mode with constant V_{DS}, i.e. using a feedback circuit which compensates for induced changes in I_{DS} by adjusting V_{GS} (6). Being the device polarized in the "linear part" of the triode region of its current-voltages characteristics (i.e. with $V_{DS} << V_{GS} - V_{TO}$), from eq. (3 b), written by neglecting the non linear term:

$$I_{DS} \sim K (V_{GS} - V_{TO}) V_{DS} \tag{6}$$

and from eq. (2) it may be inferred that:

$$V_{GS} = \frac{I_{DS}}{K V_{DS}} + V_{TO} = \frac{I_{DS}}{K V_{DS}} + E_{ref} + E_{lj} + X_e - \Psi_o(pH) + V_{TO} \tag{7}$$

Eq. (7), for I_{DS}, V_{DS} held constant, becomes

$$V_{GS} = \text{const} - \Psi_o(pH) \tag{8}$$

where the term "const" summarizes all the terms in eq. (7) which are unaffected by a change in the pH of the solution. By assuming the potential difference function $\Psi_o(pH)$ to be linear, any change ΔpH in the electrolite will induce a corresponding linearly related change ΔV_{GS} in the ISFET gate voltage V_{GS}. This results in fact to a first approximation in a shift of the I_{DS} vs V_{GS} characteristic as indicated in fig. 3.

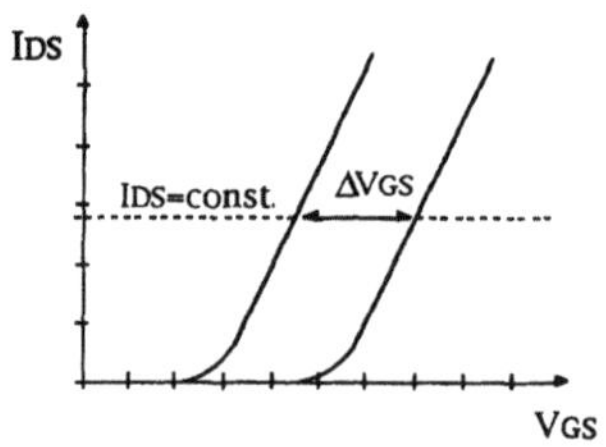

Figure 3 Shift of the ISFET gate voltage Δ V_{GS} as induced by a corresponding shift ΔpH in the pH value of the electrolyte.

It has been found that the pH-sensing properties of the ISFET greatly depend on the chemical/morphological characteristics of its insulating layer exposed to the electrolyte. This is clearly evident by data collected from the literature and summarized in fig. 4, where the pH sensitivity for the most commonly used gate insulators is represented.

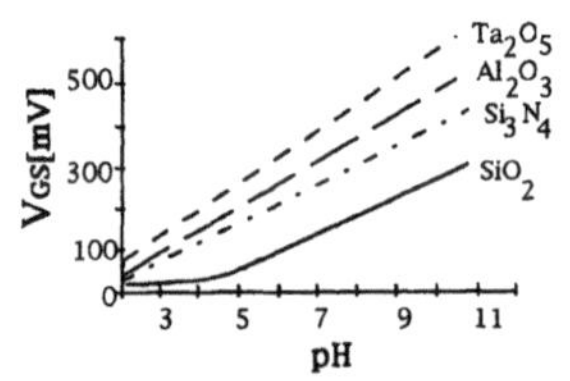

pH sensitivity:

$Si_3N_4 \approx$ 50-55 mV/pH

$Al_2O_3 \approx$ 53-57 mV/pH

$Ta_2O_5 \approx$ 55-60 mV/pH

Figure 4 pH sensitivity for different ISFET gate insulators.

From these data it may be observed that:

a) thermally grown silicon oxide (SiO_2) has the lowest, and strongly non-linear, response
b) silicon nitride (Si_3N_4) shows an almost linear pH sensitivity of the order of 50 ÷ 55 mV/pH depending on deposition technology (usually Chemical Vapor Deposition CVD)
c) aluminum oxide (Al_2O_3), also deposited by CVD techniques, shows a slightly improved pH sensitivity, of the order of 53 ÷ 57 mV/pH
d) tantalum oxide (Ta_2O_5) obtained by thermal oxidation of vacuum deposited Ta, shows about, 55 ÷ 59 mV/pH, i.e. the highest pH sensitivity.

In most of the work on pH ISFETs published so far the three gate insulators Si_3N_4, Al_2O_3, Ta_2O_5 have been used. The materials each have their specific advantages/disadvantages; in any case their sensing properties strongly depend on deposition technology, surface morphological characteristics at atomic level, thermal and electrochemical history. Other more exotic insulators, such as IrO_2, ZrO_2, Nb_2O_5 and TiO_2 have been occasionally tried [I. Lundström et al., 1991]. It is however widely recognized that more fundamental studies using surface physical-chemical characterization techniques as well as improved phenomenological modelling will be needed for a better understanding and control of these pH sensitive materials.

ISFETS FABRICATION TECHNOLOGY AND MAIN ELECTROCHEMICAL PARAMETERS

ISFET technology

The layout/cross-section of ISFET devices already developed at IRST - Microelectronics Division are shown in fig.5 and the main technological parameters characterizing the devices are summarized in the table 1, inserted at the bottom of the figure. The standard chip contains two ISFETs to be used, after appropriate encapsulation and packaging, as chemical pH sensitive transducers, while the adjacent Al-gate n-channel ALUFET transistor (together with more conventional integrated test-patterns not shown in the chip layout) is used for process monitoring and control.

The IRST-ISFET baseline technology, summarized in table 2, requires 6 photo-steps and 3 implants (p-well, guard-ring, drain-source), and allows the fabrication of good quality chemical sensors which use LP CVD-deposited silicon nitride as ion-sensing gate dielectric. Features characterizing these devices are:

1) n-channel for increased carriers mobility and device transconductance;
2) extended source/drain region to outdistance the ohmic contacts from the pH sensitive gate area: this allows an easier encapsulation of the device;
3) use of p-well with p+ guard-ring for improved device electrical isolation;
4) use of appropriate annealings to stabilize the layered-oxide/nitride dielectric and to reduce the interface charge traps density.

ISFET electrochemical main characteristics

Chemically sensitive insulators used in ISFETs are characterised by three main parameters: pH sensitivity, drift and hysteresis which are now briefly reviewed. These parameters appear to rely on the same fundamental phenomena occurring at the insulator/electrolyte interface and are therefore correlated. Their values, being influenced by the measurements procedure as well, suffer from lack of standardization, which often makes it difficult to compare quantitative results published in the scientific literature.

ISFET pH sensitivity is defined as the gate voltage variation ΔV_{GS} induced by a ΔpH change in the pH value of the electrolitic solution. It is important to observe that as pointed out in ref. [L. Bousse et al., 1990] and summarized in fig. 6, the ISFET ΔV_{GS} response following the ΔpH step variation contains a "delay" exponential term, i.e. it can be described by a first order approximation as:

$$\Delta V_{GS}(t) = (1 - \theta)\, \Delta V_{GS}(\infty) + \theta\, \Delta V_{GS}(\infty)\, [1 - \exp(-t/\tau)] \tag{9}$$

Table 1. ISFET Technological Parameters.

Gate dielectric : thermal SiO2, 100 nm / CVD Si3N4, 100 nm
Gate length (L): 25 mm
Gate width (W): 400 mm
n+source/drain junction depth: 3,5 mm
p well junction depth: 4,9 mm
p+source/drain juntion depth: 2,9 mm

Table 2. IRST-ISFET Fabrication Process Outline.

Substrate: n-type Si, (100), CZ 8 ohm.cm
Mask/Field oxide growth: T=975°; t=3h; steam; 550nm
1st photostep: p-well
1st implant: Boron; E=100KeV; D=4.5*10^{12}/cm^2 through screen oxide
p-well drive-in: T=1150°C; t=15h; dryO_2
2nd photostep: guard-ring
2nd implant: Boron E=80KeV; D=5*10^{15}/cm^2 through screen oxide
3rd photostep: n^+ Source/Drain
3rd implant: Phosphorous; E=80KeV; D=5*10^{15}/cm^2 through screen oxide
Guard-ring and n^+ Source/Drain drive-in; T=1150°C; t=1h; Nitrogen
4th photostep: gate area
gate oxide growth: T=950°C; t=4h; dry O_2; 100nm
LPCVD nitride deposition: T= 795°C; t=0.5h; 100nm
5th photostep: contacts
Al sputter deposition: 1200nm
6th photostep: Al metal
Contact sintering: T=400°C; t=5min; forming gas

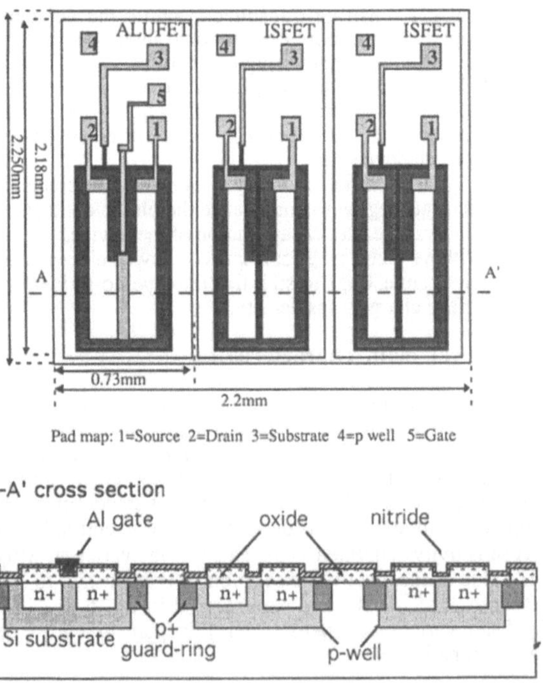

Figure 5. IRST-ISFET standard chip layout cross sectional view of the device.

being the first term of the right-hand side:

$$(1 - \theta)\,\Delta V_{GS}(\infty) \tag{10}$$

the so-called "immediate" response, while the second term represents the device "slow" response. Other terms in eq. (9) represent:

$\Delta V_{GS}(\infty)$ is the saturated gate voltage variation following the ΔpH change step. In pratical devices saturation occurs after at least 18 ÷ 24h

$(1 - \theta)$ is the "immediate" fractional response

τ is the slow response time constant.

High quality ISFET devices using silicon nitride as pH sensitive layer present $(1 - \theta) \sim 0.93 \div 0.95$ with τ of the order of 3 ÷ 4 h. This means that the immediate response, which accounts for more than 90 % of the gate voltage variation, occours almost immediately, i.e. in a time interval of less than 1 ÷ 2 min, while the slow response, i.e. the "memory" for the ΔpH change, lasts for many hours and justifies the long delay previously pointed out to reach saturation. It has been suggested that pH reactive sites (like amine and silanol) located at the surface of the insulator account for the immediate response, while protons sensitive NH groups within the nitride just below the surface (often called secondary amines) may account for the slow response of the device. ISFET drift is defined as the gate voltage V_{GS} variation rate with time at constant pH. Due to the slow response to ΔpH variation previously pointed out, this term has to be measured after an appropriate stabilization time, as shown in fig. 7, in order to minimize the memory for previous pH change and to avoid any significant overlap between the long transient response and the actual device drift. Hight quality ISFET devices using silicon nitride as pH sensitive layer show, after appropriate stabilization, drift of the order or 1 ÷ 3 mV/h. ISFET hysteresis is defined as the residual gate voltage ΔV_{GS} regression after a pH closed cycle, as shown in fig. 8. Again hysteresis appears to be an almost direct conseguence of the slow response to pH step variations, and an accurate and precise definition of time intervals and pH cycling is required to quantitatively assess this parameter.

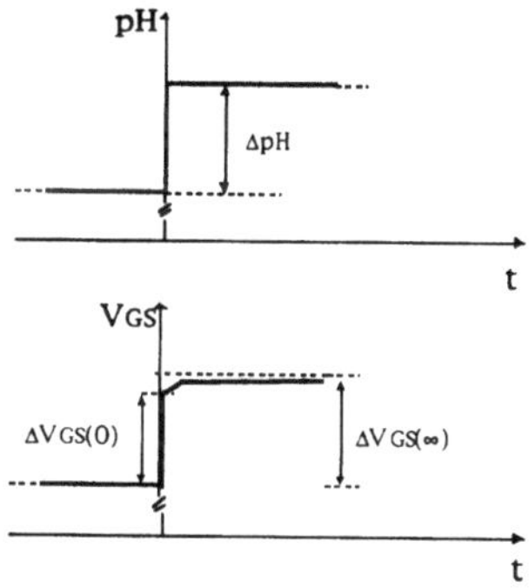

Figure 6 ISFET pH sensitivity.

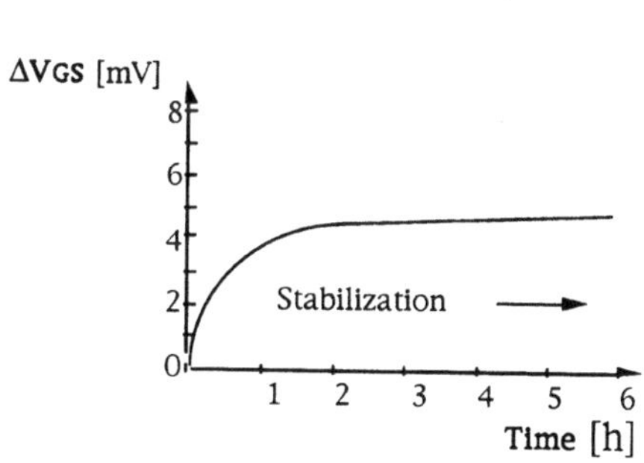

Figure 7 ISFET drift.

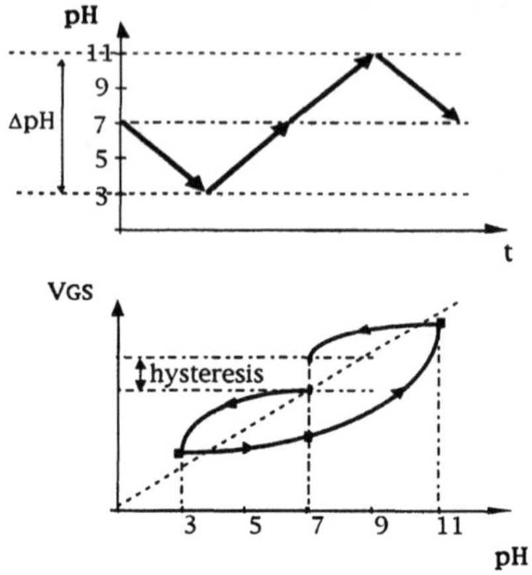

Figure 8 ISFET hysteresis.

CONCLUSIONS

An overview of ISFETs basic principles and fabrication technology, mainly with reference to devices currently under development at IRST Microelectronics Divison has been presented. It has be pointed out the key role played by the slow-response term to pH step variations in characterizing device quality and electro-chemical performance. This slow response is most probably related to NH groups (secondary amines) located within the silicon nitride insulator close enought to the surface to partecipate in the slow equilibrium exchange with protons in the electrolite. A reduced presence of hydrogen during silicon nitride deposition has been observed to have a beneficial effect on the electrochemical properties of nitride as pH sensitive material.

REFERENCES

L. Bousse et al., 1990, *Sensors and Actuators*, B1:361

J.A. Davis et al., 1978, *J. Colloid Interface Sci.*, 63:480

D.L. Harame et al., 1987, *IEEE Trans. on Electron Devices*, ED34:1700

J. Janata et al., 1980, "*Ion selective electrodes in anlytical chemistry*" Vol 2 (*H. Freiser ed.*) Plenum, New York

I. Lundström et al. in "*Sensors, a comprehensive survey*", 1991, W. Göpel et al., eds., VHC

G. Massobrio and P. Antognetti, 1993, "*Semiconductor device modelling with SPICE*" Mc Graw-Hill

R.S. Muller and T.I. Kamins, 1977, "*Device electronics for integrated circuits*" Wiley, New York

W. Smit et al., 1980, *J. Colloid Interface Sci.*, 7:1

S.M. Sze, 1981, "*Physics of semiconductor devices*" Wiley, New York

COMPUTATIONAL PROPERTIES OF RETINAL ROD PHOTORECEPTORS

G.M. Ratto*, G. Di Schino and L. Cervetto$

Istituto di Neurofisiologia CNR, Via San Zeno 51, Pisa, 56127 Italy
*Technobiochip, Marciana (LI) and
$Istituto Policattedra di Discipline Biologiche, Pisa

INTRODUCTION

Artificial neural networks

The interest for neural networks can be traced back to the early 40's with the mathematical models of neurones by McCulloch & Pitts [1943], Hebb [1949] and to continue with the work of Rosenblatt [1959] and Widrow & Posch [quoted by Lippmann, 1987], up to the more recent contributions by Hopfield [1982, 1984], Hopfield & Tank [1986], Grossberg [1986]. A renewed interest for the neural networks coincides with the introduction of new powerful optimisation methods inspired by the physics of spin-glasses [cfr. Kirkpatrick & Toulouse, 1985]; with the development of new techniques of VLSI analog implementation and with important advances in neurosciences as well. The recently renewed interest for neural networks has stimulated the convergence of goals and expertise from different fields including electrical engineering, physics, artificial intelligence and neurophysiology, generating for the first time a common language. The interaction and collaboration between groups with different cultural and technical background offers exciting perspectives in both developing new machines and getting deeper insights into the nervous system. The future applications of ideas generated by an improved knowledge of the natural neural networks rests upon the possibilities offered by a further development of microelectronics [Baccarani *et al.* 1990] and of hardware circuits possessing enough computational power to tackle optimisation problems that are not solvable by serial computations with complete polynomials. It is perhaps important to note that there are aspects of sensory functions such as vision which can be formulated as optimisation problems [Poggio *et al.* 1985].

Considering the impressive amount of biologically relevant data reaching the brain every second, it seems obvious to assume that neural information processing must occur via parallel distributed computational operations. It is true that the mechanisms of biological computation are still under something of a cloud, but there are indications suggesting that artificial networks implemented with non-linear analogic elements operating in parallel and inspired by natural neural networks can offer fairly good approximations to complex optimisation problems such as that of the travelling salesman [Hopfield & Tank, 1986, see also Bounds, 1987]. The recent wave of new computational algorithms is also accompanied by a new ferment in neurosciences, where the development of powerful techniques has produced outstanding results and generated a justified hope to quantitatively characterise the properties of neurones and of their interactions within natural networks.

From Neural Networks and Biomolecular Engineering to Bioelectronics
Edited by C. Nicolini, Plenum Press, New York, 1995

A biological neurone: the photoreceptor

Photoreceptors are the transducers feeding electrical information to the visual system. At variance with most man-made equivalent transducers (i.e. the photo multiplier tube), however, they not only convert the signal from the native form (i.e. temporally modulated light signals) to the output (synaptic activity), but they also implement the first level of computation in the visual system.

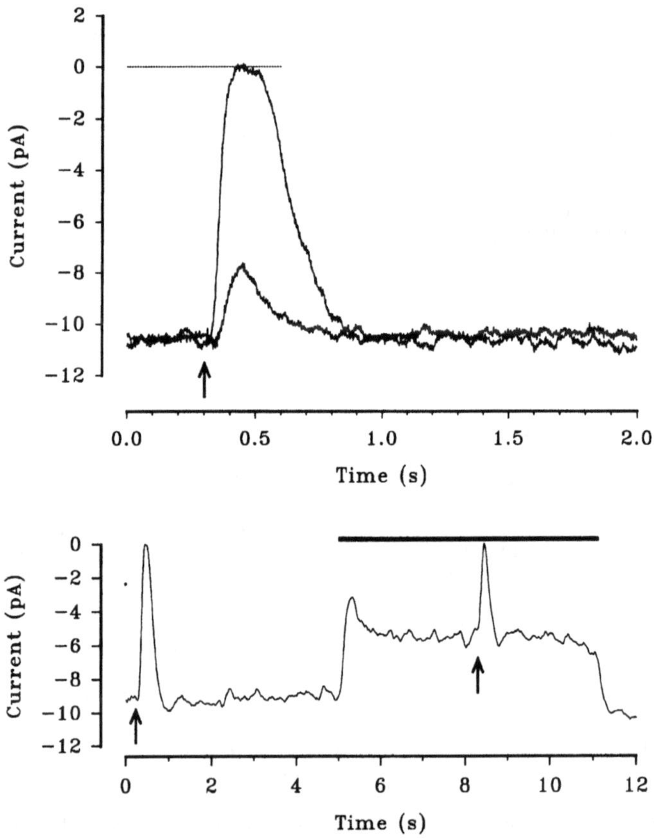

Figure 1 A mammalian photoreceptor response to stimuli of different intensities.
Upper panel: Responses to a bright (90 photons/flash) and a dim (4 Φ/flash) flashes of light of a guinea pig rod photoreceptor. The light pulse (20 ms) is delivered at the arrow.
Lower panel: Responses to a step of light of constant intensity (indicated by the thick bar) and an overimposed flash. At the onset of the light step the response is large, then it is followed by a decrease of the response. This decrease in sensitivity is characteristic of the process of light adaptation.

Photoreceptors and photo multipliers both detect single photons with high gain and low noise. Although they are physically quite different, one may identify similar constraints and strategies in the design of each. In both types of detectors, single photons absorbed by the light-sensitive material (the light-absorbing pigment, or the metallic photo cathode) elevate electrons to an excited state. In order to minimise the effect of the thermal noise which limits the detector performance, the activation energy must be high. The photo multipliers with low intrinsic noise use a cathode with a relatively high working function, which consequently requires absorption of the more energetic short wavelength photons in order to exceed the elevated work function. In a similar way, rods require photons in the blue-green region of the spectrum and are weakly excited by low energy red light.

After the incident photon has been absorbed, the signal must be amplified, and both photoreceptors and photo multipliers achieve this amplification through a cascade of amplification stages. In a photo multiplier the electrons ejected by light from the photo cathode are accelerated toward a series of cascading anodes (dynodes) and secondary electrons are ejected at each stage. The gain of each stage depends on the dynode material and on the accelerating voltage and in general this gain is fixed. In the photoreceptor the excited molecule of the visual pigment initiates a chain of chemical transformations that terminates with the inactivation of an internal messenger (cGMP) and the closure of ionic channels. This closure stops an ionic current that flows across the visual cell in the dark (dark current) and generates a voltage change (hyperpolarization), which, transmitted to the synaptic ending, sets up the light message. At each stage the light signal is amplified and, in contrast with the photo multiplier, the overall gain of this chain is regulated by the ambient light in a process called *light-adaptation*.

The efficiency of a retinal rod in converting incident photons into an electrical signal can be as high as 60%, whereas the maximum efficiency of a photo multiplier is typically 20%. The range of the light intensities over which a rod can operate is small, as the response to the absorption of a single photon is on the order of one-tenth of the maximum excursion of the response to a bright light (upper panel of figure 1). When the light level is changed more gradually, however, the process of light-adaptation extends the operating range to about five orders of magnitude by reducing the gain as the ambient illumination level rises. The ability to light adapt of retinal receptors is recognised by a partial recovery of the light-sensitive current during a prolonged exposure to steady light (see lower panel of figure 1).

Here we address the issue of how the adaptive pattern of photoreceptors might be implemented with the use of algorithms and system architecture based on parallel distributed processing. Our analysis is strongly influenced by the result of the biophysical experiments conducted in parallel with the computational approach and the network used to simulate the photoreceptor properties is trained with learning sets directly obtained from our measurements.

Computation of the photoreceptor response by using a neural network

The first level of computation in the retina is applied at the level of photoreceptors themselves. Non-linear transforms are applied to the input signal, and they determine the computational value of the photoreceptors. These nonlinearities can be roughly described as follows, discussing separately the case of short light flashes (impulsive response) and of longer stimuli of constant luminance.

Impulse response. For fast impulsive inputs (duration $\tau<150$ ms) the photoreceptor behaves roughly as a low pass filter [Baylor & Hodgkin, 1974], to the extent that the input impulse is slowed down considerably (see upper panel of figure 1). The relationship between the input and output amplitude is not linear but is modelled by a saturating exponential of the type:

$$R = 1 - e^{-I \cdot K} \tag{1}$$

where R is the response normalised to the saturating value, I is the flash intensity and K is a parameter variable from species to species and - to a lesser extent - even among different cells from the same retina. K describes the cell sensitivity: the larger is K the higher is the photoreceptor sensitivity. The lower panel of figure 2 shows examples of responses to light flashes of different intensities (continuous lines): the response amplitude increases with the stimuli intensity up to a point were complete saturation appears (trace indicated as 8). This can be well seen in the upper panel where the amplitude is plotted as a function of the intensity (continuous thick line).

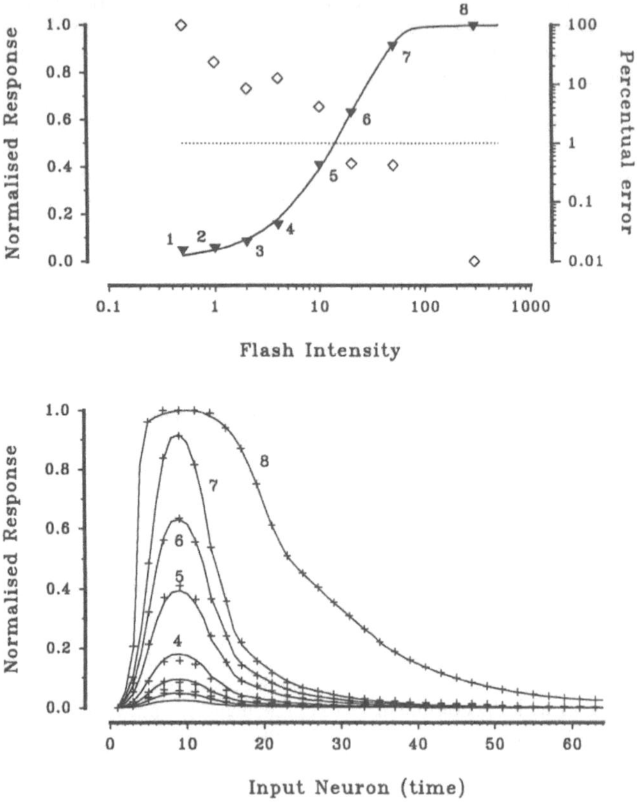

Figure 2 Network computation of a flash response.
Upper panel: Theoretical intensity-response curve for flash responses (equation 1). The flash intensity is expressed as photons/flash. The filled triangles are the amplitudes computed by the network. The empty symbols shows the percentage error for each input intensity.
Lower panel: Physiological responses (continuous curves) and responses computed by the network (crosses).

Step response. The response to a long stimulus (τ>1 sec) is shaped as a steep component followed by a relaxation to a plateau of smaller amplitude lasting throughout the duration of the light step (see lower panel of figure 1 and continuous lines in the lower panel of figure 3). After a large transient response, the photoreceptor de saturates and becomes responsive to light again, adjusting its operative range to a higher level of average luminance. This sensitivity change can be considered as an automatic gain control mechanism. The amplitude of the step response can be described by a modified saturating exponential:

$$R = 1 - e^{-m \cdot (I \cdot K)^m} \qquad (2)$$

This equation is similar to equation 1 but for the presence of the term *m*. *m* sets the gain of the photoreceptor. It is always equal to 1 for an impulsive stimulation and it gets smaller as the adaptive mechanism sets in. The smaller is *m* the smaller is the gain of the photoreceptor. The upper panel of figure 3 shows the saturating exponential for a flash response (continuous curve), for the peak of a step response (dashed line, *m*=0.95) and for the final plateau of the step response (dotted line, *m*= 0.85).

For inputs shorter than 150 ms the photoreceptor linearly adds the incoming photons and returns an impulse response with amplitude determined by the amplitude of the input signal integral, and by the exponential law. For longer duration the step-like behaviour gradually emerges. A model describing the photoreceptor behaviour must be able of correctly predicting both the step and the flash response.

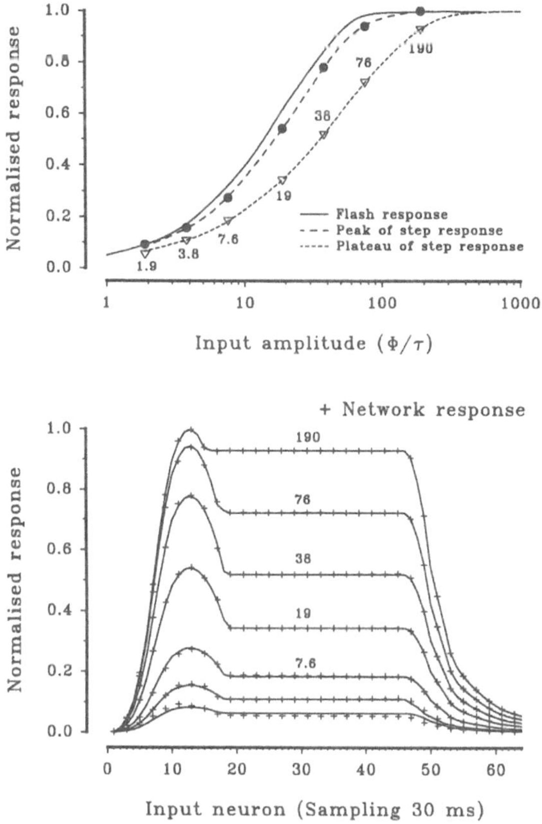

Figure 3 Network computation of a step response.
Upper panel: Theoretical intensity-response curve for flash responses (continuous curves), and to the peak and plateau of the step response (dashed and dotted lines respectively). These were obtained from equation 2 with m=1, 0.95, 0.85 respectively. The symbols represent the amplitudes computed by the network.
Lower panel: Physiological responses (continuous curves) and responses computed by the network (crosses).

In spite of a great deal of efforts there is so far no detailed quantitative knowledge of the processes underling the photoreceptors responsivity, and there is no equation set describing the response kinetic as a function of the input time course. Therefore there is no direct analytical solution to determine the photoreceptor responses to an arbitrary input. Rather than trying such an analytical method, we are following a different approach, by using a neural network to predict the input-output relationship.

The simulation We used a semilinear feed-forward net [Rumelhart, Hinton & Williams; 1986] with a back propagation learning paradigm. The learning set was obtained from light responses recorded during electro-physiological experiments. Figure 4 shows a wiring diagram of the network we employed for the first simulation. The input signal is presented at the input layer, each neurone being fed with the value of the light intensity at a temporal point. The activation function of each neurone in the network is:

$$o_j = \frac{1}{1+e^{-(a_j+\Theta_j)/\Theta_0}} \tag{3}$$

where o_j is the output of neurone j, a_j is its activation state, θ_j is the threshold and θ_0 determinies the steepness of the sigmoid.

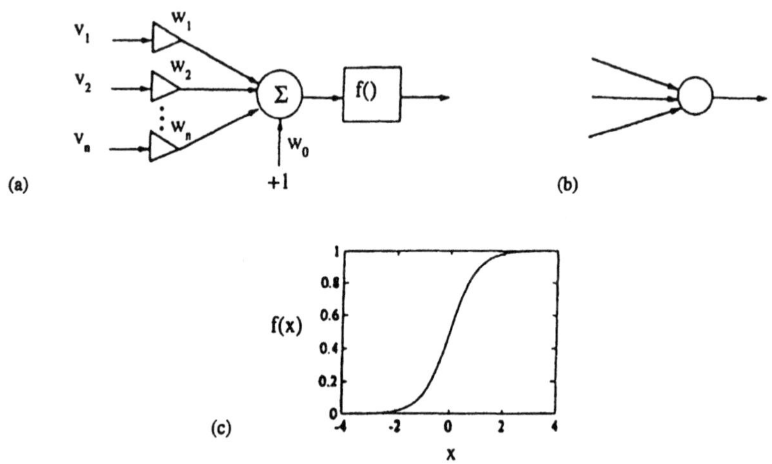

Figure 4 Schematic diagram of a network neurone.
a) the neurone sums all the inputs v_i, each one associated with a weight w_i. The result of this integration is processed by the non-linearity f (see equation 3).
b) shows how each neurone receives a number of inputs, while returning a single output.
c) shows the activation function.

The duration of the sampled sweep depends on the number of neurones and on the time resolution. This network is trained with recordings obtained from experiments performed on isolated mammalian photoreceptors. In these examples we worked with 32 neurones in the input state sampling 1.86 seconds of the input signal with a temporal resolution of 60 ms (figure 2). Fig 5 show a diagram of each neurone in the network.

In the lower panel of figure 2 is shown the learning set (continuous line) together with the network responses (crosses) obtained after the training of a net with 4 neurones in the hidden layer and 32 neurones in the input and output layers. The upper panel shows the theoretical intensity/response relationship (continuous line) together with the data obtained for the training set (filled triangles). The open symbols indicates the percentual error (scale on the right). The network successfully learned the exponential saturation after training on only a few elements of the training set.

When longer stimuli are employed the light adaptation mechanism gradually emerges, and the kinetic of the response becomes more complicated reflecting the gain adjustment. The same network has been trained to respond to longer steps of light. Figure 3 (lower panel) shows the training set (continuous lines), while the network response is given by the superimposed symbols. The data for the amplitude response characteristics to flashes and steps of light are derived from experimental data. The intensity-response curves shown in the upper panel were calculated from equation 2 using a value of $m = 0.95$ for the peak and 0.85 for the plateau measurements. The amplitudes of the network responses after training are represented with the filled circles (peak response) and empty triangles (plateau response).

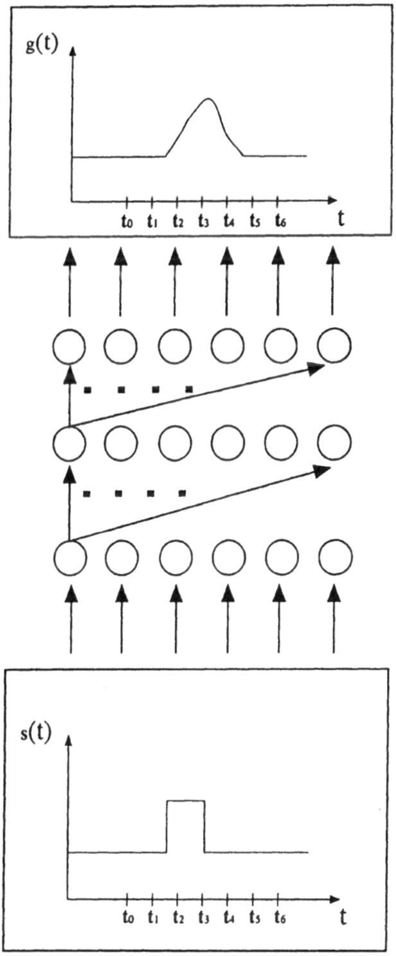

Figure 5 Diagram of the network.
The input signal (lower panel) is presented to the input layer of the network. The signal is propagated trough the hidden layer to reach finally the input layer. The output of these neurones hold the output response.

This network does not satisfy the property of temporal invariance, i.e. the response to a given stimuli must be invariant for temporal shift in the input pattern, and it must only cause a similar temporal shift in the output. As a matter of fact, because of the architecture of this network, a temporal signal is sampled and presented simultaneously to all the neurones of the input layer. Therefore stimuli differing only for their temporal location are seen by the input layer as different patterns, leading to an incorrect response. It is not practical to solve this problem by increasing the size of the learning set.

This problem can only be overcome by using a radically different architecture. Figure 6 shows the wiring diagram of the new network. This network performs computations on time-dependent signals. A temporal window of the input signal is presented to the input layer and at each clock cycle a new point enters at the left while an old point leaves the layer at the right. The layer acts as a sort of shift register. The output is present at the exit of the single neurone output.

Let s(t) be the input signal. This is sampled at the points $s(t_0)$, $s(t_1)$,...., $s(t_N)$. At time zero the first point is presented to the network, and the remaining neurones of the input layer remain silent. Now the output $g(t_0)$ depends only on $s(t_0)$. At the successive clock cycle $s(t_0)$ is shifted to neurone 2, and $s(t_1)$ enters in the network. The output $g(t_0)$ depends both on $s(t_0)$ and $s(t_1)$. After N iterations the first point $s(t_0)$ exits the network. At time t_z the output $s(t_0)$ is:

$$g(t_z) = g\{s(t_{z-N}), s(t_{z-N+1}), \ldots, s(t_z)\} \qquad (4)$$

Therefore each sampled input point gives its contribution for a time which is:

$$T = \tau N \qquad (5)$$

T being the characteristic memory of the network. Clearly the number of the input neurones, the characteristic memory **T** and the sampling period t are related. **T** is defined by the dynamic characteristic of the system under study. Let us consider a stationary input. At time zero only $s(t_0)$ will be different from zero and, at each successive iteration, one more input neurone is recruited. During this phase the input pattern is changing as each new point is added to the sequence. Therefore, the output signal is changing as well. After the characteristic time **T** is elapsed all the input neurones have the same internal state and the output will stabilise. It is easy to understand that the network response consists of a transient response occurring during the time **T** when the input neurones are activated in succession. After that, a steady-state value is finally reached. A second interpretation of **T** is found analysing the response to a very short input, lasting only one temporal unit τ. It will take N iterations in order for all input neurones to return to zero. Therefore **T** is the time when the response is non zero to a very brief flash.

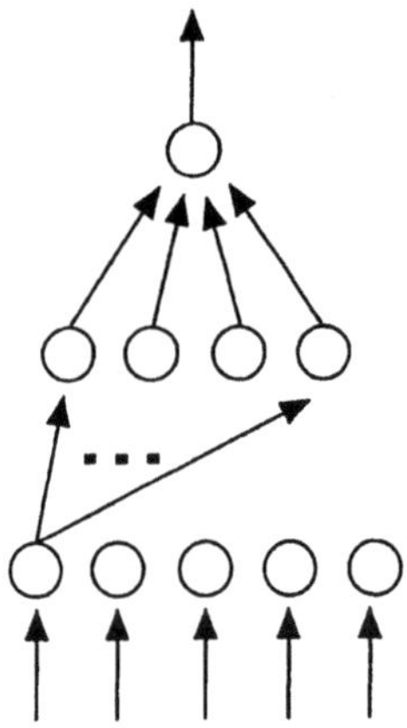

Figure 6 Diagram of the temporally invariant network.
The light stimuli is presented at the input layer, and it reaches the single output neurone through the hidden layer. The network we used had two hidden layers.

As a matter of fact the photoreceptor satisfies this condition provided the stimuli are not too bright. In this condition the response of a mammalian rod photoreceptor to a light flash extinguishes in about 500 ms. At difference with the previous network, this one can analyse signals of arbitrary length, independently from the number of neurones in the input layer. The only factors affecting the size of the input layer being the time resolution τ and the characteristic memory **T**.

Figure 7 shows some results obtained by this network. The time resolution was kept low (only 3 neurones in the input layer, two hidden layers with 5 neurones each), since we were interested just in seeing the capability of dealing with composed stimuli. The upper panel shows the network response to a light flash superimposed to a steady light. Compare this response with the lower panel of figure 1. In the lower panel the response to a single flash followed by steady illumination is shown. An additional constant stimuli is presented during part of the time.

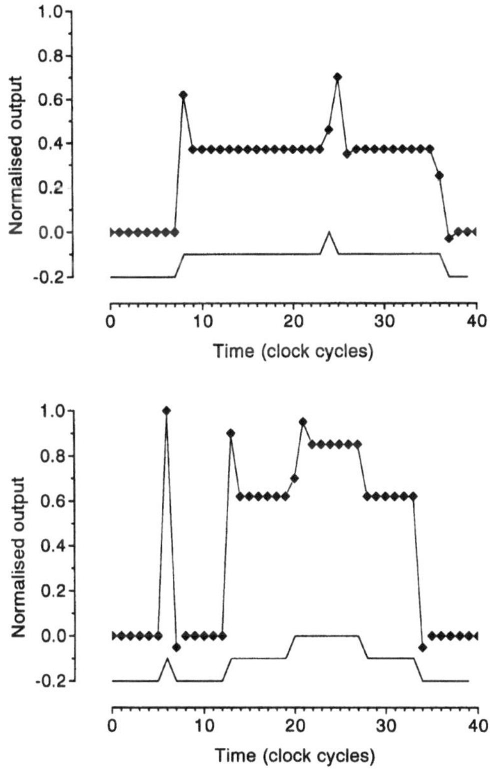

Figure 7 Responses of the temporally invariant network to overimposed stimuli. The lower trace of each panel shows the stimulation timing.

In conclusion we hope to have convincingly shown how even a simple neurone like the photoreceptor has a number of interesting computational properties, to the point that it is necessary to use a relatively complex neural network to reproduce its behaviour.

REFERENCES

Baccarani G., Guerrieri R. & Tartagni M., 1990, in *Towards the Biochip.* Nicolini ed. World Scientific, Singapore, 183-205.

Baylor D.A. & Hodgkin A.L., 1973, *J. Physiol.*, 234:163-198.

Bounds D.G., 1987, *Nature* , 329:215-219.

Grossberg S., 1986, *The Adaptive Brain.* Elsevier/North Holland, Amsterdam.

Hebb D.O., 1949, *The Organisation of Behaviour.* John Wiley and Sons, New York.

Hopfield J.J., 1982, *Proc. Natl. Acad. Sci. USA* , 79:2554-2558.

Hopfield J.J. & Tank D.W., 1986, *Science* , 233:625-633.

Kirkpatrick S. & Toulouse G.J., 1985, *J. Phys. Paris.*, 46:1277-1292.

Lippmann R.P., 1987, *IEEE Proc.*, 88:36-54.

McCulloch W.S. & Pitts W.H., 1943, *Bulletin of Mathematical Biophysics*, 5:115-133.

Poggio T., Torre V. & Koch C., 1985, *Nature*, 317:314-319.

Rosenblatt R., 1959, *Principles of Neurodynamics.*, Spartan Books, New York.

Rumelhart D.E., Hinton G.E. & Williams R.J., 1986, *Nature* , 323:533-536.

STRUCTURE-FUNCTION RELATIONSHIPS IN HUMAN GROWTH HORMONE: THEORETICAL AND GENETIC ENGINEERING STUDY

Michail P.Kirpichnikov[1,2], Andrei E.Gabrielian[2], Alexei A.Schulga[1,2], Evgeny S.Severin[3] and Konstantin G.Skryabin[1]

1- Centre "Bioengineering" Russian Academy of Sciences, Vavilov Str., 34, Moscow, 117984 Russia
2- Engelhardt Institute of Molecular Biology Russian Academy of Sciences, Vavilov Str., 32, Moscow, 117984 Russia
3- Russian Scientific Centre of Molecular Diagnostics and Treatment, Simferopolsky Blvd., 8, Moscow, 113149 Russia

The problem of searching for functional sites in protein molecules is far from solving. The straight experimental way is to isolate molecules of both receptor and ligand, co-crystallize them, obtain X-ray data of this complex, analyze it and directly find interacting sites [Burt et al. 1991]. Unfortunately in most cases this ideal could not be reached because of known difficulties in cloning, obtaining, purification, crystallization etc. of both components. This method provides full information about structures of recognition sites but not about particular biological functions of amino acids constituting these sites (e.g. binding, or signal transduction etc.). The usual way to obtain functional information is changing the structure of selected protein sites (via site-directed mutagenesis or chemical modification) followed by comparative biological testing. Step-by-step mutagenesis of the whole protein molecule remains difficult and time-consuming task. Selection of proper mutagenesis strategy makes it enormously multi-variant, although some simplistic approaches (e.g. alanine mutagenesis) could be successful. Thus it is preferential to find probable functional sites and start mutagenesis from it.

Theoretical methods of functional sites searching could be divided onto two main categories. If the protein you are interested in belongs to protein family with known (previously cloned and sequenced) members, you can perform *analysis of conservative positions* in related proteins. Usually conservative amino acid residues are either functionally, or structurally important. Hydrophobic conservative sites are frequently considered as responsible for maintaining the definite structure, characteristic for the given protein family. Hydrophilic conservative sites are mainly believed as functionally important. One could also scan the target protein using pattern libraries containing amino acids combinations (consensus sequences) responsible for known biological functions and/or spatial structures [Aitken, 1990; PATSCAN, 1990; Hodgman, 1989]. If protein of interest is not homologous to any protein known and has not evident functional sites and signatures, it is possible to analyze it using *profile analysis* methods. These methods are based on prescribing to amino acids the values of physico-chemical or statistical parameters that were found correlated with probability to be involved in certain function of parent protein [Hopp, 1993]. Plotting the value of selected parameter (usually averaged over 3-9 neighbours) *versus* amino acid number gives graph, or profile, for concrete protein. Analyzing the main extema on this profile it is possible to estimate localization of functional sites. For example, hydropathy profile analysis is based on the reasonable assumption that functional site must be exposed and for this reason should contain hydrophilic residues that tend to be on the

From Neural Networks and Biomolecular Engineering to Bioelectronics
Edited by C. Nicolini, Plenum Press, New York, 1995

surface of protein globule. Sites especially rich in hydrophilic amino acids should correspond to peaks on hydrophilicity profile. Profile analysis methods are mostly applied to searching for *sequential (linear)* functional sites, constituting of residues neighbouring in primary structure. To search for *conformational* functional sites, that are consisting of spatially close, but sequentially distant residues, more adequate is analysis of conservative positions. Both methods are supplemented each other and, if possible, should be applied together.

Every predictive method is based on the assumption that the sites responsible for some biological function must differ somehow from the sites that are not involved in this function. It is evident that success of profile analysis is caused by selection of proper parameters. One of popular characteristics is above-mentioned average hydrophilicity as a measure of site accessibility for interacting receptor (ligand) molecule. Another approach is related to information theory. Simply stated, this theory predicts that rare elements of any message contain more information, that frequent ones. In our case, we consider protein sequence as text, and could speculate that functional sites bear more information, than non-functional. If so, localization of sequential functional sites should correspond to rare (infrequent) oligomers in the structure of protein. We have developed an algorithm for information analysis of sequences and tested it on proteins with known localization of receptor- or ligand-binding sites [Gabrielian et al., 1990, 1991]. The algorithm showed high predictive ability, correctly predicting the localization of functional site in 75 % of cases. At present, there is no consensus in explanation of possible biological mechanisms of this phenomenon. Unusually high proportion of rare oligomers could decrease the probability of random, non-functional contacts of a specific ligand-receptor system with other proteins and peptides not involved in its biological activity. Thus the *uniqueness* of the site is the necessary pre-requisite for contact specificity [Gabrielian et al., 1991].

Hydrophilicity and uniqueness are two independent (non-correlated) characteristics of sequential functional sites. If protein site is characterized by high hydrophilicity and, simultaneously, by high uniqueness, the probability of this site to be involved in specific functional activity of protein is raised. We developed the new method of sequential functional sites' predictions, that shared advantages of hydrophilicity and uniqueness profile analysis: two-dimensional plot "hydrophilicity - uniqueness". Two axes set are scales of hydrophilicity and uniqueness. Protein residues could be represented as points in this two-dimensional space. For every amino acid of the protein one calculated hydrophilicity (average for 6 neighbouring residues and central, examined residue) and uniqueness (for details see [Gabrielian et al., 1991]). The relative position of point, dictated by hydrophilicity and uniqueness of amino acid, characterized its potential functional importance.

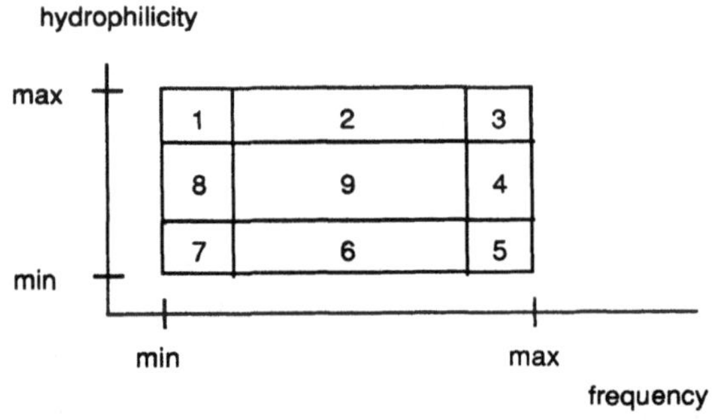

Figure 1. Principal "zone" scheme of two-dimensional plot "hydrophilicity - uniqueness".

All possible variations of hydrophilicity and uniqueness (inverted frequency) are positioned in a square limited by minimal and maximal values of these scales. This square could be dissected onto some areas (see Fig.1). Area 1 (hydrophilic and unique amino acids) included most perspective amino acids, main candidates to form functional sites. Areas 2 and 3 (hydrophilic amino acids with average and low uniqueness) are less perspective, but could include immunodominant domains. Probability of taking part in specific contact is decreased from left to right, while probability of *any* contact decreased from top to bottom. Areas 4-6 are constituted mostly by *structurally* important amino acids, responsible for maintaining of three-dimensional structure of protein. Areas 7 and 8 (average and low hydrophilicity - high uniqueness) include amino acids that could be involved in specific protein recognition, but not always in primary interaction. When analyzing short hormones, areas 7 and 8 are equivalent to area 1. And, finally, area 9 - the biggest but hardly characterizing one. It includes amino acids with average hydrophilicity and uniqueness, and interpretation of structural/functional significance is practically impossible.

We have analyzed primary structures of mammalian growth hormones and prolactins using a set of theoretical methods. Although functional topography of growth hormone family was extensively studied earlier [De Vos et al., 1992; Cunningham et al., 1989a, 1989b, 1990; Watahiki et al., 1989], some questions concerning exact localization of functional determinants remain unresolved. The characteristic features of growth hormones (GHs) are species specificity and multiple biological activity. Human GH is unique among mammalian somatotropins due to the ability to elicit both growth-promoting and lactogenic effects. These effects are initiated by binding with membrane receptors [Mathews, 1991]. Recently, using the strategy named "alanine-scanning mutagenesis" the regions responsible for binding to the prolactin and growth hormone receptors from human liver have been mapped [Cunningham et al., 1989b, 1990]. The results obtained suggest that binding determinants mainly coincide. However the coincidence of functionally important sites is unusual and seems doubtful even for homologous prolactin and growth hormone receptors. Our theoretical and experimental investigations were performed to support the hypothesis that it should be some differences between binding determinants to the different receptors. First, we have analyzed conservative residues distribution in growth hormones-prolactins superfamily. Conservative residues in primary structures of growth hormones and prolactins could be assigned to five clusters [Watahiki et al., 1989]. The hydrophilic ones (presumably functional) are located in the fourth helix and the first non-structured region. Detailed examination of conservative positions separately in growth hormones and in prolactins revealed the shift between the conservative hydrophilic clusters. It is in accordance with the fact that mammalian somatotropins and prolactins do not cross-react. At the same time, in the human growth hormone having both growth-promoting and lactogenic activity, conservative residues in specified regions correspond to both clusters. It suggests that somatogenic and lactogenic determinants in hGH molecule could overlap, but not coincide. To further investigate this hypothesis we analyzed the two- dimensional plot "hydrophilicity-uniqueness" for human and bovine growth hormones and human prolactin (Figs. 2-3). Some regions of somatotropin molecule were previously identified as responsible for biological activity of hormone. Analyzing the two- dimensional plot we could compare the theoretical results with experimental data and then classify predicted regions according either already known or probable function. The region around residues 136-147 is related to the main proteolytic site of somatotropin molecule [Yadley et al., 1973] and site of dimer formation . Site 59-65 was previously mapped as involved in binding with somatotropin [Cunningham et al., 1989a, 1989b] and prolactin [Cunningham et al., 1990] receptors, as well as fourth helix site 177 - 180 . The site around residues 31-47 correspond to the fragment whose deletion in natural variant somatotropin caused significant reduction of early insulin-like activity [Kostyo et al., 1985]. The region in the first helix (around residues 16-17 and 25) contain two histidine residues reported as essential for zinc coordination between human somatotropin and prolactin receptor [Cunningham et al., 1990]. And, finally, the region around amino acids 164-169 was the only one which function was not previously identified. This region coincides with conservative cluster of hydrophilic amino acids in fourth helix of mammalian prolactins. Based on analysis of conservative amino acids, we have supposed earlier that this site could be involved in lactogenic activity of human growth hormone. The interesting aspect of this analysis is concerned with fact that hGH is the only lactogenic somatotropin. Thus, on the plot of non-lactogenic GH the region 164-169 should not be concerned as functional. On Fig.3 it is evident that this region in bovine GH is shifted considerably to the "trivial", high frequency area of two-dimensional plot as compared to the human GH plot. Consequently, in bovine GH this region is no more predicted as functional. Taking into account both the results of experimental studies and of theoretical analysis we selected regions 164-181, 51-64 and 7-18 as potential formers of lactogenic determinant in the human growth hormone molecule.

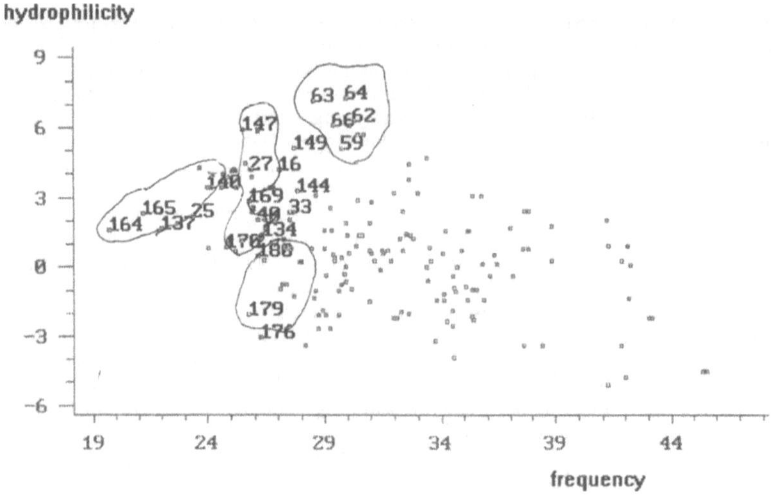

Figure 2. Two-dimensional graph "hydrophilicity - uniqueness" of human growth hormone.

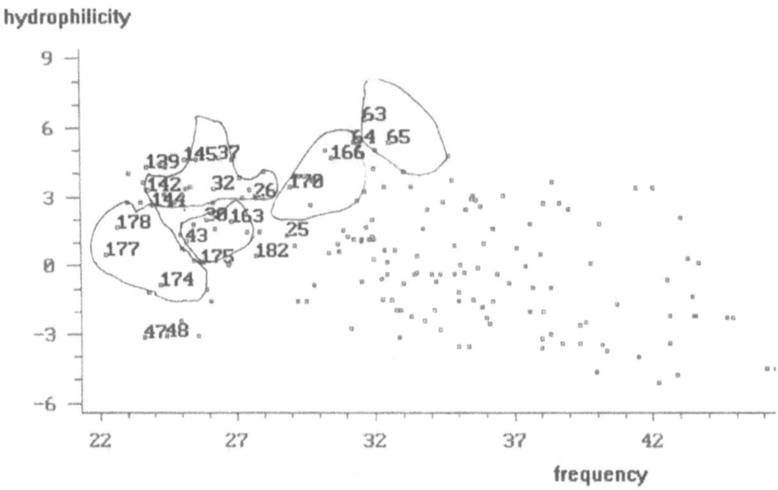

Figure 3. Two-dimensional graph "hydrophilicity - uniqueness" of bovine growth hormone.

To support experimentally this hypothesis, we used the method named "homolog-scanning mutagenesis". As was previously mentioned, porcine growth hormone in contrast to human one does not possess lactogenic activity. At the same time the primary structures of these hormones are highly homologous and their tertiary structures are similar. Consequently some mismatches in primary structures, but not in spatial structures, are causing the differences in biological functions. These regions could be mapped by substitution of hGH segments for homologous ones from porcine GH and selection of the mutants having decreased lactogenic activity. The search of the crucial mismatches was undertaken in the regions predicted by theoretical methods, and then either multiple, or single point mutations were introduced to hGH molecule. Totally six hGH analogs were produced carrying either multiple mutations: "8-18" (R8S, D10A, M13V, H17Q, R18H), "164-167" (Y164S, R167K), "164-171" (Y164S, R167K, D171H) or single point mutations S51A, S57T, R64A (Fig.5). Mutant genes were recloned to the *E.coli* expression vector. The presence of mutations was verified by direct sequencing of plasmid DNA. All the mutant hormones were effectively expressed in E.coli, and purified using "inclusion bodies" technique. To study the lactogenic activity of mutant hormones, Nb2 lymphoma assays were performed. It was shown previously that cells of this line carry homogenic population of prolactin receptors. The results of the tests indicated that at least two segments in human somatotropin molecule are involved in binding to lactogenic receptor. The significant reduction of activity was revealed in the case of hormone "164-171" with mutations in the forth α-helix of hGH (Fig.4 b, d). This variant differs from mutant hormone "164-167) by single substitution D171H. This only additional mutation totally eliminates the lactogenic activity of human somatotropin. We concluded that aspartic acid 171 is very important for exhibiting of human somatotropin lactogenic activity. The similar dramatic reduction of proliferation activity was observed in the case of analog R64A (Fig.4 d). Recently it was shown that His 17 participates in coordination of Zn atom between hGH and prolactin receptor. The alanine substitution of this amino acid residue resulted in about 50 times reduction in affinity of the hormone to the receptor. In all mammalian somatotropins as well as in the case of analog "8-18" the corresponding histidine residue is shifted to one position (His 18), that placed it on the opposite hydrophobic side of the first α-helix and not allowed to participate in Zn coordination. Our results do not agree with these data. The introduction of multiple mutations R8S, D10A, M13V, H17Q, R18H ("8-18") slightly decrease the proliferation activity of hGH (Fig.4 c). Perhaps in our case such mechanism of Zn atom coordination does not take place. Therefore two regions in hGH molecule are important for binding to the prolactin receptor: the region around R64 residue which has not regular structure and the region located in the fourth α-helix. Both these regions were successfully predicted by two-dimensional graph "hydrophilicity-uniqueness".

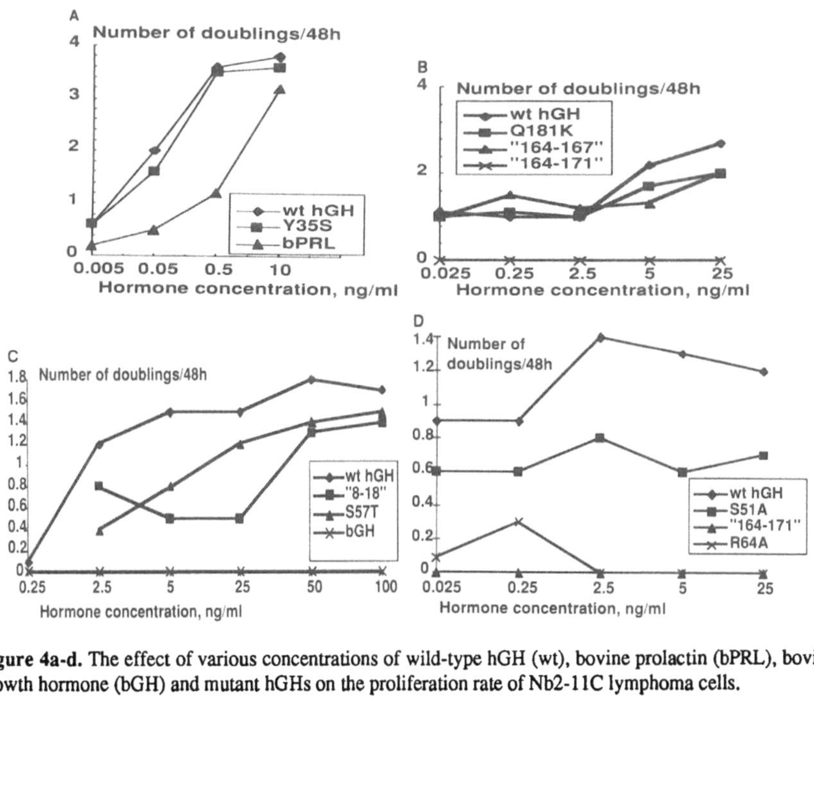

Figure 4a-d. The effect of various concentrations of wild-type hGH (wt), bovine prolactin (bPRL), bovine growth hormone (bGH) and mutant hGHs on the proliferation rate of Nb2-11C lymphoma cells.

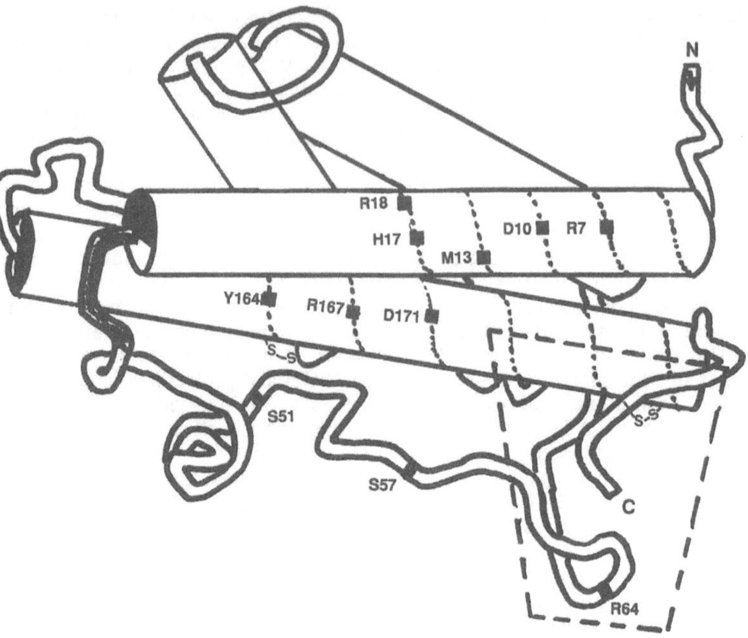

Figure 5. Structural model of hGH. Approximate locations of the mutations are designated by the black squares.

REFERENCES

Aitken A., "Identification of protein consensus sequences", 1990, *Ellis Horwood series in biochemistry and biotechnology*, Ellis Horwood Ltd., England

Burt S., Hutchins C.W., Greer J., 1991, *Curr.Opin.Struct.Biol.*,, 1:213-218

Cunningham B., Bass S., Fuh G., Wells J., 1990, *Science*, 250:1709-1712

Cunningham B., Jhurani P., Ng P., Wells J., 1989a, *Science*, 243:1330-1336

Cunningham B., Wells J., 1989b, *Science*, 244:1081-1085

De Vos A., Ultsch M., Kossiakoff A.A., 1992, *Science*, 255:306-312

Gabrielian A.E., Ivanov V.S., Kozhich A.T., 1990, *Comp.Appl. in the Biosci.*, 6:1-2

Gabrielian A.E., Nekrasov A.N., Kirpichnikov M.P., 1991, *Biomed.Science*, 2:481-484

Hodgman,T., 1989, *Comp. Appl. in the Biosci.*, 5:1-13

Hopp T.P., 1993, *Peptide Research*, 6:183-190

Kostyo J., Cameron C., Olson K., Jones A., Pai R., 1985, *Proc.Natnl.Acad.Sci.USA*, 82:4250-4253

Mathews L.S., 1991, *Trends Endocrinol. and Metabol.*, 2:176-180

PATSCAN, a new method for characterizing proteins. 1990, DNASTAR users manual. DNASTAR, Inc. Madison, Wisconsin

Watahiki M., Yamamoto M., Yamakawa M., Tanaka M., Nakashima K., 1989, *J.Biol.Chem.*, 264:312-316

Yadley R., Chrambach A., 1973, *Endocrinology*, 93:858-865

PROTEINS AND ENZYMES FROM EXTREMOPHILES: ACADEMICAL AND INDUSTRIAL PROSPECTS

Mosè Rossi

Istituto di Biochimica delle Proteine ed Enzimologia, CNR, Via Marconi 10, 80125 Napoli, Italy
and
Dipartimento di Chimica Organica e Biologica, Università di Napoli, Via Mezzocannone 16, 80134 Napoli, Italy

In the recent years the general taxonomy of living organisms has been revised and now there is a general agreement in dividing living organisms in three groups: Bacteria, Archaea and Eukarya. The new evolutionary line of Archaea is formed of three major phenotypes: halophiles, methanogens and thermophiles, each of which lives in environments that would normally kill all other known organisms [Woese G.R. et al, 1977; Stetter K.O., 1986]. The discovery of many microrganisms living optimally over 100°C, at extreme pHs, in saturated saline solutions has generated a lot of interest both academical and commercial on these organisms and on their enzymes and has prompted studies on their utilization for industrial applications.

Since the potential exploitation of enzymes for practical purposes is one of the main goal in many biotechnological processes, the use of enzymes and the availability of extremophiles as a source of enzymes stable and active at high temperature and in adverse chemical environments are important developments in this area. Thermophiles, either Bacteria or Archaea, seem to be the most interesting and promising organisms for biotechnological exploitation [Rossi M. et al, 1993; Peek K. et al, 1992].

We focused our attention on proteins that could be extracted from extreme thermophilic microorganisms, such as Archaea because:

a) they represent one of the most ancient form of life and, because of they early divergence from Bacteria and Eukarya have evolved molecular mechanisms, metabolic pathways and compounds that are peculiar to this kingdom;

b) organisms surviving and growing in such extreme environmental conditions must have macromolecules with high thermostability;

c) while thermophilic eubacteria evolved presumably from mesophilic lines by devising different ways to turn mesophilic macromolecules into more thermophilic ones, the extreme thermophilic bacteria may have invented and further transmitted an unique, most successfull principle to construct thermostable proteins allowing them to grow even in boiling water [Kandler O., 1984].

In fact enzymes isolated from extremophiles are thermophilic and thermostable and exhibit an enhanced general stability if compared with mesophilic counterpart. They are also "natural" examples of thermostable, thermophilic and solvent resistant proteins and, after determining their structure, can be used as natural models for designing and constructing proteins with new properties. In Naples we are working on several enzymatic activities and proteins isolated from extremophiles such as a β-glycosidase [Pisani F.M. et al, 1990], an alcohol dehydrogenase [Ammendola S. et al, 1992], a DNA polymerase [Pisani F.M. et al, 1992], an esterase and proteins involved in oxidoreductions reactions and protein folding such as thioredoxyns and chaperones [Guagliardi A.M. et al, 1992]. Our interest is pointed out to the study of the molecular mechanisms of protein thermostability and thermophilicity, in the protein folding at high temperature, in the production of these proteins in reasonable amount to study their properties and in the evaluation of their industrial potentiality.

Production of thermophilic biomasses.

Though it not very difficult to grow extremophiles in the gram scale using small glass fermentors, it is not an easy task to scale up at industrial level the production of biomasses. The high temperature and corrosive nature of the media used for growing many thermophiles represent engineering problems that cannot be solved at a reasonable cost. In Naples it has been designed and built with good results a new fully automatic 150 liter fermentor for high temperature and extreme pHs made with special stainless steel. But even if the fermentation could be scaled up to hundred liters, this would hardly be sufficient even for scientific use since the yield of enzymes of extremophiles is far lower than that of mesophiles. However, an increased production can be obtained by better knowledge of growth requirements, by the control of genic expression and by mutagenic tecniques.

RESULTS

Production and utilization of enzymes

Enzymes from extremophiles can be very similar to mesophiles except for their general stability to heat, to organic solvents, to detergents and to the common protein's denaturing agents. There is an explanation, at least in principle, of the molecular basis of these properties. In fact, it must be pointed out that the folded state of a protein is stabilized against unfolded state (conformational entropy) by weak interactions, such as hydrophobic, electrostatic and hydrogen bonding. For a typical enzyme ΔG of unfolding is of the order of 20-50 Kj/mole and derives from energy differences of the order of ~ 1000 Kj/mole. Since each weak interaction can contribute ~10 Kj/mole, two or three of such bonds would be enough to change the half life of an enzyme from 1 hour at 50°C to 1 hour at 85°C. Attempts to stabilize enzymes by generating such extra interactions have not been particularly successful, underlining the need for a better understanding of the stabilizing network of forces present, and the pathways of denaturation.

However the general stability of enzymes from extremophiles should confer a number of advantages in terms of industrial applications. In fact, in addition to longer lifetime, the possibility to operate with enzyme reactors at high temperature will increase substrate solubility and diffusion rates and reduce viscosity of the medium and bacterial contamination. In the case of reactors and membrane reactors it would be easier to clean the enzyme from fowling by using, for example, urea at a certain concentration (2-4 M) and acetone. Proteinases, cellulases, amylases, glycosidases, esterases, DNA polymerases, alcohol dehydrogenases, are the thermophilic enzymes more suitable for potential biotechnological exploitation. As shown in Table 1, proteinases from thermophiles are, with no doubts, more stable; in addition, since mesophilic proteins are the substrates, the high reaction temperature and the possible addition of detergents result in the denaturation of these proteins, which, as demonstrated, are more susceptible to proteolysis under these conditions. Consequently thermophilic proteases result to have higher specific activity than analogous mesophilic enzymes. Cellulases, hemicellulases and amylases are another group with several potential industrial applications since cellulose, hemicellulose and starch represent a large renewable resource from which many chemicals, including fuel and food, can be derived.

Several processes may employ thermostable cellulases, hemicellulases and amylases in the future. These include the production of liquid fuels from renewable biomass, food and fruit juice processing, production of high fructose syrup, paper manufacture and laundry detergency.

However, though enzymes from thermophiles seem well-suited for industrial applications, until now only a few of them have found application. Exceptions are the use of thermolysin in the industrial synthesis of the sweetener aspartame, and of DNA polymerase from *Termus aquaticus* in the polymerase chain reaction. This last enzyme, reasonably stable at 94°C, has permitted the automated amplification of DNA using specific oligonucleotide primers and a template DNA. The polymerase chain reaction allows the detection of one or few gene copies of viruses, and is useful to study inherited genetic disorders, and for the identification of populations and organisms in anthropological studies. Furthermore Taq polymerase has made more easy DNA sequencing procedures and this will be important in the sequencing of human genomes.

Table 1. Thermal stability of some thermophilic enzymes [Peek et al., 1992].

Enzyme	Organism	Half Life
Proteinase	*Thermus spRt41A*	13,5 h 80°C
Proteinase	*Thermococcus celer*	45' 95°C
Proteinase	*Desulfurococcus sp*	1.5 h 95°C
Proteinase	*Pyrococcus furiosus*	33 h 98°C
Endocellulase	*Clostridium thermocellum*	5 h 70°C
Endocellulase	*Acidothermus cellulolyticus*	2 h 85°C
Endocellulase	*Caldocellum saccharolyticum*	1.5 h 85°C
Exocellulase	*Thermotoga sp.*	11 h 95°C
Glycosidase	*Sulfolobus solfataricus*	2 h 95°C
Amylase	*Sulfolobus solfataricus*	> 1h 95°C
Amylase	*Desulfurococcus sp.*	> 1h 95°C
DNA polymerase	*Thermus aquaticus*	40' 95°C
DNA polymerase	*Thermococcus litoralis*	1 h 100°C
DNA polymerase	*Sulfolobus solfataricus*	30' 85°C
Esterase	*Sulfolobus acidocaldarius*	1 h 100°C
Esterase	*Bacillus acidocaldarius*	30' 85°C
Alcohol dehydrogenase	*Thermoanaerobium brokii*	2 h 85°C
Alcohol deydrogenase	*Sulfolobus solfataricus*	2 h 85°C

The reluctance to change existing processes and the difficulty to economically produce thermophilic enzymes on industrial scale are the keys reasons of the few applications. Same of these enzymes are still scientific curiosity and often it is a hard task to obtain the amount needed to try a crystallization.

The only methodology for mass production of these enzymes lies in the cloning and overexpression of their genes in mesophilic hosts, which should be generally recognized as safe (GRAS) and characterized by high biomass yield and optimized fermentation conditions.

In our laboratory, in the attempt to set up a general methodology for the production of archaebacterial enzymes, several enzymatic activities from *Sulfolobus solfataricus*, such as a β-glycosidase [Pisani F.M. et al, 1990; Cubellis V. et al, 1990; Moracci M. et al, 1992], a DNA polymerase [Pisani F.M. et al, 1992], an alcohol dehydrogenase [Ammendola S. et al, 1992] have been cloned and expressed either in *E.coli*, yeast and human cells. This methodology offers several advantages, also regarding the down stream processing of enzyme purification. The results can be summarized as follows:

a) Archaea utilize the universal genetic code and their genes are translated by Eukarya and Bacteria;
b) the expression product mantains its properties such as thermophilicity and thermostability;
c) thermophilic proteins are, in general, inactive at the host optimal growth temperature and, consequently, an accumulation of a higher quantity of enzyme may occur;
d) thermophilic proteins have a rigid structure at low temperature and are thus resistant and not degraded by the host proteases;
e) the thermostability of the expressed protein makes possible its purification from mesophilic proteins by using thermal denaturation steps, which can be easily scaled up in an industrial down stream process;

f) the degree of purity needed for industrial applications of thermophilic enzymes expressed in mesophilic hosts is lower since, after a thermal denaturation step, the only remaining enzymatic activity is the thermophilic protein being host enzymes contaminants inactivated at high temperature.

β-glycosidase from *Sulfolobus solfataricus*

I will discuss here, as example, the results obtained with the β-glycosidase (Sβ-gly) isolated from *Sulfolobus solfataricus*.. This enzyme was at first characterized as a β-galactosidase, because purified by using o-nitrophenil-β-D-galactopyranoside as substrate. However the homogenous enzyme was discovered to possess a broad specificity for β-D-gluco-, fuco- and galactosides; hydrolyzed a large number of β-linked glycoside oligomers, and had a remarkable exoglucosidase activity [Nucci et al., 1993]. This enzyme, purified at first time [Pisani et al., 1990] from strain MT-4. Sβ-gly, has a molecular mass of 60 KDa, migrates as a tetramer in gel filtration studies and has a half life of about 10 hours at 80 °C but is quite rapidly denatured at 100 °C. The *S. solfataricus* enzyme is significantly different from β-galactosidases such as the LacZ enzyme of *E. coli*; its molecular weight is much lower, it does require divalent cations for activity and it is not inhibited by sulphydryl blocking reagents. Recently a new purification procedure and substrate specificity study of the enzyme [Nucci et al., 1993] have permitted new insights into the protein properties and opened new perspectives for its utilization. Results of this study are summarized in Table 2: the enzyme is active on o-glycosides in the β-anomeric form with a wider tolerance regarding structural variations of aglycon moiety for aryl and alkyl-glucosides. All the β-linked glucose dimers were substrate of the enzyme, the preference being β 1-3 > β 1-4 > β 1-6. In addition the incubation of the enzyme with cellotetraose produces glucose and cellotriose, trace amount of cellobiose being detected only after few minutes from hydrolysis of cellotriose [Nucci et al., 1993].

These results indicate that the *S. solfataricus* enzyme is a glycosylhydrolase with remarkable exoglucosidase activity. In addition, inhibition studies suggest a common site for most of the tested substrates, as found for similar enzymes from other sources [Grogan, 1991]. In fact several substrates were strong competitive inhibitors (i.e. cellobiose on β-galactosidase activity) and a combination of two substrates gave the typical behaviour of two substrates competing for the same site [Nucci et al., 1993]. The enzyme is active over 95 °C. However the Arrhenius plot of Np-Gal hydrolysis reaction shows a break at about 65 °C, with values of activation energy of 75 Kj/mole between 35 and 65 °C and 30 Kj/mole between 65 and 95 °C. In addition the enzyme activity is enhanced in the presence of low concentrations of SDS (0.025%). In an attempt to correlate these properties with the conformation of Sβ-gly upon heating or in presence of SDS, circular dicroism measurements, were carried out. The results obtained indicate only a subtle if any temperature dependent conformational change of the protein and a more pronounced one in the presence of SDS.

The data are interpreted as indicating that the rigid structure of the thermostable and thermophilic β-glycosidase has an adverse effect on enzyme catalysis and that mobility and thus catalytic efficiency can be enhanced by slightly perturbating the enzyme structure upon heating or in the presence of moderate concentrations of SDS.

Expression in mesophilic hosts

Expression of genes coding for enzymes of thermophilic organisms in different mesophilic hosts allows the selection of appropriate hosts, in line with safety rules and requirements for industrial fermentation processes. The choice of a particular expression system depends upon a number of parameters, such as the nature of the enzyme itself, the availability of activity assays, the purification procedure, the type of purity required, the field of its possible industrial applications. Although some of these parameters can be controlled, it is often difficult to predict *a priori* which is the best expression system, but it has to be empirically determined for each different enzyme. The Sβ gly has been expressed in *E. coli* and in the yeast *Saccharmomyces cerevisiae*, and both systems have been found convenient for different applications. Expression in *E. coli* allows production of proteins essentially useful for structural and biochemical analysis (crystallization and X-ray crystallography, analysis of amino-acid modifications such as methylated lysines, enzymological studies) by simple growing in laboratory scale. Expression in yeast, besides the advantages of high biomass production, particularly with the use of fed-batch systems, is a pilotsystem for production of the enzyme for the food industry, since yeast is a GRAS organism.

Table 2. Substrate specificity [adapted from Nucci et al., 1993]

Substrate	Hydrolysis[a]	Substrate		Hydrolysis[b]
p-Nph-β-D-Glc*p*	+	cellobiose	(Glc*p* β1-4Glc)	+
p-Nph-β-D-Fuc*p*	+	b-gentiobiose	(Glc*p* β1-6Glc)	+
p-Nph-β-D-Gal*p*	+	laminaribiose	(Glc*p* β1-3Glc)	+
p-Nph-β-D-Xyl*p*	+	cellotriose	(tri(Glc*p* β1-4Glc))	+
p-Nph-β-D-Man*p*	tr	cellotetraose	(tetra(Glc*p* β1-4Glc))	+
o-Nph-β-D-Glc*p*	+	cellopentaose	(penta(Glc*p* β1-4Glc))	+
o-Nph-β-D-Gal*p*	+	lactose	(Gal*p* β1-4Glc)	+
p-Nph-β-D-Cell[b]	+	maltose	(Glc*p* α1-4Glc)	-
p-Nph-α-L-Ara	tr			
o-Nph-β-D-Gal*p*-6-P	tr			
p-Nph-β-D-GlcU	-			
p-Nph-β-D-GalU	-			
p-Nph-α-L-Fuc*p*	-			
p-Nph-β-L-Fuc*p*	-			
p-Nph-α-D-Gal*p*	-			
p-Nph-α-D-Glc*p*	-			
p-Nph-β-D-GlcNAc	-			
p-Nph-β-D-GalNAc	-			

[a] Hydrolysis products were detected as glucose (for oligosaccharides) and as *o*- or *p*-nitrophenol (for arylglycosides) as described in Materials and Methods.
Products were present (+), not detectable (-), or present only in traces (tr).

[b] The hydrolysis of this substrate was detected either as glucose or as *p*-nitrophenol production.

Two genomic libraries from *S. solfataricus* in λGT_{11} and $\lambda EMBL_3$ vectors were screened either with antibodies and oligonucleotide probes and a DNA fragment (pD 22) containing the gene coding for the enzyme and its 5' and 3' flanking regions was isolated and sequenced. Starting with ATG at position 229 an aminoacid sequence is encoded which is identical to that obtained from the purified βS-gly suggesting that the first residue of the mature protein is also the first translated aminoacid [Nucci R. et al, 1993]. The open reading frame between position 230 and position 1696 encodes for a protein of 489 aminoacids with a predicted molecular weight 56650 Da. The molecular weight and the aminoacid composition determined directly from the purified protein and inferred from the nucleotide sequence are in good agreement the aminoacids immediately preciding the termination codon TTTTCTTTT correspond to the carboxyterminal residues of βS-gly determined by carboxypeptidase hydrolysis.

The expression in *E. coli* was achieved by subcloning the 3.0 Kb Xba1 fragment in pEMBL18 plasmid in the two possible orientations that originates two different construct called pD 22 and pD 23. In the overnight culture of a β-gal minus *E. coli* strain (JM 109) transformed with these plasmids a β-galactosidase activity was found using an enzymatic assay at high temperature. Remarkably the activity was found regardless of DNA orientation in the vectors suggesting that the expression of this archaebacterial gene was not initiated from a vector promoter but from its own regulatory sequences and the activity was not induced by IPTG or lactose. However to obtain higher expression the Sβ-gly gene was cloned in the plasmid p T 7-7 after removal of the upstream non coding sequence. The *E. coli* cells having the new vector showed the Sβ-gly activity increased by more than 30 fold over that of *S. solfataricus* [Unpublished results]. The gene coding for this β-glycosidase shows a good homology with enzymes (BGA family) including β-glucosidases, β-galactosidases and phospho-β-glucosidases both from eubacteria and eukarya [Grabnitz F. et al, 1991]. This family includes the human enzyme lactase/phlorizin hydrolase that is required for utilization of lactose in milk.

The enzymes of BGA family show a relevant sequence similarity with enzymes of cellulose family A, including microbial enzymes with β 1-4 glucanase activity which are involved in the degradation of celulose. As shown in Fig. 1 the thermostability and thermophilicity of native and recombinant enzyme are similar, as are similar kinetic characteristics, optimal pH, isoelectric point and molecular weight.

A 20 fold purification (Table 3) of the expressed enzyme with about 100% yield was obtained from the recombinant *E. coli* strain by an heating step performed on a crude cell extract followed by an isoelectric precipitation [Moracci M. et al, 1990]. In contrast only 10 fold purification with about 45% of recovery of the initial enzyme activity was obtained from *S. solfataricus* homogenate using a DEAE column and a pH precipitation of protein contaminants [Pisani F.M. et al, 1991].

The expression of the Sβ-gly gene in yeast was obtained by using two different inducible promoters [Moracci et al., 1992], the classical galactose inducible promoter (UAS gal) and a new heat-inducible promoter constructed by using a repetitive sequence from the nematode *Caenorhabditis elegans* (poly-HSE). The poly HSE induction is obtained rapidly (45 min) just by temperature shift without particular growth medium, while the induction of the GAL-promoters occurs in media with galactose as sole carbon source.

Table 3.

	ml	Protein mg	U* tot	U*/mg	U° tot	U°/mg	Yield %	Purific. Fold
Crude extract	45	305	1,8	0,006	-	-	-	-
75 °C Supernatant	33	23	2,0	0,09	32	1,4	100	15
pH 3.0 pellet	8	16	1,9	0,12	31	1,95	95	20

U* = units at 40 °C; U° = units at 75 °C.

Furthermore the induction by poly-HSE sequence is very rapid and efficient since, after 45 min of induction, the level of protein that are comparable with those obtained with GAL-promoter after a 24 hours induction. Transformed yeast colony were screened very efficiently by using the synthetic substrate of the enzyme X-GAL at 37 °C or 65 °C. Colonies producing β-glycosidase appeared as a dark blue colonies on plates incubates at 65 °C.

The enzyme was purified as described before and also in this case its characterization indicates that the enzyme has properties similar to those of the native enzyme. This system allows the possibility of obtaining mutants of the protein by random mutagenesis on the transformed *E. coli* or yeast or on the plasmids. Studies are in progress on the attempt to modify, for example, the thermophilicity of the enzyme without impairing stability to heat and other agents. For this purpose the expression of the *S. solfataricus* in yeast is a convenient model system because it lacks endogenous activity and production of heterologous enzyme can be assayed in individual colonies, allowing screening for mutant affecting thermostability and thermophilicity of the enzyme.

Crystallographic studies

Since the properties of proteins from extremophiles cannot be attributed to a common factor but are rather the consequence of a variety of cooperative stabilizing effects, much more structural work is needed.

Until now the primary structure of only about a dozen proteins from extremophiles have been analyzed, only a few extremophilic proteins have been crystallized and only one structure reported [Day M.W. et al, 1992]. The resolution of tridimensional structures are needed which might allow insight into the construction principle of these proteins. On this line by using native S-β-gly over 50 different crystallization conditions were explored using microbatch tecniques, with a protein at a concentration of 10 mg/ml [Pearl L.H. et al, 1993]. The conditions of crystallization refined by using hanging drops give a protocol in which large bipyramidal crystals grow in 3-4 days from 14% polyethylene glycol 4000, 170 mM ammonium acetate and 100 mM acetate buffer pH 4.6. The protein crystallizes in $P3_121$ or $P3_221$ with unit cell dimension of a=b=169.4 Å and c=98.3 Å. Calculations based on standard values of protein specific volumes suggested that the crystals either contained a dimer of the protein in the asymmetric unit with a solvent content higher than 60% (v/v) or a tetramer (the biologically observed form) with a solvent content lower than 30% (v/v). However the value of the density of 1.28 g/cm^3, measured in a bromobenzene/xylene gradient fits well with the value of 1.26 g/cm^3 for a tetramer in the asymmetric unit and with a solvent content of 27% (v/v) which is one of the lowest observed for a protein crystal.

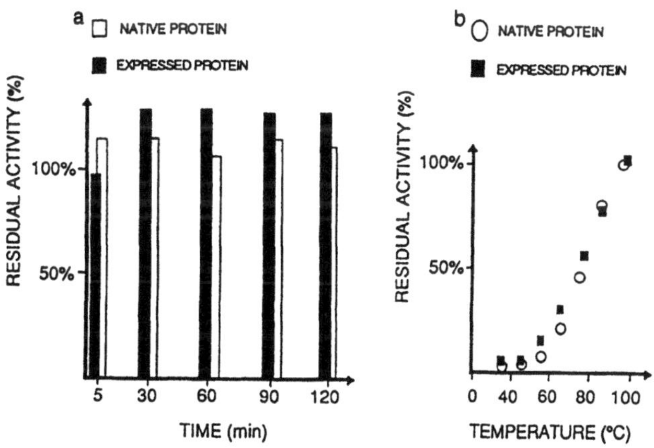

Figure 1. Thermostability and thermophilicity of native and expressed β-galactosidase
a. Thermal stability at 75°C
b. Dependence of the activity on temperature [Pisani et al., 1990]

Crystals are reproducible in size, quality and yield, withstand X-ray exposure remarkably well and are very robust to handling and heavy soaking. Useful diffraction to 2.5 Å resolution has been observed for native crystals using Synchroton radiation and 2.8 Å resolution can be obtained on laboratory X-ray sources. Since Sβ-gly is the first member of the family to be crystallized work is in progress to solve the structure of the enzyme with multiple isomorphous replacements with heavy atoms.

REFERENCES

Ammendola S., Raia C.A., Caruso C., Camardella L., D' Auria S., De Rosa M. and Rossi M., 1992, *Biochemistry*, 31:12514-12523

Colbear T., Daniel R.M. and Morgan H.W., 1992, in "*Advances in Biochemical Engineering/Biotechnology*", Fiechter Ed., Springer Verlag, Berlin, 45:57-98

Cowan D.H., 1992, in "*The Archaebacteria: Biochemistry and Biotechnology*", Bioch. Soc. Symp. n. 5

Cubellis V., Rozzo C., Montecucchi P. and Rossi M., 1990, *Gene*, 94: 89-94

Day M.W., HSU B.T., Joshua-Tor L., Park J.B., Zhou Z.H., Adams M.W.W., Rees D.C., 1992, *Protein Science*, 1:1494-1507

Grabnitz F., Seiss M., Ruknagel K.P. and Staudenbauer W.L., 1991, *Eur. J. Biochem.*, 200:301-309

Grogan D.W., 1991, *Appl. Environ. Microbiol.*, 57:1644-1649

Guagliardi A.M., Cerchia L., De Rosa M., Rossi M. and Bartolucci S., 1992, *FEBS Letters*, 1:27-30

Kandler O., 1984, *Archaeabacteria: biotechnological Implication*, Proc. 3rd Eur. Congr. on Biotechnology, Mucnhen, IV:551-560

Moracci M., De Rosa M., Rossi M., 1990, in "*Separation for Biotechnology*" D.L. Pyle Ed., Sci. by Elsevier Applied Science, 557-582

Moracci M., La Volpe A., Pulitzer J.F., Rossi M. and Ciaramella M., 1992, *J. of Bact.*, 174:873-882

Nucci R., Moracci M., Vaccaro C., Vespa N. and Rossi M., 1993, *Biotechnol. Appl. Bioch.*, 17:239-250

Pearl L.H., Hemmings A.M., Nucci R., Rossi M., 1993, J. *Mol. Biol.*, 229:558-560.

Peek K., Ruttersmith L.D., Daniel R.M., Morgan H.V. and Berquist P.L., 1992, *BFE* 466-470

Pisani F.M., De Martino C. and Rossi M., 1992, *Nucleic Acid Res.*, 20:2711

Pisani F.M., Rella R., Raia C.A., Rozzo C., Nucci R., Gambacorta A., De Rosa M. and Rossi M., 1990, *Europ. J. Bioch.*, 187:321-328

Rossi M. and De Rosa M., 1993, *Extremophiles in Biotechnology*, 6th Europ. Congr. Biotechn., Florence (In press)

Rossi M., 1992, Int. *Conference: Science and Thermophiles*, Reykjavik, Iceland, 11-12

Stetter K.O., 1986, in "*Thermophiles: General, Molecular and Applied Microbiology*", Brock T.D. Ed., pp. 33-74, J. Wiley and Sons, New York

Woese G.R., Fox G.E., 1977, *Proc. Natl. Acad. Sci. USA* , 74:5088-5090

ELECTRON TRANSFER FROM BACTERIAL DEHYDROGENASES

J.A. Duine

Department of Microbiology & Enzymology
Delft University of Technology
Julianalaan 67, 2628 BC Delft
The Netherlands

INTRODUCTION

In the canonical scheme of the respiratory chain, a number of flavoprotein dehydrogenases, embedded in the membrane, transfer their electrons to the chain at the level of ubiquinone or cytochrome b. Overviews on the situation in bacteria also show such a picture, as exemplified for *Escherichia coli* [Spiro and Guest, 1991]. However, additional pathways exist in these organisms: some dehydrogenases are loosely attached to the outside of the cytoplasmic membrane to different sites of the chain, either directly or via small blue copper proteins or cytochromes. Since these electron carriers are soluble redox proteins, electron transfer can be studied of systems consisting of natural components. An additional factor explaining the popularity of these systems is that, in a number of cases, the gene has been cloned and the 3-dimensional structure is known, enabling modification of sites presumed to be essential in the pathways of electron transfer or for interaction between the components. Needless to say that this approach will increase knowledge about long range electron transfer mechanisms, important not only to understand the functioning of the respiratory chain but also to develop artificial systems in which the electron flow from the cofactor in a dehydrogenase is directed towards an artificial electron acceptor. At a practical level, basic knowledge about these systems is required in the attempts to construct amperometric biosensors or electrode devices for electrochemical cofactor regeneration in biocatalytic reactors. For this development, the availability of a large reservoir of diverse dehydrogenases and mediating redox proteins is a very important aspect. Therefore, the overview presented here focuses on our work on the different types of dehydrogenases, and soluble blue copper proteins and cytochromes showing good electron acceptor activity for them *in vitro*.

DIFFERENT TYPES OF DEHYDROGENASES

Flavoproteins

Flavoprotein dehydrogenases have either FMN or FAD as cofactor. Normally the cofactor firmly sticks to the protein and in some cases it is even covalently attached. The mode of attachment varies not only with respect to the nature of the amino acid residue involved but also to the site of the flavin ring structure [Singer and McIntire, 1984]. Besides flavin, the enzyme may also contain other redox cofactors, for instance a haem, as illustrated by yeast L-lactate dehydrogenase (EC 1.1.2.3), containing FMN and haem b, and p-cresol dehydrogenase (EC 1.17.99.1), containing a flavoprotein as well as a haem c-containing subunit [Hopper and Taylor, 1977].

Table 1. Microbial amine oxidoreductases.

Enzyme	Cofactor	Source	References
Methylamine dehydrogenase	TTQ	Gram-negative methylotrophic bacteria	1.4.99.3
Aromatic amine dehydrogenase	TTQ	*Pseudomonas* sp.	1.4.99.4
Methylamine dehydrogenase	haem c/quinone	Methylotrophic bacteria	
Methylamine oxidase	TTQ/Cu^{2+}	Methylotrophic bacteria	1.4.3.6
Aromatic amine oxidases	TPQ/Cu^{2+}	*Escherichia coli*	
Lysyl oxidase	$quinone?/Cu^{2+}$	yeast	1.4.3.13
Amine oxidase	FAD	fungi/bacteria	1.4.3.4

Quinoproteins

Quinoproteins are enzymes containing the quinone cofactor prroloquinoline quinone (PQQ), tryptophanyl tryptophanquinone (TTQ) or topaquinone (TPQ) (Fig. 1) [Duine, 1991]. The latter two are integrated in the protein chain and they are in fact modified aromatic amino acids of that chain. TPQ has been found in copper-containing amine oxidases (EC 1.4.3.6), enzymes present from bacteria to man (Table 1). PQQ and TTQ only occur in bacterial dehydrogenases, of which TTQ exclusively in amine dehydrogenases (EC 1.4.99.3). PQQ is the cofactor of several different types of alcohol and sugar dehydrogenases (Tables 2 and 3, respectively). Combinations with other prosthetic groups also occur, e.g. quinohaemoprotein ethanol dehydrogenase containing haem c besides PQQ, either in the same protein chain (type I) [Groen et al., 1986] or in a special subunit (type II) [Ameyama and Adachi, 1982].

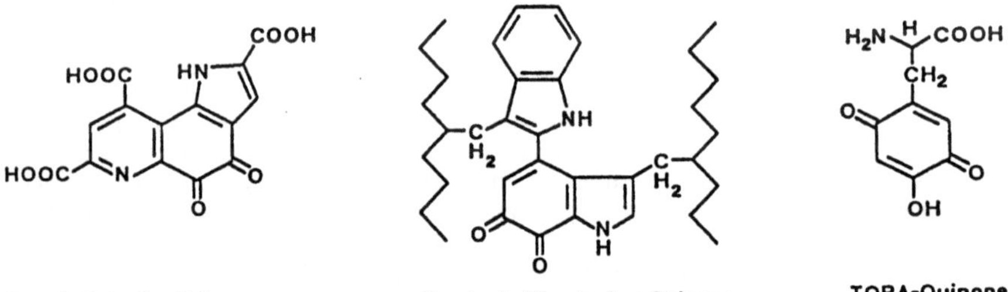

PyrroloQuinolineQuinone (PQQ)

TryptophylTryptophanQuinone (TTQ)

TOPA-Quinone (TPQ)

Figure 1. The quinoid cofactors.

Molybdoproteins

Molybdoproteins contain a whole array of cofactors: Fe-S clusters, FAD (some molydoproteins do not contain the flavin), and molybdenum attached to a pterin dinucleotide cofactor via S-bridges [Meyer et al., 1993] (also "tungstoproteins" exist like carboxylate reductase (EC 1.2.99.6) [White et al., 1989] and aldehyde dehydrogenase [Mukund and Adams, 1991]]. Several molybdoprotein dehydrogenases have been found in bacteria: carbon-monoxide dehydrogenase (EC 1.2.3.10), aldehyde dehydrogenase [Poels et al.,

1987; van Ophem et al., 1993], xanthine dehydrogenase, quinoline dehydrogenase, nicotine dehydrogenase (EC 1.5.99.4), and formylmethanofuran dehydrogenase (EC 1.2.99.5).

Table 2. Bacterial alcohol dehydrogenases (dye-linked)

Enzyme	Cofactor	Source	References
Quinoproteins			
Methanol dehydrogenase	PQQ	methylotrophic bacteria	1.1.99.8
Ethanol dehydrogenase	PQQ	*Pseudomonas aeruginosa*	Schrover et al., 1993
Ethanol dehydrogenase (quinohaemoprotein, I)	PQQ, haem c	*Comamonas testosteroni*	Groen et al., 1986
Ethanol dehydrogenase (quinohaemoprotein, II)	PQQ, cytochrome c	Acetic acid bacteria	Ameyama and Adachi, 1982
Polyvinylalcohol dehydrogenase	PQQ, haem c	*Pseudomonas* spec.	1.1.99.23
Polyethyleneglycol dehydrogenase	Quinone	*Flavobacterium* spec.	1.1.99.20
Nicotinoproteins			
Formaldehyde dismutase	NAD	*Pseudomonas* sp.	1.2.99.4
Methanol/formaldehyde oxidoreductase	NADP	*Amycolatopsis methanolica*	Bystrykh et al., 1993
p-nitrosodimethyl aniline/ alcohol dehydrogenase	NAD	*Amycolatopsis methanolica*	van Ophem et al., 1993

Nicotinoproteins

Nicotinoproteins contain either firmly bound NAD or NADP [van Ophem et al., 1993; Bystrykh et al., 1993]. In contrast to normal NAD(P)-dependent dehydrogenases, the NAD(P)H formed after oxidation of the substrate is not released but remains sticked to the protein. Free NAD does not exchange with it and transfer of reduction equivalents occurs neither. From this it is clear that nicotinoproteins cannot be detected by reduction of added NAD but require a special assay in which p-nitroso-N,N-dimethylaniline is used as artificial electron acceptor. Regeneration of the bound NAD(P)H *in vivo* occurs in a variety of ways: in a homocatalytic reaction (e.g. formaldehyde dismutase (EC 1.2.99.4), oxidizing formaldehyde to formate after which the next molecule is reduced to methanol, in a heterocatalytic reaction (e.g. glucose/fructose oxidoreductase, coupling oxidation of glucose to gluconic acid with reduction of fructose to sorbitol [Zachariou and Scopes, 1986]), via the respiratory chain (most probably in the case of a number of alcohol dehydrogenases (vide infra)).

Table 3. Sugar oxidoreductases oxidizing (at the C_1-position).

Enzyme	Cofactor	Source	References
Quinoproteins			
Glucose dehydrogenases (soluble type)	PQQ	*Acinetobacter calcoaceticus*	1.1.99.17
Glucose dehydrogenase (membrane-bound type)	PQQ	Gram-negative bacteria	Duine, 1991
Nicotinoprotein			
Glucose/fructose oxidoreductase	NADP		Zachariou and Scopes, 1986
Flavoprotein			
Glucose dehydrogenase	FAD	*Aspergillus oryzae*	1.1.99.10
Cellobiose dehydrogenase	FAD	fungi	1.1.5.1
Glucose oxidase	FAD	fungi	1.1.3.4
Hexose oxidase	Cu^{2+}?	seaweed	1.1.3.5

COUPLING TO THE RESPIRATORY CHAIN

Direct

The classes of dehydrogenases discussed have been found to couple at nearly all redox levels to the respiratory chain. Contrary to expectation, this is not strictly related to the redox potential of the free cofactor (compare PQQ/PQQH2(+90 mV) with NAD/NADH (-320 mV)). Apparently, the protein in which the cofactor is embedded modifies the potential to a level compatible with that of the natural electron acceptor. Some examples of the diversity will be presented in the following.

NADH-Dehydrogenase. Indications exist [N. Govorukhina, L. Bystrykh, L. Dijkhuizen and J.A. Duine, unpublished results] for a multi-enzyme complex in which at least an NADH-dehydrogenase and the nicotinoprotein "methanol/formaldehyde oxidoreductase" play a role in dehydrogenation of alcohols. Although conclusive evidence is still lacking, the complex could form part of the respiratory chain and chanelling of reduction equivalents could occur in this since oxidation of alcohol via the complex can occur only in the presence of DCPIP or a special tetrazolium dye [van Ophem et al., 1991].

Ubiquinone. Membrane-bound, PQQ-containing glucose dehydrogenase, occurring in many Gram-negative bacteria, including *Escherichia coli*, transfers its electrons to ubiquinone [Matsushita et al., 1989].

Quinohaemoprotein ethanol dehydrogenase (type II) is coupled to cytochrome o in Acetic acid bacteria [Matsushita et al., 1990]. This forms the shortest respiratory chain possible, perhaps related to the task required in this organism when producing acetic acid (vinegar): generating a very efficient proton pump in order to cope with the substantial amounts of acid penetrating the membrane due to the high outer concentrations of vinegar.

Via Blue Copper Proteins

TTQ-containing methylamine dehydrogenase uses amicyanin as electron acceptor (Fig. 2). Much is known already of these two components, especially of those from *Thiobacillus versutus*: of the dehydrogenase [Vellieux et al., 1986], the structural [Ubbink et al., 1991] as well as the genes for TTQ-biosynthesis (or maturation) [Chistoserdov et al., 1991], the 3-dimensional structure [Vellieux et al., 1989; Huizinga et al., 1992], the electrochemistry [Burrows et al., 1991] and the kinetics with methylamine and amicyanin [A.C.F. Gorren and J.A. Duine, unpublished results] are known; of the amicyanin [van

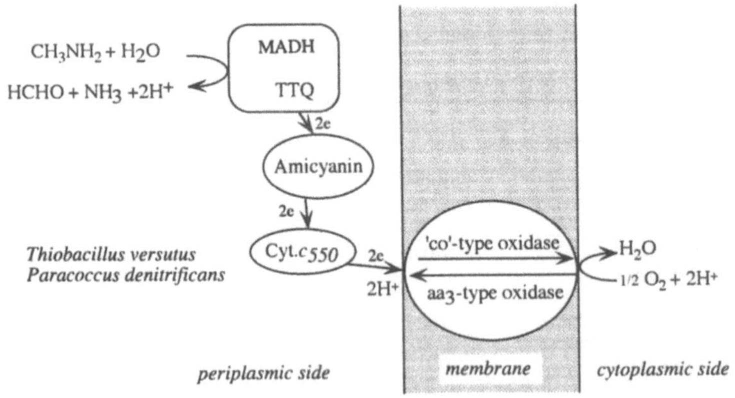

Figure 2. The respiratory chain for methylamine oxidation in *Paracoccus denitrificans*.

Houwelingen et al., 1985], the gene [Ubbink et al., 1991], the 3-dimensional structure (also of a co-crystal with the dehydrogenase [Chen et al., 1992]) is known and production of variant proteins is under way.

Other well known bacterial blue copper protein are azurin and pseudoazurin. Although their structure predicts that they are also electron carrier proteins, their precise role in electron transport chains is less well known. The following suggestions have been made regarding a function: as electron donor for nitrite reductase in nitrate respiration [Zumft et al., 1987]; as electron acceptor for methylamine dehydrogenase of *Methylomonas* spec. J [Ambler and Tobari, 1989]; as electron acceptor for amicyanin in the chain for methylamine oxidation [Auton and Anthony, 1989]. Clear evidence has now been obtained for a role of azurin in an aerobic electron transfer chain of *Pseudomonas aeruginosa* (Fig. 3). Variants of azurin prepared by protein engineering are available [Pascher et al., 1989; den Blaauwen and Canters, 1993].

Via Soluble Cytochromes

B-Type Cytochromes. The soluble type of PQQ-containing glucose dehydrogenase has been specifically found in *Acinetobacter calcoaceticus* [Dokter et al., 1986]. The sole redox protein in the cell-free extract which oxidized the reduced dehydrogenase appeared to be the small, soluble, basic cytochrome b_{561} [Dokter et al., 1988].

C-Type Cytochromes. PQQ-containing ethanol dehydrogenase has been discovered in alcohol-grown Pseudomonads. As was recently demonstrated for the enzyme from *Pseudomonas aeruginosa*, the electron acceptor is the small, soluble cytochrome c_{EDH} [Schrover et al., 1993]. Although the cytochrome has a maximum at 551 nm, the other properties are quite different from those of the already known cytochrome c_{551} of this organism. The only redox protein in the extract accepting electrons from cytochrome c_{EDH} appeared to be azurin [J. Schrover, S. de Vries and J.A. Duine, unpublished results]. In the case of the related quinoprotein methanol dehydrogenase, two soluble cytochromes c function in the chain: cytochrome c_L as electron acceptor for the dehydrogenase and cytochrome c_H as electron acceptor for cytochrome c_L [Anthony, 1992; Frank and Duine, 1990].

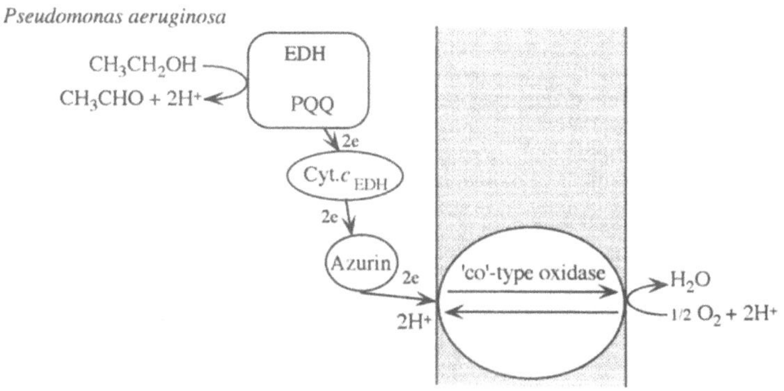

Figure 3. The ethanol oxidation chain via quinoprotein ethanol dehydrogenase of *Pseudomonas aeruginosa.*

ELECTRODE-DIRECTED ELECTRON TRANSFER

Redox proteins have been modelled by evolution to perfect catalysts for oxidizing/reducing substrates and transferring electrons in a particular organism under certain conditions, not for carrying out redox reactions in a reactor in a chemical plant or donating electrons to a solid surface like an electrode. Nevertheless, many attempts are nowadays undertaken to achieve the construction of amperometric biosensors. Furthermore, although cofactor regeneration occurs very efficiently in the cell, the use of the enzyme in vitro has certain attractive aspects (*vide infra*). Coenzyme regeneration has been achieved in several ways, e.g. in a membrane reactor using NAD attached to a polymer, preventing escape, and reoxidizing the NADH formed with a second dehydrogenase (reductase) [Hummel and Kula, 1989]. However, cofactor regeneration (i.e. reoxidation of reduced quinones or flavins in the proteins) is more problematic. Since one of the physiological roles of dehydrogenases is to transfer the electrons specifically to their natural electron acceptors in an ordered way (otherwise the bioenergetics of the organism would become in disorder if the dehydrogenase reacts with every redox compound having a suitable redox potential), special constraints exist in order to achieve this specificity (buried active site, affinity to the natural electron acceptor, pathways for electron transfer between the redox centra through the proteins). Thus, although certain low molecular weight, man-made dyes (but also natural dyes, being inhibitors for competitive organisms by acting as short cut or shunts for natural electron flow in their respiratory chain [Haraguchi et al., 1992] are rather good electron acceptors, in general the electron transfer process to foreign compounds is rather inefficient. However, the use of artificial electron acceptors is unattractive since they may contaminate the final product. Therefore, electrochemical cofactor regeneration seems an interesting option to circumvent this problem. Moreover, in our studies on kinetic resolution of the enantiomers of solketal and glycidol with quinohaemoprotein ethanol dehydrogenases [Geerlof et al., 1990; Geerlof and Duine, 1991], an additional reason exists. These C_3-alcohols are interesting building blocks for the synthesis of many pharmaceuticals. It appeared that the enzymes *in vitro* had very high enantioselectivity but that of whole cells was significantly less [Geerlof et al., 1994]. Since no interfering other alcohol dehydrogenases were present, it was concluded that substrate diffusion limitation caused the lower enantioselectivity of the cells. Thus, the use of free enzyme is obvious in these cases [work in collaboration with W. Somers].

Electron transfer from a redox protein to a bare electrode has been observed in a number of cases. Although the efficiency of this can be increased by addition of a promotor [Burrows et al., 1991], the current densities obtained and the scarcity of information on the stability and fixation of the enzyme seem not attractive for application. In this connection, it would be interesting to see whether the recently discovered method of depositing enzymes in a very stable form to a surface [Prof. Nicolini, personal communication] could be applied here.

Indirect electron transfer can be achieved by using a mediator. The mediator can be either a low molecular weight compound, cycling between the dehydrogenase in solution and the (modified) electrode, or a conducting polymer network, bridging the active site of the dehydrogenase with the electrode surface. Both possibilities have been investigated by us with respect to PQQ-containing dehydrogenases. In the case of conducting polymers, the "wired soluble glucose dehydrogenase" gave the highest electron flow per surface area ever achieved [Ye et al., 1993]. Work is in progress to attain efficient electron transfer between quinoprotein dehydrogenases and conducting polymers by investigating different electron conducting polymer systems, and by incorporating PQQ in the network.

CONCLUSIONS AND PROSPECTS

A multitude of bacterial dehydrogenases, catalyzing one and the same reaction but having different protein structure or cofactor identity, exists. Striking examples are found in direct, non-phosphorylative oxidative degradation of amines, alcohols, and sugars.

The diversity is not restricted to dehydrogenases but also occurs at the level of the natural electron acceptors, i.e. the redox compounds embedded in the cytoplasmic membrane as well as soluble cytochromes and blue copper proteins functioning as electron carriers. With these dehydrogenases and carriers, long range electron transfer studies can be conducted of a "soluble chain", with all its inherent advantages for the experimental set ups. Since mutant proteins and 3-dimensional structures of dehydrogenases and carriers are or will soon become available, effective tools exist now to probe putative amino acid residues involved in contact between the proteins and in electron transfer pathways through the proteins.

Fundamental studies indicated above are not only relevant for understanding the mechanisms operating in respiratory chains but also in artificial systems consisting of a dehydrogenase and an unnatural electron acceptor (e.g. a man-made dye in a test strip, the solid surface of an electrode or a conducting synthetic polymer attached to it). Progress on these aspects will have its impact on development of diagnostics, amperometric biosensors, and electroenzymology connected with application of oxidoreductases in synthesis of fine chemicals.

ACKNOWLEDGEMENTS

I thank Jaap Jongejan and Simon de Vries for preparing the figures.

REFERENCES

Ambler, R.P. and Tobari, J., 1989, *Methylomonas* J., *Biochem. J.* 261:495.
Ameyama, M. and Adachi, O., 1982, *Meth. Enzymol.* 89:450.
Anthony, C., 1992, *Biochim. Biophys. Acta* 1099:1.
Auton, K.A. and Anthony, C., 1989, *J. Gen. Microbiol.* 135:1923.
Barata, B.A.S., Legall, J. and Moura, J.J.G., 1993, *Biochemistry* 32:11559.
Blaauwen, T. den and Canters, G.W., 1993, *J. Am. Chem. Soc.*115:1121.
Burrows, A.L., Hill, H.A.O., Leese, T.A., McIntire, W.S., Nakayama, H. and Sanghera, G.S., 1991, *Eur. J. Biochem.* 199:73.
Bystrykh, L.V., Vonck, J., Bruggen, E.F.J. van, Beeumen, J. van, Samyn, B., Govurokhina, N.I., Arfman, N., Duine, J.A. and Dijkhuizen, L., 1993, *J. Bacteriol.* 175:1814.
Chen, L., Durley, R., Poliks, B.J., Hamada, K., Chen, Z., Mathews, F.S., Davidson, V.L., Satow, Y., Huizinga, E., Vellieux, F.M.D. and Hol, W.G.J., 1992, *Biochemistry* 31:4959.
Chistoserdov, A.Y., Tsygankov, Y.D. and Lidstrom, M.E., 191, *J. Bacteriol.* 173:5901.
Dokter, P., Frank, J. and Duine, J.A., 1986, *Biochem. J.* 239:163.
Dokter, P., Wielink, J.E. van, Kleef, M.A.G. van and Duine, J.A., 1988, *Biochem. J.* 254:131.
Duine, J.A., 1991, *Eur. J. Biochem.* 200:271.
Frank, J., and Duine, J.A., 1990, *Methods Emzymol.* 188:303.
Geerlof, A., Tol, J.B.A. van, Jongejan, J.A. and Duine, J.A., 1994, *Biosci. Biotech. Biochem.*, in press.
Geerlof, A., Groen, B.W. and Duine, J.A. 1990, *Eur. Pat.* 90201651; USA Pat. 07/541,910.
Geerlof, A. and Duine, J.A., 1991, *Eur. Pat.* 91201576.
Groen, B.W., Kleef, M.A.G. van and Duine, J.A., 1986, *Biochem. J.* 234:611.
Haraguchi, H., Abo, T., Hashimoto, K. and Yagi, A., 1992, *Biosci. Biotech. Biochem.* 56:1221.
Hopper, D.J. and Taylor, D.G., 197, *Biochem. J.* 167:162.
Houwelingen, T. van, Canters, G.W., Stobbelaar, G., Duine, J.A. and Tsugita, A., 1985, *Eur. J. Biochem.* 153:75.
Huitema, J., Beeumen, J. van, Driessche, G. van, Duine, J.A. and Canters, G.W., 1993, *J. Bacteriol.* 175:6254.
Huizinga, E.G., Zanten, B.A.M. van, Duine, J.A., Jongejan, J.A., Huitema, F., Wilson, K.S. and Hol, W.G.J., 1992, *Biochemistry* 31:9789.
Hummel, W. and Kula, M-R., 1989, *Eur. J. Biochem.*, 184:1.
Matsushita, K., Shinagawa, E., Adachi, O. and Ameyama, M., 1989, *J. Biochem.*, 105:633.
Matsushita, K., Shinagawa, E., Adachi, O. and Ameyama, M. 1990, *Proc. Natl. Acad. Sci. USA* 87:9863.
Meyer, O., Frunke, K. and Mörsdorf, G., 1993, *in*: "Microbial growth on *C1*-compounds," J.C. Murrell and D.P. Kelly, eds., Intercept, Andover (1993) pp. 433-459.
Mukund, A. and Adams, M.W.W., 1991, *J. Biol. Chem.* 266:14208.
Ophem, P.W. van, Euverink, G.J., Dijkhuizen, L. and Duine, J.A., 1991, *FEMS Microbiol. Lett.* 80:57.
Ophem, P.W. van, Beeumen, J. van and Duine, J.A., 1993, *Eur. J. Biochem.* 212:819.
Pascher, T., Bergström, Malmström, B.G., Vänngärd, T. and Lundberg, L.G., 1989, *FEBS Lett.* 258:266.
Poels, P.A., Groen, B.W. and Duine, J.A., 1987, *Eur. J. Biochem.* 166:575.
Schrover, J.M.J., Frank, J., Wielink, J.E. van and Duine, J.A., 1993, *Biochem. J.* 290:123.
Singer, T.P. and McIntire, W.S., 1984, *Meth. Enzymol.* 106:369.
Spiro, S. and Guest, J.R., 1991, *Trends Biochem.* 16:310.
Ubbink, M., Kleef, M.A.G. van, Kleinjan, D-J., Hoitink, C.W.G., Huitema, F., Beintema, J.J., Duine, J.A., Canters, G.W., 1991, *Eur. J. Biochem.* 202:1003.
Vellieux, F.M.D., Huitema, F., Groendijk, H. , Kalk, K.H., Frank, J., Jongejan, J.A., Duine, J.A., Petratos, K., Drenth, J. and Hol, W.G.J., 1989, *EMBO J.*8:2171.
Vellieux, F.M.D., Frank, J., Swarte, M.B.A., Groenendijk, H., Duine, J.A., Drenth, J. and Hol, W.G.J., 1986, *Eur. J. Biochem.* 154:383.
White, H., Strobl, G., Feicht, R. and Simon, H., 1989, *Eur. J. Biochem.* 184:89.

Ye, L., Haemmerle, M., Olsthoorn, A.J.J., Schuhmann, W. Schmidt, H-L., Duine, J.A. and Heller. A., 1993, *Anal. Chem.* 5:238.
Zachariou, M. and Scopes, R.K., 1986, *J. Bacteriol.* 167:863.
Zumft, W.G., Gotzmann, D.J., and Kroneck, P.H.M., 1987, *Eur. J. Biochem.* 168:301.

THE USE OF PARAMAGNETIC PROBES FOR NMR INVESTIGATIONS OF BIOMOLECULAR STRUCTURES AND INTERACTIONS

M. Scarselli [a], G. Esposito [b], H. Molinari [c], M. Pegna [d], L. Zetta [e] and N. Niccolai [a]

a) Dipartimento di Biologia Molecolare, Università di Siena, Pian dei Mantellini 44, 53100 Siena, Italy
b) Dipartimento di Scienze e Tecnologie Biomediche, Università di Udine, Via Gervasutta 48, 33100 Udine, Italy
c) Istituto di Chimica Biologica, Facoltà di Farmacia, Università di Sassari, 09200 Sassari, Italy
d) ITALFARMACO c/o ICM-CNR, Via Ampere 56, 20131 Milano, Italy
e) ICM-CNR, Via Ampere 56, 20131 Milano, Italy

The structure/function relationship in protein is a fundamental topic of investigation in every branch of biotechnology. The possibility of *de novo* design of transelectronase enzymes to be used in bioelectronic devices implies a good knowledge of folding mechanism and solution structures. Furthermore, the enzyme surface accessibity to substrates is of particular interest. Nuclear magnetic resonance (NMR) is nowadays a well established technique for the investigation of the solution conformation of complex molecules (Wüthrich; Dweck *et al.*, Campbell *et al.*) and of their mutual interaction processes.

The strong improvement of this spectroscopy for the study of biological systems at the atomic resolution results in a very favourable synergism among different scientific fields, as described in Fig.1. Recently, NMR has been proved to be a particularly powerful technique for elucidating the molecular conformation of protein molecules, provided that their size is in the range of 20 KD of molecular weight. The twenty aminoacid residues commonly occurring in proteins give rise to characteristic spectral patterns which overlap degree is strongly dependent on the overall structure of the protein. Random coil sequences always show infact a severe degeneracy of resonances, irrespective of the specific primary structure. On the contrary, the presence of a native conformation can partially remove the superposition of identical spin systems, leading to a major spread of the chemical shifts that can be considered a first hint of the presence of a structured protein. However, when the number of residues in the sequence increases, the spectrum becomes very crowded, despite the secondary shifts experienced by the single resonances. The complexity of protein NMR spectra has been a major limitation in previous studies, where only resonances occurring outside the main spectral envelopes could be studied in detail.

In order to unravel the information contained in complex 1D NMR spectra, a complete identification of individual resonance lines is needed. Their overlap may be at least partially circumvented by spreading the resonances over several frequency dimensions. By choosing suitable instrumental parameters, it is possible to investigate the behaviour of the nuclear magnetisation, spread over two (Aue *et al.*; Nagayama *et al.* 1977; Nagayama *et al.* 1979; Wüthrich K *et al.*; Wider *et al.*) or more time variable domains (Griesenger *et al.*; Kay *et al.*; Ikura *et al.*), thus correlating, in the corresponding multiple frequency dimensions, different NMR parameters. The correlation is visualised by the presence of a cross peak, the

From Neural Networks and Biomolecular Engineering to Bioelectronics
Edited by C. Nicolini, Plenum Press, New York, 1995

frequency coordinates of which are related to the parameters represented in the 2D spectral cross-sections.

Thus, information on structure, dynamics and biological mechanisms may be obtained from multidimensional NMR techniques which provide efficient tools to simplify the wealth of dipolar and/or scalar connectivities encoded in a complex spectrum (Bertini et al., 1990) . Multiple quantum and X filters are typical examples of efficient strategies to reduce the spectral complexity which can often inhibit a detailed analysis of the NMR parameters. Thus, simplification may be achieved on the basis of the specific spin coupling patterns in a way which is controlled only by the physical behaviour of the investigated spin system, regardless of any stereochemical feature of the protein in solution.

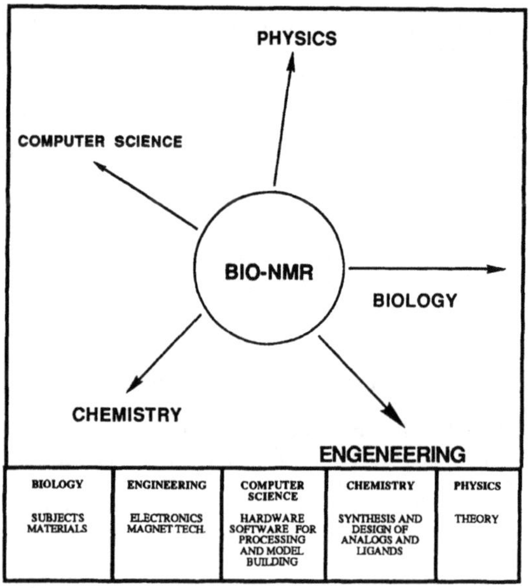

Figure 1. Synergism between NMR of biological systems and other scientific fields.

A different type of simplification could arise from the individuation of limited ensemble of resonances corresponding to molecular moieties involved in recognition processes. The presence of nitroxide stable spin labels bound to proteins or nucleic acids has been proposed proposed by several authors (Schmidt et al; Moonen *et al.*; Kosen *et al.*; De Jong *et al.*) to study the environment of the unpaired electron, allowing the mapping of the ligand binding sites through 1D or nD methodologies.

Alternatively, soluble free radicals such as TEMPO or its derivatives, have been used (Niccolai *et al.* 1982; Niccolai *et al.* 1984; Zhou *et al.* 1985) for spin-labelling of the solvent. In this case information can be obtained on the nitroxide and solvent molecules accessibility of proton nuclei of small, but structured, molecules from 1D spectra: the formation of weak collisional adducts between nitroxide and diamagnetic solute determines a significant perturbation on the relaxation rates of nuclei localised on the accessible area of the molecule. More recently, as reported below in this review, the use of nitroxides for solvent spin-labelling has been proposed as a paramagnetic relaxation filter of signal intensities of multidimensional NMR spectra of peptides (Niccolai et al., 1991) and proteins (Esposito et al. 1992).

MECHANISMS OF THE SOLUTE-SPIN LABEL INTERACTION

The paramagnetic effects on nuclear relaxation, induced by the presence in solution of transition metal ions or free radicals, has been extensively reviewed either in 1D or 2D investigations (Dweck R. A.; Bertini I. *et al.1987*; Banci L. *et al.,1991*). Since the structural analysis of the spin lattice perturbations induced by the spin labelling of the solvent upon the addition of soluble nitroxides was first proposed (Niccolai *et al.*, 1982) several papers appeared on the mechanisms of interaction between paramagnetic probes and diamagnetic solutes (Niccolai *et al.*, 1984, N. Zhou *et al.*). In all these cases, the nature of the observed enhancement of the proton relaxation was fully interpreted in terms of electron - nucleus dipolar interactions. Scalar contributions could be ruled out as the ESR spectrum of the free radical was unaffected by the presence of the biomolecules in solution and no significant chemical shift changes were detected. The paramagnetic relaxation rates, R_p, measured from 1D Inversion Recovery experiments and calculated by the difference of the proton spin lattice relaxation rate of a given nucleus in the presence, R_{exp}, and in the absence, R_0, of the nitroxide are a measure of the relaxation contribution given by the nitroxide to the total proton relaxation for each nucleus.

From the study of these effects on molecules with well defined structures, differential accelerations were observed and, in particular, inner nuclei had R_p's much smaller than the ones of surface exposed protons, suggesting that the dipolar interaction of the electron with the nuclear spins is mainly controlled by the molecular conformation and that the dipolar coupling is modulated by the self-diffusion.

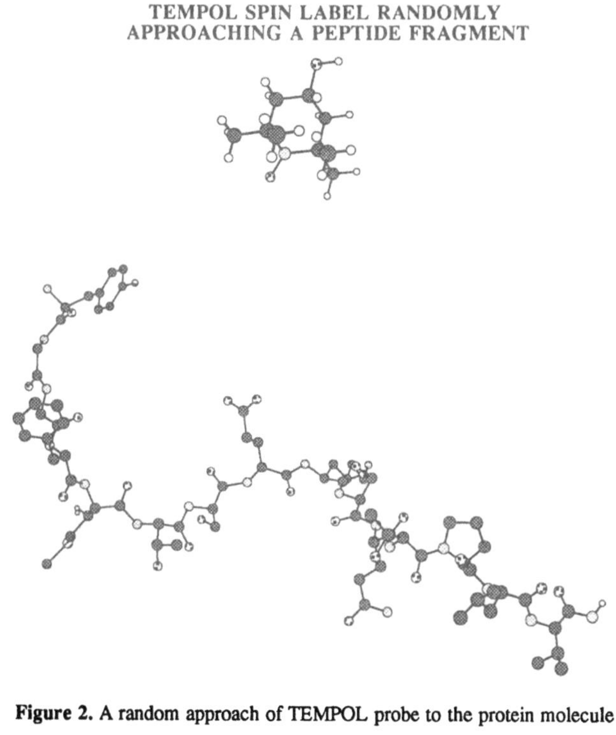

Figure 2. A random approach of TEMPOL probe to the protein molecule

From the analysis of the paramagnetic effects induced by soluble spin labels on the molecular systems so far investigated by the authors of the present work some insight on the dynamics of the interaction of the diamagnetic and paramagnetic species can be given. A mainly random approach between the protein and the probe, governed by the conformation of the biomolecule, as shown in Fig. 2, is suggested by several observations: i) no or small J and chemical shift changes are induced on proton protein signal upon the addition of the paramagnetic molecule, indicating that no strong intermolecular interaction between TEMPOL and the investigated molecule occurs; ii) a statistical nature of the interaction is suggested by the dilution studies performed on the gramicidin S - DMSO - TEMPOL system, being the observed paramagnetic effects strongly and linearly dependent on the concentration of the two solutes; iii) the role of the solvent molarity is an addition clue for supporting a diffusion controlled formation of collisional bimolecular adducts, as comparable signal attenuations of protein or of peptide resonances are measured at very different molar ratios of the two solutes, depending on the used solvent. In fact, in water solutions a spin label concentration five to ten times higher than the one of the investigated molecule is typical for yielding sizeable reductions of cross peak intensities, while in organic solvents like DMSO or $CHCl_3$, which are almost five time more diluted than the aqueous solvent, similar conditions are reached at equimolar concentrations.

The role of TEMPOL hydrogen bonding capability with the polypeptide backbone amide protons has been analysed in different molecular systems and solvents. In the case of gramicidin S in DMSO solution, diagonal peaks of amide protons were twice as much attenuated than the similarly surface exposed Hα's suggesting that the nitroxide N-oxyl group efficiently compete with the sulphoxyl moiety of the solvent for the formation of intermolecular H bonds at the exposed NH protons.

The hydrogen acceptor behaviour of the TEMPOL N-oxyl group was better elucidated in a study where the interaction of the chemical probe with a cyclic tetrapeptide in chloroform solution was investigated (Scarselli *et al.*). In this case, the various surface accessibilities to solvent and spin label molecules of NH proton signals was revealed by differentials resonance attenuations and shifts, suggesting that hydrogen bonding in the low polarity environment play a relevant role.

In a study of the TEMPOL induced paramagnetic filters on b endorphin proton resonances in water solution, the spin label preferential interaction with the peptide through H bonds donor/ acceptor groups or hydrophobic sites was investigated. In the presence of the large variety of different side chains offered by that peptide, the unordered nature of the molecular conformation yielded such similar paramagnetic filters to exclude the possibility of strong solute-solute interactions (Esposito *et al.*, 1993).

Thus, some conclusions can be drawn: i) each class of hydrogens interact dipolarly with the unpaired electron of TEMPOL in a way which is primarily determined by its solvation behaviour; ii) within each class of hydrogens, the different paramagnetic relaxivities reflect the location in the molecule; iii) intramolecular hydrogen bonding of amide protons is efficiently recognised; iv) the 1D spin-lattice relaxation study is limited by the spectral dispersion of the NMR signals and, therefore, the structural analysis through this method is confined to small molecules. Then, the possibility to exploit the nitroxide perturbation to discriminate between surface and core nuclei in more complex molecules implies the study of the information flow of the paramagnetic relaxation in multidimensional NMR spectroscopies.

THE PHYSICAL BASIS OF THE PARAMAGNETIC PERTURBATION OF SOLUBLE SPIN LABELS ON PROTEIN MULTIDIMENSIONAL SPECTRA

The proposed correlation between Rp's and conformational features, as summarised above, is limited by the signal dispersion in the proton 1D spectrum and possible applications are confined to molecules of low complexity. Hence, for obtaining structural information on spin systems as complex as proteins, the extension of this methodology for the analysis of the paramagnetic effects in multidimensional spectra was needed.

In principle, in the presence of the paramagnetic probe, a sizeable intensity decrease of the cross-peak intensity, $\underline{a}$, had to be expected, as the relaxation processes control its extent, as shown in the Fig. 3 for a 2D NOESY experiment where A and B nuclei relax according to their characteristic spin- spin relaxation times, T_2, in the t_1 and t_2 time domains. The expected differential perturbations, induced by the TEMPOL nitroxide, were found in the 2D spectra of gramicidin S (Niccolai *et al.*, 1991) and lysozyme (Esposito *et al.*, 1992). Since the observed spectral changes were consistent with the well known conformational features of the two molecules, a general strategy for the determination of biomolecular structures in spin labelled solvents has been proposed, based on the delineation of a inner/outer profile, according to the different attenuation degree for each cross-peak intensity. This multidimensional NMR approach can be successful adopted, provided that suitable experimental conditions are chosen. A problem derived from the line broadening dependent cancellation typical of antiphase multiplets of COSY cross-peaks can be circumvented by using NOESY or TOCSY types of pulse sequences, so that spurious J modulated effects are avoided. In the case that one or more of the J or dipolar connectivities in the nD spectrum arise from nuclei which have different solvent/spin label exposures, non-square data matrices, *i.e.* with a strong difference in the acquisition times t_1, t_2 .... t_n, should be used. In this way, the ambiguity of the cross-peaks of nuclear pairs with a large difference in nitroxide exposures may be explored, as the T2 contribution to the cross-peak intensity of each nucleus may be differently propagated in the two or more frequency dimensions.

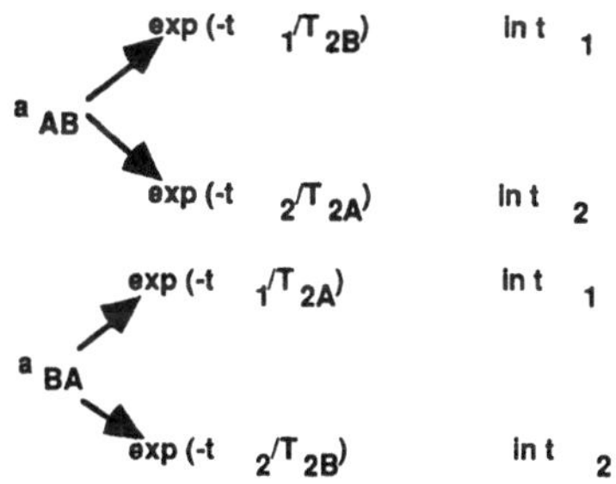

Figure 3. Relaxation effects in NOESY cross-peaks.

Different local mobility of molecular moieties could explain the ambiguous results obtained by others (Petros *et al.*) in analysing the nitroxide exposure of protein side chains. The structural interpretation of paramagnetic filters seems, therefore, to be reliable for backbone nuclei, which are generally involved in analogous rotational motions.

QUANTITATIVE ANALYSIS OF PARAMAGNETIC EFFECTS

A quantitative evaluation of the paramagnetic filter effects can be performed by measuring the cross-peak volumes of the NH-Hα correlations in the nD spectrum obtained in the diamagnetic, V_d, and paramagnetic, V_p, solutions.
The spin label induced attenuations, Ap's, defined as

$$A_p = 1 - V_p / V_d \qquad (1)$$

can be autoscaled, using the mean value as a scaling factor. The absolute reduction of intensity is not very informative on the surface accessibility of a given nucleus, since, as discussed in the previous section, the signal attenuation depends not only on the nitroxide concentration, but also on the solvent molarity and on the polypeptide concentration. Thus, the relative fluctuations of the A_p's, *i.e.* the variances, should be discussed for a structural interpretation of the spectrum changes rather than the absolute extents of the signal attenuations.

In the Figures shown below, obtained for lysozyme, gramicidin S and β endorphin the variance around the mean attenuation, for each resolved cross peak, is reported. In all the three cases, the cross peak attenuation variances were fully consistent with the structural features of the molecules in solution.

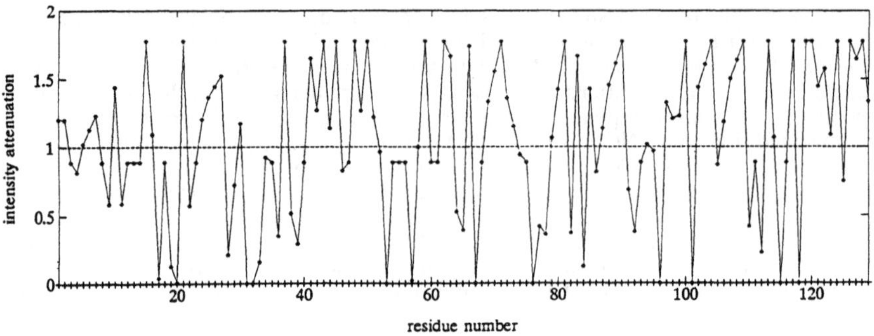

Figure 4. The cross peak attenuation, reported as the auto-scaled value A_{pi} of Eq. 1, induced by the presence of TEMPOL for the NH-Hα correlations of lysozyme protons in water solution. Numbers in abscissas refer to the residue sequence position.

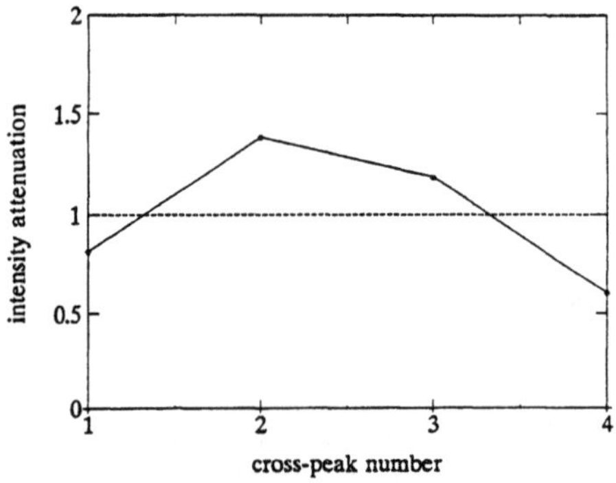

Figure 5. The cross peak attenuation induced by the presence of TEMPOL for the NH-Hα correlations of gramicidin S protons in DMSO solution. Numbers in abscissas refer to valine, phenilalanine, ornithine and leucine respectively .

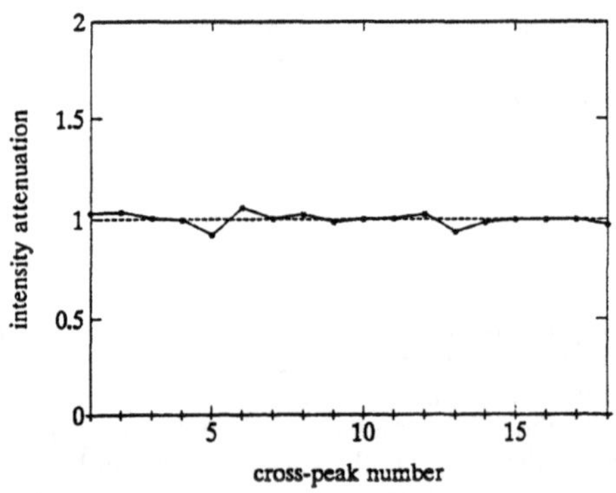

Figure 6. The cross peak attenuation induced by the presence of TEMPOL for the NH-Hα correlations of the 1-27 β-endorphin fragment. Numbers in abscissas refer to the cross peak labelling of Fig. 1 of the β-endorphin investigation (Esposito *et al.*, 1993)

LIMITS AND PERSPECTIVES OF THE USE OF THE SOLVENT SPIN LABELLING FOR STRUCTURAL DETERMINATIONS

The use of chemical probes which cause a paramagnetic environment for the nuclear relaxation of complex molecular systems seems to be very promising. Proton relaxation itself, in diamagnetic solution, is not very informative of specific structures, since each nucleus relaxes similarly due to similar magnetic environments for each class of hydrogens. In the studies summarised in the present report, it has been confirmed that solvent spin-labelling can remove the latter degeneracy in the proton relaxation process and that the paramagnetic effects induced on the spin-lattice relaxation rates and on cross peak attenuation by the presence in solution of soluble nitroxides are governed by the molecular solution structure. Soluble nitroxides such as TEMPOL, appeared good candidates to be used in this kind of experimental procedure, due to i) their chemical stability, ii) the similarity with some solvent molecules and iii) the low affinity with the chemical groups present in the naturally occurring amino acids. In the case that the investigated biopolymer presents on its surface chemical groups with a potential reducing activity (*e.g.*, free SH's of cystein) some concern should be given on the possibility that the TEMPOL N oxyl can be changed into the diamagnetic N hydroxyl species. Then, alternative chemical probes should be chosen, such as DOTA or DPTA Ga(III) complexes (Aime *et al.*) or other types of paramagnetic systems. Interaction processes should be also investigated with this NMR approach, since the molecular surface accessibility, modulated by the complex formation, should yield paramagnetic filter profiles which reflect the sterical aspects of the intermolecular binding. Spectral editing of inner/outer nuclei should be also possible by comparing nD spectra obtained in the presence and in the absence of the paramagnetic probe.

REFERENCES

Aime S., Botta M., Panero M., Grandi M. and Uggeri F, 1991, *Magn. Res. in Chem.*, **29**:923-927.

Aue W.P.,Bartholdi E. and Ernst R.R., 1976, *J.Chem. Phys.*, **64**:2229.

Banci L., Bertini I., Luchinat C. and Piccioli M, 1991, in "*NMR and Biomolecular Structure*" VCH Weinheim, 31- 58.

Bertini I and Luchinat C., 1986, "*NMR of Paramagnetic Molecules in Biological Systems*" Benjamin/Cummings, Menlo Park, California.

Bertini I., Molinari H. and Niccolai N., 1990, in "*NMR and Biomolecular Structure*", VHC, Weinheim.

Campbell I.D. and Dobson C.M., 1979, in '*Methods of Biochemical Analysis*', Glick D. ed., Wiley Interscience, New York, **25**:1-133.

Dweck R. A., 1973, "*Nuclear Magnetic Resonance in Biochemistry*" Oxford Univ. Press, New York ().

Dwek R.A., Campbell I.D., Richards R.E. and Williams R.J.P., 1977, "*NMR in Biology*", Accademic Press, London.

De Jong E.A.M., Claesen C.A.A., Daemen C.J.M., Harmsen B.J.M., Konings R.N.H., Tesser G.I. and Hilbers C.W., 1988, *J. Magn. Reson.*, **80**:197.

Esposito G., Lesk A. M., Molinari H., Motta A., Niccolai N. and Pastore A., 1992, *J. Mol. Biol.*, **224**:659-670.

Esposito G., Molinari H., Niccolai N., Pegna M. and Zetta L., 1994, *J. Chem. Soc. PERKIN II*, in press.

Griesenger C., Sorensen O.W. and Ernst R.R., 1987, *J. Magn. Res.*, **73**:574-579.

Ikura M., Kay L.E. and Bax A., 1990, *Biochemistry*, **29**:4659-4667.

Kay L.E., Clore G.M., Bax A. and Gronenborn A.M., 1991, *Science*, **249**:411-414.

Kosen P.A., Scheek R. M., Naderi H., Basus V. J.,Manogaran S., Schmidt P. G., Oppenheimer N. J. and Kuntz I. D., 1986, *Biochemistry*, **25**:2356.

Moonen C.T.W., Scheek R. M., Boelens R. and Muller F., 1984, *Eur. J. Biochem.*, **141**:323.

Niccolai N. , Valensin G., Rossi C. and Gibbons W. A., 1982, *J. Am. Chem. Soc.*, **104**:1534-37.

Niccolai N., C. Rossi, G. Valensin, P. Mascagni and W.A.Gibbons, 1984, *J. Phys. Chem*, **88**:5689-92.

Niccolai N. , Esposito G., Mascagni P., Motta A., Bonci A., Rustici M., Scarselli M., Neri P. and Molinari H., 1991, *J. Chem.. Soc.-PERKIN II*, 1453-1457..

Nagayama K., Wüthrich K., Bachrann P. and Ernst R.R., 1977, *Biochem. Biophys. Res. Commun.*, **78**:99.

Nagayama K., Wüthrich K. and Ernst R.R., 1979, *Biochem. Biophys. Res. Commun.*, **90**:305.

Petros A.M., Mueller L. and Kopple K. D., 1990, *Biochemistry*, **29**:10041-10048.

Scarselli M., Bonci M , Butini L., Vasco A., Mascagni P. and Niccolai N., 1993, *Spectroscopy Letters*, submitted.
Schmidt P.G. and Kuntz, I. D. 1984, *Biochemistry*, **23**:4260.
Wüthrich K., Nagayama K. and Ernst R.R., 1979, *Trends Biochem. Sci.*, **4**, N178.
Wüthrich K., 1986, *NMR of Proteins and Nucleic Acids*, Wiley, New York.
Wider G., Macura S., Kumar A. and Wüthrich K., 1984, *J. Magn. Res.*, **56**:207.
Zhou N., Mascagni P., Gibbons W.A., Niccolai N., Rossi C and Wyssbrod H., 1985, *J. Chem. Soc. PERKIN II*, 581-87.

USE OF MOLECULAR DYNAMICS FOR THE RECONSTRUCTION OF COMPLETE STRUCTURES: A STUDY ON THIOREDOXIN

E.A. Carrara*, A. Anselmino, C. Gavotti, S. Vakula and C. Nicolini

*Cibernia Srl, Sassari, Italy
Institute of Biophysics, University of Genoa, Genoa, Italy

ABSTRACT

A new molecular mechanics protocol for the reconstruction of complete structure of proteins from the positions of the α-carbons is presented. The method can efficiently avoid local minima, resulting in a good ability to reconstruct sidechain dihedral angles.

The structural features predicted in our computations are in satisfactory agreement with the full crystallographic structure resolved at 1.68 Å, but the secondary structures shows an even better agreement with the 3D atomic structure of thioredoxin from 2DFTNMR studies in solution.

An analysis of the relevance of the structural variability of minimizations produced by different potential functions is made, and its results discussed. The congruence between structures produced minimizing the same starting structure with different potentials increases with decreasing degree of refinement of the starting structures.

INTRODUCTION

The tertiary structure of proteins plays an important role in many of the protein functions, such as catalysis, ligand binding, antibody-antigen interactions. Therefore the possibility to predict theoretically certain structural features might be extremely useful for the elucidation of the functional behaviour. Most structure prediction algorithms (with the notable exception of those based on homology) do not predict the conformation of side chains, but give only an estimate of local backbone topology (e.g. position of secondary structure or geometry of loops). Furthermore X-ray studies of big complexes at the initial refinement stage are often limited to the position of C_α. Nevertheless, the backbone topology gives little or no information on protein properties and function, as they are usually determined by side chain packing inside the protein, by the nature and position of prosthetic groups. Most of the properties are individuated by the side chain configuration, and therefore require a full structure to be estimated. The first part of this study describes a method based on molecular mechanics for the reconstruction of full protein structures from C_α coordinates. The use and results of the methodology are shown on thioderoxin with realistic data. Different energy minimizations were then performed to refine structures of oxidized thioredoxin available in Protein Data Bank, utilizing different forcefields and different strategies, in order to investigate the consistency and the accuracy of calculation based on minimized structures.

From Neural Networks and Biomolecular Engineering to Bioelectronics
Edited by C. Nicolini, Plenum Press, New York, 1995

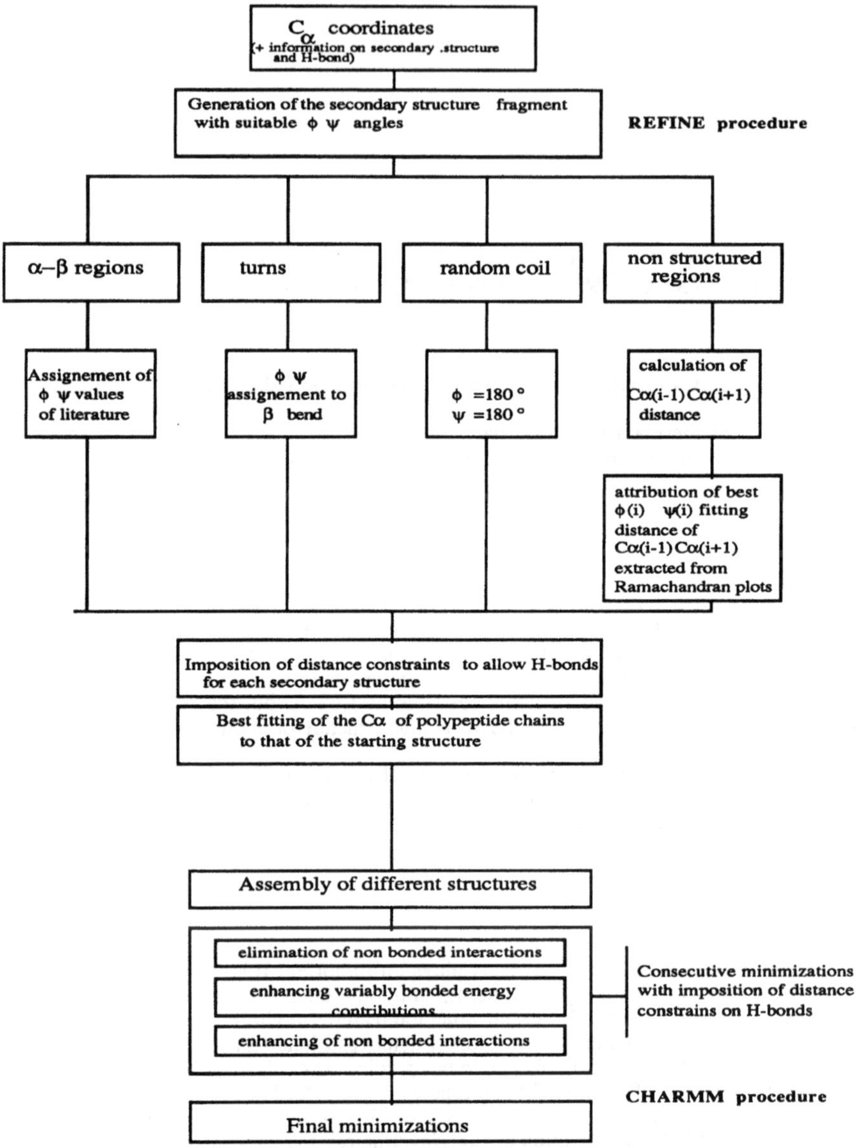

Figure 1. Procedure to reconstruct three dimensional structure from Cα coordinates.
In the REFINE procedure calculations were performed utilizing the forcefield for non-bonded interactions described by Nelson and Hermans (1973) and by Nomany et al. (1975) and topology residue information, charge distributions, force constant for bonded interactions described by Poland and Sheraga (1967).
The ϕ ψ angles are set according to the secondary structure type as described in the following: α helices ϕ = -62° ψ= -41°. This values represent mean values observed for α helix in 57 resolved proteins (Barlow and Thornton, 1988); β strand ϕ = -140° ψ= 135°; turns are considered as β turns ϕ = -60° ψ= -30° and ϕ = -90° ψ= 0° respectively for (i*1) and (i+2) residues (Smith, J.A., Pease, L.G., 1980). Random coil segments have been generated starting from the end of a structured region with dihedral angles values of 180°. Non-structured regions : the best ϕ(i) and ψ(i) values for each given crystallographic distances between Cα(i-1) and Cα(i+1) were extrapolated from the Ramachandran plots. In order to disentangle the side chains we performed different procedures in consecutive steps. Distance constraints (NOE routine) and harmonic force to crystallographic target have been imposed.

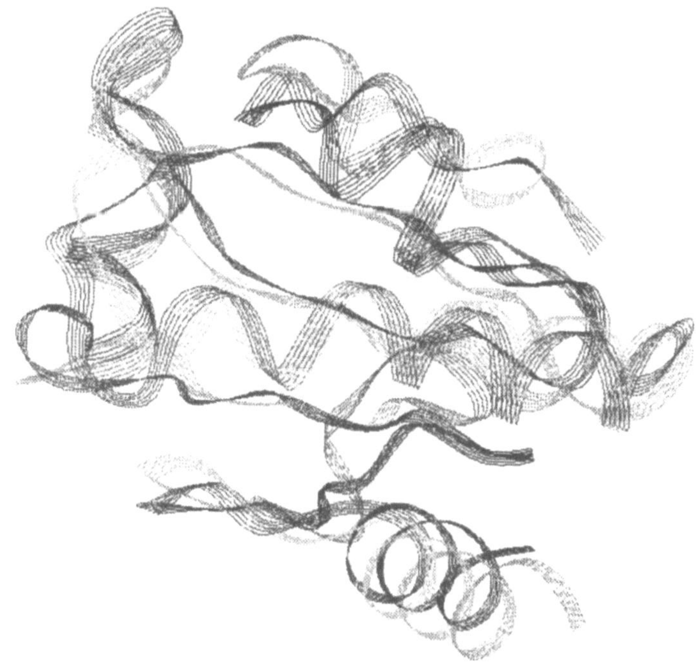

Figure 2. Superimposed structures obtained by our procedure of coordinates reconstruction on crystallographic (1.68 Å) structure. Only ribbons are visualized. Dark lines are crystallographic structure, light lines - predicted structure. The α helix at the bottom of the figure is at the C-terminal.

METHOD

Procedure of reconstruction of the 3D structure from only Cα atomic coordinates

The flow-chart of the procedure utilized is shown in fig.1. The procedure starts generating a polypeptide chain with suitable and angles for each secondary structure. After that, the segments are superimposed to the original structure by least-squares fitting of the Cα coordinates of corresponding residues and the resulting structure is minimized with increasingly precise potentials. Finally, a full minimization of the structure is performed. Two different molecular mechanics programs have been used to implement the simulation protocol. The program REFINE (Ferro et al., 1980, Allinger and Yuh, 1980) is used to determine the overall structure for the whole aminoacidic chain, assigning initial positions to backbone and to side chains atoms. The final structure refinement is performed using the CHARMM (Brooks et al., 1983) routines of minimization and molecular mechanics.

To test our reconstruction method on a realistic case we have used it to rebuild the full structure of thioredoxin from the original data on Cα positions from Holmgren et al. (1975) (PDB data set 1srx 2.8 Å resolution), and compared the structure obtained with two more recent thioredoxins from *E.Coli*, one obtained from X-ray data (Katti et al., 1990, chain A of PDB data set 2trx, 1.68 Å resolution) and the other from Nuclear Magnetic Resonance data (Dyson et al., 1989. PDB data set 1trx).

Thioredoxin has been isolated from many different organisms and shows a very ubiquitous behaviour, being involved in many reactions within the cell. In particular this protein plays an important role in thiol-dependent redox reactions (see Holmgren, 1985, for a review). The active site of thioredoxin, CYS-GLY-PRO-CYS, in the oxidized form contains a disulphide bridge which is usually reduced by NADPH and FAD-containing enzyme thioredoxin reductase. Reduced thioredoxin is a powerful general protein disulphide reductase and also an hydrogen donor for ribonucleotide sulphoxide reductase and sulphate reductase.

Minimizations with different potentials

Energy minimizations and molecular dynamics (MD) simulations have become a standard tool for determining protein structures with constraints from X-ray crystallography or NMR spectroscopy. MD simulations are also used in modelling new proteins or protein structures from homology, and can be a help in interpreting data from different spectroscopic methods. Anyway several problems still remain in the evaluation of simulations results, mainly due to (1) the short length of the simulations compared to the experimental time scale (2) the accuracy of the potentials in describing the motion and the energy (3) the partial or full neglect of solvent effects and (4) trapping in local energy minima.

Escherichia coli thioredoxin has been recently used for an analysis of how consistent molecular dynamics simulations are (Elofsson and Nilsson, 1993). In a complementary way, in our study we investigate the impact of different potentials on the results of energy minimizations to evaluate the influence of problem (2) above on structure determination.

The structure produced by our 3D-coordinates assignment procedure is, from an energetical point of view, far from a minimum. So it can be an appropriate starting point to examine different forcefields effects occurring during a minimization procedure. We have performed several energy minimizations with different force fields and strategies. Starting from a reconstructed structure generated from a Cα-only structure of thioredoxin as described in section (a) above (tref), various steps of steepest descent and conjugate gradients have been alternated to molecular dynamics runs at 300 °K, utilizing both CVFF (Dauber-Osguthorpe et al., 1980) and AMBER (Wiener et al., 1986) forcefields. To use a strategy comparable with the process of refinement performed with CHARMM, we have imposed the same NOE constraints on distance between atoms. The results are to compare with CHARMM refinement structures and with themselves.

Furthermore other accurate energy minimization refinement of oxidized thioredoxin has been performed starting from the PDB structure 2trx mentioned above. In the following we will refer to this structure as TRX. To simulate the behaviour of the protein in solution (the main difference between X-ray and NMR, as methods of determining 3D structure of proteins, is the different environment of the molecule analyzed: in crystal for X-ray and in solution for NMR), the thioredoxin molecule has been surrounded by a water layer of 5Å. Then the energy of the assembly has been minimized utilizing the CVFF forcefields as implemented by the package DISCOVER by BIOSYM. Minimizations with steepest descent and conjugate gradients have been alternated with simulation of dynamics at temperatures ranging from 300 °K to 600 °K, to reduce the problems of possible local minima.

After 35000 steps by both algorithms, energy had decreased from 6000 kcal/mol to -15600 Kcal/mol. Further on we will refer to the monomer A of this resulting structure as TWAT. All minimizations and dynamics were carried on with standard DISCOVER forcefields (CVFF). Nevertheless other potentials can be considered and can be useful to understand dependence of results on different choices. Both structures (2TRX chain A and TWAT) have been minimized by 10000 steps of steepest descent and conjugate gradients. This strategies of minimization has been carried on each structure with two different forcefields (Amber and CVFF).

COMPARISONS

The most popular index of similarity between different structures of the same molecule is surely the root-mean square deviation (RMS), defined by

$$RMS = \sqrt{\frac{1}{N}\sum_{i=1}^{N}(r_i^a - r_i^b)^2}$$

where r_i denotes the Cartesian coordinates of atom i in the structures a and b, respectively. In this work RMS differences have been calculated over all atoms in the structures or only on the backbone atoms or trace atoms (Cα), to localize differences between different structures.

The conservation of secondary structure segments in different 3D structures has been evaluated both by visual inspection and by utilizing the DSSP procedure for the attribution of the secondary structure from atomic coordinates (Kabsh and Sanders (1983).

RESULTS

Reconstruction of full 3D structure from Cα positions

A summary of the RMS difference among the reconstructed thioredoxin structure and the crystallographic and NMR structures is reported in table I.

Table I. RMS analysis on 3D-coordinates assignment procedure results

RMS	1SRX/2TRX	1SRX/TFIN	2TRX/TFIN	1TRX/TFIN	TREF/1SRX	TREF/2TRX	TREF/TFIN
trace	1.97	1.17	2.04	2.09	0.51	2.09	1.35
backbone	---	---	2.06	2.11	---	2.25	1.39
all atoms	---	---	3.14	2.75	---	4.19	2.70

(a) 1SRX: crystallographic structure at 2.8 Å (only C_α positions).
(b) 2TRX: crystallographic structure at 1.68 Å.
(c) 1TRX: NMR structure.
(d) TREF: 3D assignment with REFINE
(e) TFIN: final structure after CHARMM refinement.

We have to note that the RMS difference on only the C_α between the structure 1SRX and the X-ray structure 2TRX is 1.97 Å, that is comparable with RMS between 1SRX and reconstructed structure. So we can deduce that no uncertainty has been added to the C positions and that the backbone coordinates are well generated. The positioning of side chains is obviously the source of the major differences. In fact, evaluating the results of only the first part of the procedure, concerning only atoms coordinates assignment, (TREF), we can see that essentially CHARMM procedure has improved positions of backbone by a great adjustment of side chains positions.

Table II. Position of secondary structures after 3D coordinates assignment procedure .

	1SRX(a)	2TRX(b)	1TRX(c)	TREF(d)	TFIN(e)
HELIX α1	11-18	11-17	11-19	12-18	11-18
α2	34-49	32-49	33-48	39-48	35-49
α3	59-63	59-63	---	---	---(f)
α4	95-107	95-107	96-107	96-107	95-107
3_{10}	66-70	66-70	---	66-69	66-69
β-sheet β1	2-8	3-8	4-7	---	---
β2	22-29	21-28	22-29	25-27	23-28
β3	53-58	53-59	53-60	56-58	54-58
β4	77-81	76-82	77-82	77-81	77-81
β5	88-91	86-92	82-92	88-91	88-90

(a) 1SRX: crystallographic structure at 2.8 Å.
(b) 2TRX: crystallographic structure at 1.68 Å.
(c) 1TRX: NMR structure.
(d) TREF: 3D assignment with REFINE
(e) TFIN: final structure after CHARMM refinement.
(f) The short 3_{10} helix has not been observed. The presence of (i,i+3), (i,i+4) and (i,i+5) hydrogen bonds indicates that α -turn could be present.

Fig. 2 shows the ribbon representation of the rebuilt structure superimposed to that of the crystallographic one. Table II shows a comparison between the locations of helices and sheet for the 3D structures generated by our method and some crystal X-ray and NMR-determined structures. Although the characteristic large central β–sheet of thioredoxin was well recovered with its most peculiar features, some hydrogen bonds on the border of the sheet were rather weak.

In particular, as a result of reconstruction, the strand β1 has been distorted and hydrogen bond connections to the β–sheet network have been lost, if compared with both the 2.8Å and 1.68Å crystallographic structure. In 2DFTNMR experimental data on the same thioredoxin carried out in solution, this β–sheet is shifted of some residues and some H-bond are weakened showing a partial distortion of the strand β1 similar to that predicted by our method. The strand β2 shows a quite big displacement with respect to the X-ray and NMR structure, greater than the mean value over all secondary structure (tab. III).

Table III. RMS analysis on different secondary structures.

RMS		TREF/2TRX	TFIN/2TRX	TFIN/1TRX
HELIX α1	11-17	1.37	1.20	1.44
α2	32-49	1.98	1.94	2.07
α4	95-107	1.47	1.42	1.74
STRAND β2	22-28	2.89	2.71	2.85
β3	54-58	1.37	1.43	1.68
β4	77-81	1.44	1.30	1.05
β5	88-90	1.31	1.32	1.11

(a) 2TRX : crystallographic structure at 1.68 Å.
(b) 1TRX : NMR structure.
(c) TREF : 3D assignment with REFINE
(d) TFIN : final structure after CHARMM refinement.

As we have outlined in the introduction, problems of such procedures of prediction of 3D structure lie in the correct assignment of side chains position. The analysis of χ_1 and χ_2 angles ca be useful to evaluate the accuracy of the positioning of the side chains, In Reid and Thornton (1989), where a procedure similar to ours has been used to reconstruct positions of all atoms of flavodoxin from extracted positions of $C\alpha$, the average error on χ_1 was 58° ± 53°, with respect to the same X-ray structure from whom starting $C\alpha$ positions were extracted, while for our procedure the average error was 48° ± 45 °, with respect to the crystallographic structure resolved at 1.68 Å, that is a structure much more refined than the starting one. This improvement might be due to the differences between the protocol of minimization used in our work and the one used by Reid and Thornton. In fact using incrementally detailed potentials makes it more difficult to get trapped in local minima increasing the chances to find optimal structure.

Table IV. RMS analysis after different forcefields minimizations starting from not refined structure.

RMS	NOECH/ NOEAM	NOECH/ NOECV	NOECV/ NOEAMB	SCVFF/ SAMBER
trace	2.04	2.23	1.88	2.35
backbone	2.17	2.32	1.92	2.42
all atoms	2.63	2.86	2.42	2.99

(a) NOECH : structure minimized with CHARMM forcefields plus NOE constrains
(b) NOEAM : structure minimized with AMBER forcefields plus NOE constrains
(c) NOECV : structure minimized with CVFF forcefields plus NOE constrains.
(d) SCVFF: structure minimized with CVFF forcefields without NOE constrains
(e) SAMBER : structure minimized with AMBER forcefields without NOE constrains

Minimizations with different potentials

The secondary structures were preserved by all minimizations.

Comparisons of the results of minimization procedures performed with CVFF and Amber forcefields from the same structures are reported in table IV. We can see that starting from structures far away from an energetic minimum, the final conformations of minimization procedures with different forcefields can show notable differences. This is strictly related to the empirical nature of utilized potential functions. The necessity of further work on forcefields determination is underlined by several works in the literature (Rashin (1993) and ref. quoted thereby, DISCOVER manual (1991)). Until all aspects could not be incorporated in the model and this is shown to be rigorous and complete, all the conclusions based only on energy calculations should be considered only a way to model and to represent complex systems rather than a rigorous thermodynamic study of them. Another point is the numerical approach of MM calculations: to reach convergence from a structure far away from a minimum a great number of steps are needed and so the different numerical approximation required by different potentials have more chances to build up and result in structural differences. The imposing of the same NOE constrains can improve convergence to the same structures, but this process is limited to atoms interested by the distance constrains, and so is related to the number and accuracy of these. Simulation in water environment has produced a structure (TWAT) very similar to the starting crystallographic structure 2TRX (RMS of 0.70 Å on all atoms). It is known that when an MD simulation is run on a protein structure starting from its known X-ray structure the RMS deviation of some conformations produced from initial one can reach 3Å for a vacuum simulation, while for simulations including water, this value lowers to 1Å (Jahnig and Edholm, 1992). Our results are confirming this general behaviour. Oxidized thioredoxin resolved with X-ray crystallography or with NMR spectroscopy, shows one main difference in the $\alpha 3$ helix (residues 59-63), that is present only in the crystal structure. Structures minimized with the same strategies from slightly different structures, but utilizing different potentials do not differ much, but form a cluster of very similar structures. RMS analysis results are shown in table V.

Table V. RMS analysis after different forcefields minimizations.

	2TRX/ XAMBER	2TRX/ XCVFF	XCVFF/ XAMBER	TWAT/ WAMBER	TWAT/ WCVFF	WCVFF/ WAMBER
trace	0.31	0.50	0.48	0.48	0.44	0.40
backbone	0.39	0.58	0.56	0.52	0.49	0.45.
all atoms	0.46	0.66	0.66	0.68	0.62	0.57

(a) 2TRX: crystallographic structure at 1.68 Å.
(b) XAMBER: structure minimized with AMBER forcefields starting from 2trx.
(c) XCVFF: structure minimized with DISCOVER forcefields starting from 2trx.
(d) TWAT: water environment simulation.
(e) WAMBER: structure minimized with AMBER forcefields starting from twat
(f) WCVFF: structure minimized with DISCOVER forcefields starting from twat.

To evaluate the variability of the results of such minimization processes, we looked at the displacement of each atom in the final structures from the starting position. Only a very small fraction of the atoms had moved more than 2Å (3 atoms in both cases), while less than 10% of atoms had moved more than 1Å. All the atoms that had moved more than 1Å were on the surface of the protein. This prompts to think that, when accurate energy parameters are used, minimizations with different semiempirical potentials are equally well suited for optimizing packing of the protein interior.

DISCUSSION

The 3D structure of thioredoxin at atomic resolution obtained in this work shows a fairly good correlation with the experimental secondary structure from X-ray diffraction pattern of oxidized thioredoxin crystals. In conclusion, it appears that our reconstruction and optimization procedure of the original set of X-ray coordinates only from Cα is able to produce a tridimensional atomic structure with a satisfactory degree of accuracy. The main difference reported in literature between the structure resolved by X-ray or NMR is the presence in crystallographic structure of the little 3 helix from 59 to 63. The folding of this helix has been lost during our procedure of coordinate assignment and refinement, giving a final structure more similar to the NMR structure than to the X-ray one. The other secondary structures are very well conserved and attributed. The greater similarity to the NMR structure despite the fact that the starting coordinates came from a crystal structure might be due to the greater similarity of the boundary conditions of the minimization to those available in solution than to those imposed by the presence of spatially confined neighbouring molecules. The helix 3 is probably stabilised by intermolecular interactions that were not represented in the potential used in our single-molecule minimizations. In our water simulation, we have seen only a slight relaxation of this helix, but it is still present, showing that, to be sure to simulate differences between structures in crystal and in solution, we need longer dynamic simulations to be performed. The analysis of the dependence of the procedures of energy minimization from different terms used in the expression of the molecular potentials shows that energy minimization of the same conformations with different potentials could lead to final conformations whose resemblance to each other varies from acceptable to highly unsatisfactory with notable differences in structure and energy. The agreement between minimized structures of the same static conformation is high if the starting structure is already well-minimized, but decreases for high-energy starting structures. In the former case, the main differences are concentrated on the surface of the protein. In fact sterical hindrance of neighbouring atoms provides additional constraints that are stronger than the differences in energy contribution due to the different potentials. Such constraints lack in exposed region of a protein, so here small differences in the potential can lead to minor but significant adjustments. When a treatment of proteins with exposed active sites is sought, the choice of the potential should be carefully evaluated.

REFERENCES

Allinger N.L., Yuh Y.H., 1980, *QCPE* 12:395

Barlow D.J., Thornton J.M., 1988, *J.Mol.Biol.*, 201:601-619

Brooks B.R., Bruccoleri R.E., Olafson B.D., States D.J., Swaminathan S., Karplus M., 1983, *J.Comput.Chem.*, 4:187-217

Cantor C.R., Schimmel P.R., 1980, in *'Biophysical Chemistry'*, Vol.1, W.H. Freeman & Co., S.Francisco

Cohen F.E., Sternberg M.J.E., 1980, *J.Mol.Biol.*, 138:321

Dauber-Osguthorpe P., Roberts V.A., Osguthorpe D.J., Wolff J., Geuest M., Hagler T., 1980, *Protein: Structure, Function, Genetics*, 4:31-47

DISCOVER manual, 1989, BIOSYM Inc.

Dyson H.J., Holmgren A., Wrught P.E., 1989, *Biochem.*, 28:7074-7087

Elofsson A., Nilsson L., 1993, *J.Mol.Biol.* , 233:766-780

Ferro D.J., McQueen J.E., McCown J.T., Hermans, 1980, *J.Mol.Biol.*, 136:1

Hagler A.T., Honig B., 1978, *Proc.Natl.Ac.Sci.U.S.A.*, 75:554

Holmgren A., 1968, *Eur.J.Biochem.*, 6:475-484

Holmgren A., 1985, *Ann.Rev.Biochem.*, 54:237-271

Holmgren A., Södeberg B.O., Eklund H., Bränden C.-I., 1975, *Proc.Natl.Ac.Sci.U.S.A.*, 72:2305-2309

Jahnig F., Edohlm O., 1992, *J.Mol.Biol.* 226:837-850

Kabsch W., Sander C., 1983, *Biopolymers*, 22:577-2633

Katti S.K., Le Master D.M., Eklund H., 1990, *J.Mol.Biol.*, 212:167-184

Levitt M., Warshel A., 1975, *Nature*, 253:694

Nemethy G., Pottle M.S., Scheraga H.A., 1983, *J.Phys.Chem.*, 87:1883

Nemethy G., Scheraga H.A., 1977, *Q.Rev.Biophys.*, 3:239

Nelson D.J., Hermans J., 1973, *Biopolym.*, 12:1269

Nomany F.A., McGuire R.F., Burges A.W., Scheraga H.A., 1975, *J.Phys.Chem.* 79(22):2361

Novotny J., Bruccoleri R.E., Karplus M., 1984, *J.Mol.Biol.*, 177:787

Poland D., Scheraga H.A., 1967, *Biochem.*, 6:3791

Rashin A.A., 1993, *Prog.Biophys.Mol.Biol,.* 60:73-200

Ried L.S., Thornton J.M., 1989, *Proteins: Structure, Function, Genetics*, 5:170-183

Smith J.A., Pease L.G., 1980, *CRC Crit.Rev.Biochem.* 8:315-400

Wiener S.J., et al., 1984, *J.Am.Chem.Soc.*, 106:765

Wiener S.J., Kollman P.A., Nguyen D.T., Case D.A., 1986, *J.Comp.Chem.*, 7:230-252

ENZYME BIOAMPLIFICATION

A. Rigo[1] and M. Scarpa[2]

[1]Biophysics Laboratory, Department of Biological Chemistry
University of Padova, Padova, Italy
[2]Department of Physics, University of Trento, Trento, Italy

INTRODUCTION

The restless quest for increased sensitivity in measurement has led to the development of a variety of devices by which the full sensitivity of measurement may be obtained by a detector followed by an amplification step. Amplification is usually carried out by means of electronics. However, amplification may be performed utilizing biological components, provided by nature in the form of enzymes, antibodies, receptors, in the light of the outstanding characteristics of these biostructures, in particular: selectivity, specificity, modularity and self-assembling. Essentially, bioamplification is based on the interaction of a substance with a bioelement through a catalytic, cycling, or multiplication mechanism to generate a relatively large signal, or in order to increase the signal to noise ratio. Bioelectronics devices are among the possible fields of application of bioamplification. In particular the introduction of a bioamplification step in biosensors permits to push down the detection limits of these devices by one or more orders of magnitude and therefore to measure very low concentrations of analytes. In this way the concentration and purification steps, often required by other analytical techniques, such as mass spectroscopy, chromatography, etc., can be avoided. In fact the detection of traces of compounds by typical analytical techniques, based on very expensive and bulky instrumentation, usually requires an enrichment step (to concentrate the sample) and a purification step (to minimize matrix effects). Furthermore these operations are time consuming and should be carried out in specialized laboratories. Therefore, at difference from biosensors, the majority of the analytical techniques, cannot be applied to real time and in loco measurements.
Bioamplification methods can be classified as follows:

Enzyme amplification	- Amplification by catalysis	
	- Amplification by cycling	*Cycling reactions
		* Electrode cycling
	- Amplification by cascade	
Amplification by multiplication	- Bacterial systems	
	- Liposome systems	
	- DNA polymerase chain reaction	

This paper deals with the description and the possible applications of bioamplification methods using enzymes.

From Neural Networks and Biomolecular Engineering to Bioelectronics
Edited by C. Nicolini, Plenum Press, New York, 1995

ENZYME AMPLIFICATION

Amplification by catalysis

Enzymes are proteins which catalyze reactions occurring in living systems according to the general scheme:

$$S \xrightarrow{E} P \qquad (1)$$

where E is the enzyme and S and P are the substrate and the product, respectively.

The most general application of the enzyme catalysis is the measurement of the S concentration, which does not involve bioamplification. In fact, if the enzyme E is used for the determination of a substrate S, the concentration changes of S or P are stoichiometrically limited by the concentration of S itself. Therefore no inherent gain in the sensitivity, for measuring S, is achieved. Bioamplification by catalysis is strictly related to the measurement of the concentration of E or of a molecule which activates or inhibits the enzyme E. In fact amplification is achieved since a molecule of E catalyzes the formation of many molecules of P. Therefore under controlled conditions the rate of P formation is related to the E activity, and may be taken as a measure of the enzyme concentration provided that P or S are measurable. Theoretically the enzyme concentration may be measured with infinite sensitivity. However practical limitations are:

i) Time of reaction. For very low concentrations of E a long time may be taken to permit the product P to accumulate up to a measurable level.

ii) Uncatalyzed reactions of S to P, which may occur in parallel to the reaction 1.

iii) Presence of catalytic impurities.

Measurement of E concentration. Conventionally the amount of P generated over a fixed amount of time is measured and taken as an indicator of E activity or concentration. The gain is given by the number of molecules of P produced by each molecule of E and, under optimal conditions, the gain is given by the turnover number multiplied by time. In Table 1 the turnover numbers of various enzymes are reported. From this Table it appears that the turnover numbers of many enzymes are of the order of 10^4 -10^5 cycles per s. Therefore gains in the 10^3 - 10^6 range may be achieved by the use of these enzymes as amplification factors, and as a consequence, if the transducer sensitivity is in the µM range, analyte concentrations in n-pM range can be detected by biosensors. These gains are similar to those of operational amplifiers which are typically of the order of 10^5 - 10^6.

Table 1 Examples of turnover numbers of various enzymes

Enzyme	Substrate	Turnover number (s^{-1})
Carbonic anhydrase	CO_2	1.0×10^6
Catalase	H_2O_2	4.0×10^5
Acetylcholine esterase	acetylcholine	2.5×10^4
Ascobate oxidase	ascorbate	1.2×10^4
Urease	urea	1.0×10^4
Alkaline phosphatase	phosphate ester	2.0×10^3
lactate dehydrogenase	lactate	1.0×10^3

In Table 2 the detection limits of various transducers are reported. If enzyme bioamplification is coupled to one of these transducers the detection limits can be pushed down by the factor 1/(turnover number x time). Taking into account Table 1 and 2, it appears that for a turnover number of 10^5, E concentrations in the range of 10^{-8} -10^{-12} M can be measured by some of the transducers reported in Table 2.

Table 2 Detection limits of various transducers

Transducers		Minimum volume[a)], mL	Detection limits M	mol
Optical	- absorption	100	10^{-6}	10^{-11}
	- emission	10	10^{-8}	10^{-13}
Electrochemical	- potentiometry	10	10^{-6}	10^{-11}
	- amperometry	100	10^{-8}	10^{-12}
	- conductometry	100	10^{-4}	10^{-8}
Piezoelectric		100	10^{-5}	

a) these values are strongly dependent on the set-up assembly.

An example is the measurement of the ascorbate oxidase concentration by an UV transducer. In fact ascorbate can be measured in the UV region with a sensitivity of 10^{-6} M. Since the turnover number of ascorbate oxidase is about 10^4 s^{-1}, enzyme concentrations of the order 10^{-11} M can be measured with an integration time of 1 min [Stevanato et al., 1985].

Extension of the catalytic properties to molecules other than enzymes. The versatility of enzyme amplification may be increased by introducing an enzyme function into a variety of molecules. For example, enzymes can be covalently bound to antibodies (Ab) or antigen (Ag) to monitor biomolecules, drugs, viruses etc, or to locate subcellular structures in tissues. By this technique molecules with an extended spectrum of detection have been produced: virtually any substance (antigen) against which antibodies can be raised, can be monitored. The labelling of a biomolecule by an enzyme is the base of the enzyme-immunoassay (EIA). In order to determine the concentration of the analyte X, this technique requires the covalent binding of E to the molecule X to give the conjugate X-E. In the simpler form of EIA, the analyte X competes with the conjugate X-E for a limited number of binding sites on an antibody specific for X. At the equilibrium X-E is present as Ab-bound form and as free form, that is:

$$X + Ab \cdots X - E \Leftrightarrow X - E + Ab \cdots X \qquad (2)$$

The requirement that the activity of the free (X-E) and bound (Ab···X-E) forms are greatly different and the low values of affinity constants of Ab···EX and Ab···X complexes are practical limitations to the sensitivity that can be achieved by EIA.

Enzyme amplification: measurement of effectors or inhibitors. The amplifying nature of enzymes offers the possibility of detection of very low levels of inhibitors or effectors. Examples are the measurement of cofactors or metal ions by apoenzymes, that is by enzymes deprived of these factors. For instance the apoenzyme of tyrosine decarboxylase was incubated with pyridoxal-5'-phosphate, which is the enzyme cofactor, and the initial rate of CO_2 production in the reaction between L-tyrosine and the regenerated holoenzyme was measured [Hassan et al., 1981]. Amplification factors up to 10^5 were achieved and pyridoxal-5'-phosphate level as low as 1 nM were measured with good precision and analysis time of only 20 min. A further example is the measurement of Cu^{2+} ion by apoCuSOD [Stevanato et al., 1981]. This metalloenzyme requires Cu^{2+} ions for the catalysis and a linear increase of the activity with the ratio [Cu^{++} added]/[apoenzyme] has been observed for values of this ratio in the range 0-2. The sensitivity and the gain achieved by this method were 10^{-10} M and 10^6, respectively.

The modulation of the enzyme activity by a specific inhibitor is the base for the measurement of the inhibitor itself as in the detection of cyanide by horseradish peroxidase (HRP) [Smit and Cass, 1990]. In this case the measurement of cyanide in submicromolar concentrations by HRP amplification is carried out by a dual working electrode system. In fact horseradish peroxidase, which is very sensitive to CN^-, is oxidized by H_2O_2 generated from molecular oxygen, at the primary electrode, and then it is reduced at a secondary electrode by mediated electron transfer using ferrocene as an electron carrier, according to the scheme reported in Fig. 1. Since the inhibition of HRP increases increasing the cyanide concentration, the cell current decreases raising the CN^- concentration.

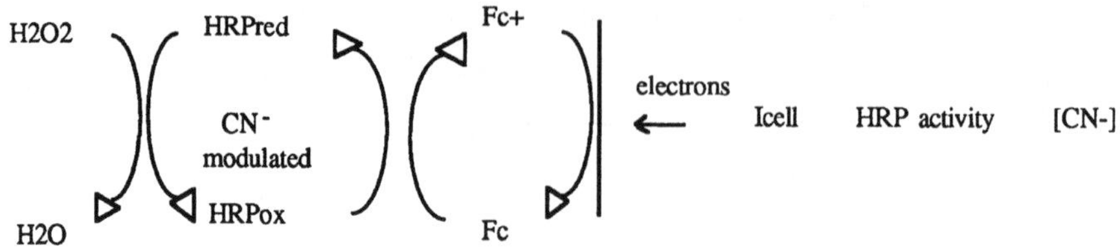

Fig. 1 Detection of cyanide by enzyme amplification. The couple ferrocene-ferricinium ion (Fc-Fc^+) is used as electron shuttle.

Amplification by cycling

Cycling enzymatic reactions. The cyclic amplification, in principle is based on two coupled enzymatic processes which cycle the substrate S, and cause a relatively large change of the concentration of a measurable species. In other words S is recycled by choosing the enzyme 1 (E1) in the way of the reaction product P is in turn the substrate for the enzyme 2 (E2). This second enzyme must convert P back to S, as reported in the following scheme, where A and C are the reagents, and B and D are the products :

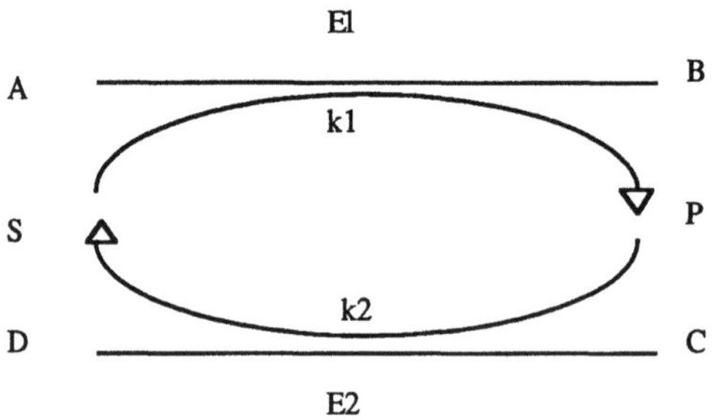

(3)

Therefore the use of the enzymes E1 and E2 permits a consumption-regeneration cycle for the species S and P and gives an increased concentration change, beyond stoichiometric limitations of the products B and D. This increased change of concentration of the reaction products provides the gain in the sensitivity for measuring the species S.

In general, if k_1 and k_2 are the apparent first-order kinetic rate constants (that is k_{cat}x[enzyme]) of the two enzymatic reactions, at the steady-state the overall cycling rate constant is:

$$k = \frac{k_1 \times k_2}{k_1 + k_2} \tag{4}$$

This means that the overall rate constant for the amplification cycle, see scheme 3, is practically controlled by the slowest reaction. It must be remarked that, if a cycling procedure is applied to the measure of S concentration, the reaction rate must be proportional to [S], and therefore it should be $[S] << Km_S$, where Km_S is the Michaelis constant for the substrate S. Large signal amplification can be obtained in this way. The first cycling method was described by Lowry [1973] using lactic dehydrogenase and glutamic dehydrogenase. In Table 3 working examples of developed amplification systems have been reported, keeping in mind that the number of potential enzyme pairs, for cycling amplification of cofactors or metal ions, is very high [Lowry, 1973; Kato et al., 1973; Cox et al., 1982; Yao et al., 1990; Scheller et al., 1985; Breckenridge, 1964; McNeil et al., 1989; Schubert et al., 1985; Carrico et al, 1976]

Table 3 Enzyme amplification systems.

Enzyme 1	Enzyme 2	Substrate	Gain
Lactic dehydrogenase	Glutamic dehydrogenase	NADP	8,000 cycle/h [Lowry, 1973]
Glyceraldehyde 3-phosphate dehydrogenase	Glutamic dehydrogenase		20,000 cycle/h [Lowry, 1973]
Alcohol dehydrogenase	Malic dehydrogenase	NAD	30,000 cycle/h [Kato et al., 1973]
Glucose 6-phosphate dehydrogenase	NAD peroxidase	NAD	4,000 cycle/h [Cox et al., 1982]
Glutamate pyruvate transaminase	Glutamate oxidase	L-glutamate	1,000 cycles [Yao et al., 1990]
Lactate dehydrogenase	Lactate oxidase	Lactate	1,000 cycles [Scheller et al., 1985]
Hexokinase	Pyruvate kinase	3,5-AMP	5,000 cycles [Breckenridge, 1964]
NADH oxidase	Alcohol dehydrogenase	NAD	15 cycles [McNeil et al., 1989]
Glucose dehydrogenase	Glucose oxidase	Glucose	8 cycles [Schubert et al., 1985]
Lactate dehydrogenase	Diaphorase	NAD	20 cycles [Carrico et al, 1976]

If cycling is efficient, without loss of S to side reactions, and if the measurable product is stable enough to permit accumulation, the amplification achieved by this approach may be very large. Under optimal conditions, with soluble enzymes making 20,000-50,000 cycles/h, detection limits as low as 10^{-15} mol have been achieved in the case of some coenzymes.

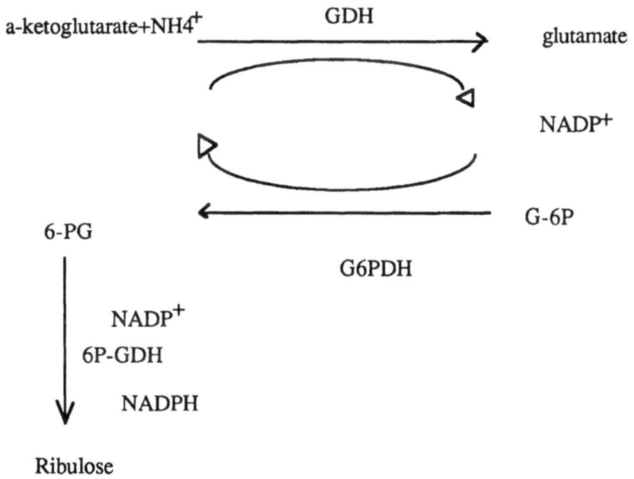

Fig. 2 Cofactor amplification by cycling. GDH, glutamic dehydrogenase; G6PDH, glucose-6-phosphate dehydrogenase; 6P-GDH, 6-phosphogluconate dehydrogenase; G6P, glucose-6-phopsphate; 6-PG, phosphogluconolactone.

A typical example of cofactor amplification is the determination of NADPH by cycling between two enzymes (glutamic dehydrogenase and glucose-6-phosphate dehydrogenase) in the presence of nonlimiting concentrations of α-ketoglutarate, ammonium ion and glucose-6-phosphate, see Fig 2. In this case cycling rates as high as 20,000 cycles per hour have been obtained [Lowry, 1973]. After stopping the cycling, 6-phosphogluconolactone is converted into NADPH with the aid of 6-phosphogluconate dehydrogenase and extra amount of $NADP^+$. With the cycle step repeated, using the NADPH formed by the first stage, and destroying the excess of $NADP^+$ with alkali, another factor of 20,000 can be obtained, to give an overall amplification of more than 10^8. In principle, two-stage determinations should permit the assay of 10^{-19} mol of NADPH. A further development of the cycling systems, proposed by McNeil et al. [1989] and Stanley et al. [1985], is a two-stage system consisting of an enzyme label, "the preamplifier", which activates a cycling system: "the power amplifier". A schematic diagram of the preamplifier and power amplifier system has been reported in Fig. 3.

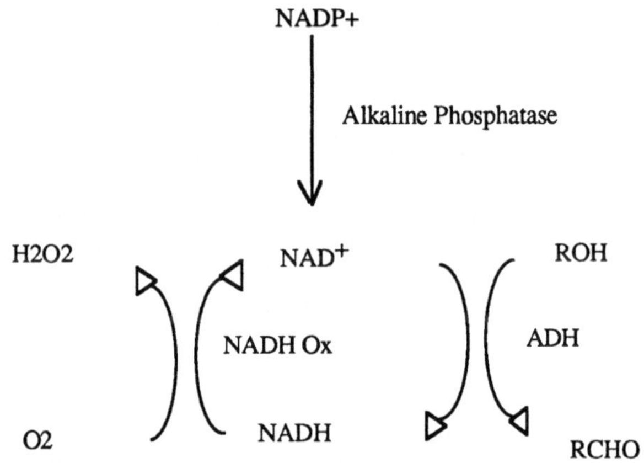

Fig. 3 Schematic outline of a bio-preamplifier-power amplifier system. Alkaline phosphatase catalyzes the dephosphorylation of $NADP^+$ to NAD^+ (biopreamplification) that enters a cycle in which NADH oxidase catalyzed oxidation of NADH produces H_2O_2 (power amplification), which is detected amperometrically.

Alkaline phosphatase (enzyme label) catalyzes the dephosphorylation of $NADP^+$ to NAD^+. NAD^+ then takes part in a cycle driven by two enzymes, alcohol dehydrogenase and NADH oxidase (power amplifier). The hydrogen peroxide produced by the reduction of oxygen can be measured electrochemically. The enzyme power amplifier is comparable to an electronic amplifier, the gain of which may be adjusted to provide the necessary sensitivity. The ability to amplify cofactors by cycling has led to a novel use of cofactors as labels in competitive binding assays [Carrico et al., 1976]. A NAD derivative was bound to the analyte X. The obtained conjugate X-NAD, in which NAD is still active as cofactor, was employed in the lactate dehydrogenase-diaphorase cycling system. In the presence of an antibody specific for the analyte, the resulting complex Ab···X-NAD was inactive as NAD cofactor in the cycling system. The inhibition can be relieved in a competitive manner by the addition to the reaction system of the free analyte X in nanomolar range.

Cycling electrode enzymatic reactions. The cycling enzyme methods are not widely used because they require a considerable effort in manipulation and long incubation times. Furthermore the electrochemical methods, compared with the spectrophotometric ones, show in general many advantages: in fact they can be used with solutions characterized by high turbidity or absorbance, show high sensitivities, due to low detection limit for products, and are relatively easy and inexpensive to operate. Therefore the combination of an enzymatic cycling reaction with an electrode system is particularly advantageous. In a typical cycling electrode enzyme reaction, an electron transfer process, occurring at the electrode, is followed by an enzyme reaction that regenerates the electroactive material. In this way a limiting current higher than that obtained from the electron transfer reaction alone is generated. Such current is called catalytic current. In general, under given conditions, the gain can be evaluated by the ratio of the steady-state currents measured in the presence and in the absence of the enzyme E. A scheme of an electrode cycling enzyme reaction is reported in Fig. 4.

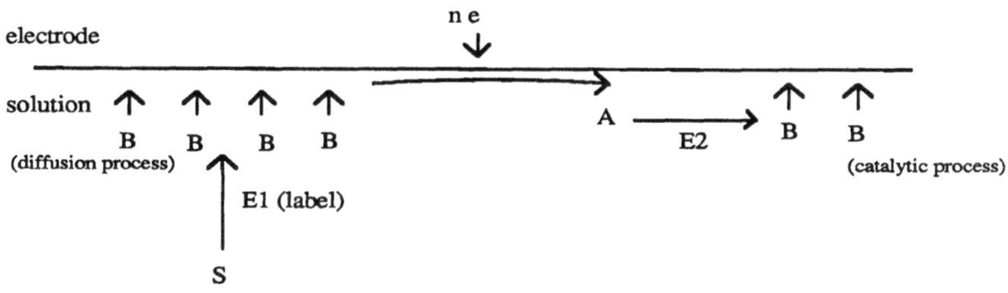

Fig. 4 Scheme of a cycling enzymatic electrode reaction

In this example to obtain a catalytic current, the working potential of the electrode must be set sufficiently cathodic to reduce B, but not to reduce S. Under these conditions, B, generated by the enzyme label E1, is converted by the electrode reaction in the product A and the cell current is proportional to B concentration (diffusion current). After addition of a second enzyme, able to convert the electrode product A back to B, an instantaneous increase of the current (catalytic current) is observed. This catalytic current permits to increase the sensitivity by orders of magnitude. Examples of the potentiality of this method are the systems:

- hydroquinone-benzoquinone and glucose oxidase;
- molecular oxygen- superoxide ion and superoxide dismutase.

In the first case hydroquinone (H2Q) is consumed by the electrode reaction to form benzoquinone but it is regenerated by the oxidase reaction according to the following scheme:

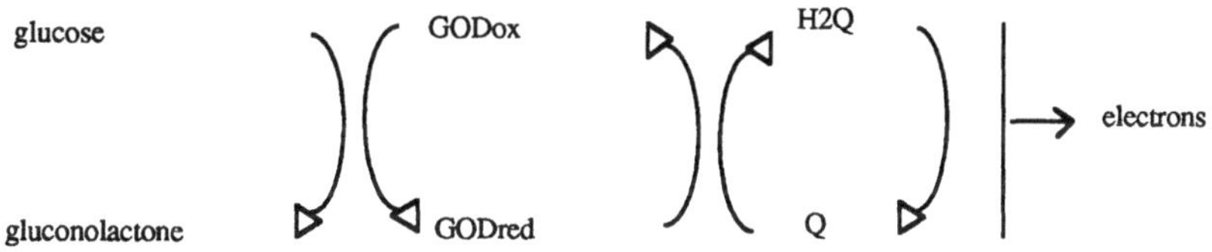

(5)

where GODox and GODred are the oxidized and reduced forms of GOD, respectively. The consumption/regeneration cycle for H2Q generates an enhanced electrode current, see Fig. 5.

In the case of the measurement of the concentration of superoxide dismutase (SOD) the electroreduction of molecular oxygen in aqueous solutions, in the presence of a surfactant such as triphenylphosphin oxide, is a monoelectronic process which generates very unstable superoxide ion, O_2^- [Rigo et al., 1975; Argese et al., 1984]. Under these conditions the value of the electrode current is controlled by the diffusion of the molecular oxygen, O_2. The addition of a compound, such as superoxide dismutase, able to catalyze the dismutation of superoxide ion into H_2O_2 and O_2, converts part of the superoxide ion present in the reaction layer to molecular oxygen which, adding to the diffusive stream of O_2 from the solution to the electrode, increases the electrode current. Therefore the electrode is at the same time a source of O_2^- and a very sensible detector of its dismutation. The electrode process can be described, in a simplified form, by the following differential equations:

$$\frac{d[O_2^-]}{dt} = D_{O_2^-}\frac{d^2[O_2^-]}{dx^2} + \frac{2xd[O_2^-]}{3tdx} - k_1[SOD][O_2^-] - k_2[O_2^-]^n \tag{6}$$

$$\frac{d[O_2]}{dt} = D_{O_2}\frac{d^2[O_2]}{dx^2} + \frac{2xd[O_2]}{3tdx} - \frac{k_1}{2}[SOD][O_2^-] - \frac{k_2}{2}[O_2^-]^n \tag{7}$$

where x is the normal distance from the electrode surface, k2 and n are the kinetic rate constant and the reaction order of the spontaneous dismutation process of superoxide ion with respect to O_2^- itself, D_{O_2} and $D_{O_2^-}$ are the diffusion coefficients of O_2 and O_2^-, respectively. Solving the differential system for n=1 (first-order condition), the following approximate solution is obtained:

$$f(R) = 7.42R / (1 - 1.25R) = t_g k_1 [SOD] t_g + k_2 t_g \qquad (8)$$

where t_g is an electrode constant. Taking $k_1 = 3 \times 10^9\ M^{-1}s^{-1}$ and $t_g = 3$ s, SOD concentrations between 10^{-11} and 10^{-8} M give a linear increase of the left side of equation 8.

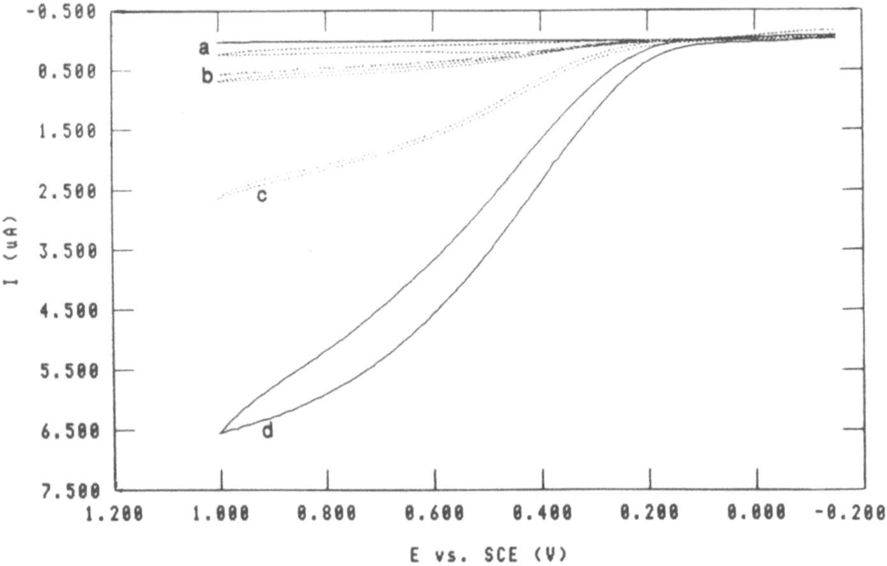

Fig. 5 Scheme of hydroquinone amplification via enzyme electrode reaction. Hydroquinone (H2Q) is consumed by the electrode reaction to form benzoquinone (Q) but it is regenerated by the glucose oxidase reaction (see text): the consumption-regeneration cycle for H2Q-Q gives an enhanced electrode current. Cyclic voltammograms of 30 mM hydroquinone in the presence of 10 mM glucose, at increasing concentrations of glucose oxidase (GOD). a) 0 mM GOD; b) 0.3 mM GOD; c) 1.5 mM GOD; d) 3.8 mM GOD.

CASCADE AMPLIFICATION SYSTEM

An enzyme cascade system provided by nature can be used as amplification mechanism. This type of amplification requires a system of enzymes such as that involved in the clotting of the blood. In this way the primary signal can be amplified by several orders of magnitude through a cascade amplification mechanism. Enzyme cascade systems, such that involved in limulus amoebocyte, have been proposed [Seki et al., 1990; Blake et al., 1984]. The reaction scheme for the limulus clotting system is reported in Fig. 6.

In this amplification system three proenzymes and an artificial substrate for the clotting enzyme are utilized: a lipopolysaccharide, LPS, (the label) activates the proenzyme C. The activated enzyme cleaves many molecules of the next proenzyme in the chain, and so on. In this way the primary signal can be amplified by several orders of magnitudes. In the reported example the clotting enzyme catalyzes the hydrolysis of the substrate and the produced p-nitroaniline is detected at 400 nm. In this way LPS can be detected in the pg ml^{-1} range in a sample volume of 20 μL.

In conclusion, gains up to 10^6 can be achieved in periods of times ranging from ms to hours by bioamplification using enzymes. These gains permit to obtain biosensors which can be used to detect concentrations of analytes in the n-pM, sensitivities which cannot be achieved directly by more sophisticated and expensive instrumentations. Finally it must be remarked that while the classic solid state devices have approached their physical limits, bioelectronic devices such as biosensors, with their characteristics of selectivity, amplification and self-assembling, are at early stages of development and a future improvement can be expected.

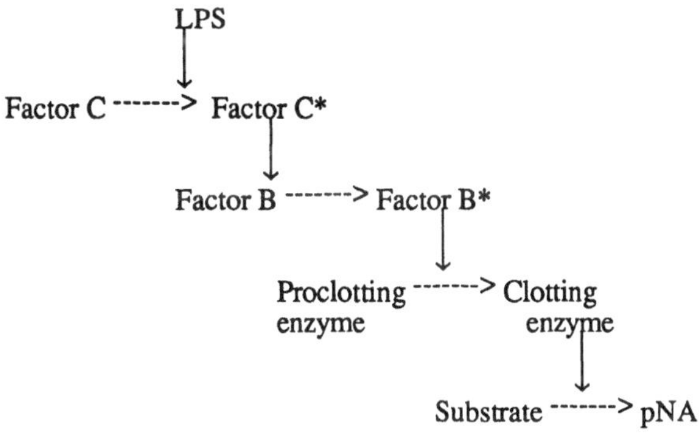

Fig. 6 Cascade amplification system (from *Limulus amoebocyte*). Lipopolysaccharide (LPS) activates four proezymes, (factors C , C', B and B'). Factor B' activates the proclotting enzyme to clotting enzyme which catalyzes the hydrolysis of the substrate. The generated p-nitroaniline (pNA) is detected by spectrophotometry. Substrate = Boc-Leu-Gly-Arg-pNA

REFERENCES

Argese.E., Orsega E.F., Carli B., Scarpa M. and Rigo A., 1984, *Bioelectrochem. Bioenerg.*, 13:385-392

Blake D. A., Skarstedt M. T., Shultz J. L. and Wilson D. P., 1984, *Clin. Chem.* 30:1452-1457

Breckenridge McL., 1964, *Proc. Acad. Nat. Science* , 52:1580-1586

Carrico R., Christner J.E., Boguslaski R.C. and Yeung K.K., 1976, *Anal. Biochem.*, 72:271-280

Cox C., Camus P., Buret J. and Duvidier J., 1982, *Anal. Biochem.*, 119:185-193

Hassan S.S. and Rechnitz G.A., 1981, *Anal. Chem.*, 53:512-515

Kato T., Berger S.J., Carter J.A. and Lowry O.L., 1973, *Anal. Biochem.*, 53:86-97

Lowry O.L., 1973, *Acc. Chem. Res.*, 6:289-293

McNeil C.J., Spoors J.A., Cocco D., Cooper J.A. and Bannister J., 1989, *Anal. Chem.*, 61:25-29

Rigo A., Viglino P. and Rotilio G., 1975, *Anal. Biochem.*, 68:1-8

Scheller F., Siegahn N., Danielsson B. and Mosbach K., 1985, *Anal. Chem.*, 57:1740-1743

Schubert F., Kirstein D., Schroder K.L. and Scheller F.W., 1985, *Anal. Chim. Acta* , 169:391-396

Seki.A., Tamiya E. and Karube I., 1990, *Anal. Chim. Acta* , 232:267-271

Smit M.H. and Cass A.E., 1990, *Anal. Chem..*, 62:2429-2436

Stanley.C.J., Johansson A. and Self C.H., 1985, *J. Immunol. Methods* , 241:571-575

Stevanato R., Avigliano L., Finazzi-Agrò A.and Rigo A., 1985, *Anal. Biochem.*, 149:537-542

Stevanato R., Viglino P., Rigo A. and Cocco D., 1981, *Bioinorg. Chem. Letters, Inorg. Chim. Acta* , 56:39-40

Yao T., Yamamoto H. and Wasa T., 1990, *Anal. Chim. Acta* , 236:437-440

ELECTRIC "CONTACTS" BETWEEN CONDUCTORS AND PROTEIN ACTIVE SITES

S.D. Varfolomeyev and A.J. Yaropolov#

Moscow State University, 119899, Moscow
#Institute of Biochemistry, Russian Academy of Science

INTRODUCTION

The problems related to the mechanisms of electron transfer between electron-conducting structures (metals, semiconductors or organic conductors) and biopolymer molecules are central and most important in biolelectronics. Our interest in this area was to study the enzyme behavior at the interface "ionic conductor (electrolyte solution or solid electrolyte) - electron conductor (metal, carbon or semiconductor)". In the experiments with various enzymes, we have shown that the direct electron transport between a conductor and the active site of redox enzyme is feasible. The rates of electron transfer, which may well proceed by the tunneling mechanism, can exceed the rates of subsequent enzymatic steps. In this case, we can think of an electric "contact" between semiconductor and enzyme active site.Experimentally, the processes of electron transfer between enzyme active site and conductor are most convenient to observe in electrochemical systems upon enzyme immobilization on electron conductor surface. In this case, the enzyme can act as electrocatalyst "pumping out" the electrons from conductor or donoring them into the conductor. This phenomenon was first revealed for a blue copper-containing oxidase (laccase) reducing molecular oxygen to water; the source of electrons was the conductor matrix [Berezin I.V et al. 1978; Berezin I.V et al., 1980]. The phenomenon of acceleration of electrode processes in direct "contact" of the enzyme and conductor was termed bioelectrocatalysis. Later, this phenomenon was observed for a number of other enzymes and for the enzymes entrapped in the matrices of organic semiconductors and organic metals [Varfolomeev S.D. et al, 1978; Hill H.A.O. et al, 1981]. This review is an endeavour to sum up these studies and to demonstrate the applicability of direct electron transport for creation of reagentless biosensors.

MECHANISMS OF TRANSFER OF ELECTRONS BETWEEN THE PROTEIN ACTIVE SITE AND THE CONDUCTOR

The electrocatalytic transport of electrons can involve two fundamentally different mechanisms [Berezin I.V. et al, 1978; Varfolomeev S.D. et al, 1988].
(1) The transfer of electrons can be performed with the help of a diffusively mobile intermediate low-molecular carrier of electrons (mediator). The scheme of the process in this case can be presented as follows:

From Neural Networks and Biomolecular Engineering to Bioelectronics
Edited by C. Nicolini, Plenum Press, New York, 1995

$$S + E \Rightarrow P + E^0$$
$$E^0 + M \Rightarrow E + M^0 \qquad (1)$$
$$M^0 \xrightarrow{electrode} M \pm e^-$$

where S and P are the substrate and the product of the reaction, E and E^0 are the oxidised and the reduced forms of the active site of the enzyme M and M^0 are the oxidised and the reduced forms of the mediator. The mediator should be a sufficiently specific substrate of the enzyme used and be electrochemically active on the electrode made of the given material. The redox potential of the mediator should be close to that of the fuel employed or oxidising agent. It is important that the mediator be stable against degradation processes The mediator mechanism of electron transport is fairly widely used in electrochemical enzymatic reactions .
(2) A direct electrocatalytic transfer of electrons between the electrode and the active site of an enzyme is possible. The direct" electron transfer process in electrochemical reactions was demonstrated with some enzymes (see below). The phenomen of direct electron transfer from protein active site to conductor (or reverse process) can be determed as DET- effect. Below are analysed the DET-effect on an example of some enzymatic reactions.

BLUE COPPER OXIDASE (LACCASE) THE ENZYMATIC ELECTROCHEMICAL REDUCTION OF OXYGEN

Studies of the enzymatic electrocatalysis of oxygen reduction are of fundamental interest. In classical electrochemistry this is one of the most complicated problems. The equilibrium redox potential of the O_2/H_2O couple (1.23 V) is known to be established only on specially pretreated platinum and in extra pure solutions. Oxygen exchange currents on platinum are very low, 10^{-11} A/cm. On the other hand, enzymes are known which actively reduce oxygen to water by the four-electron mechanism without the intermediate formation of hydrogen peroxide in the solution. Laccase is a copper-containing enzyme performing the four- electron reduction of oxygen, with different aromatic amines and phenols used as donors. The active site of the enzyme consists of four copper ions involved in a coordinated oxygen reduction. Laccase is a "typical" enzyme as regards its catalytic efficiency $k_{cat} = 200$ s^{-1}, Km with respect to oxygen is 10^{-5} M) Oxygen electroreduction in neutral or weakly acidic solutions on carbonaceous materials is known to proceed at high overvoltage. It was found however that laccase in minor quantities (10^{-9} M) strongly shifts the stationary potential towards the zone of positive values and accelerates oxygen electroreduction [Berezin I.V. et al, 1978; Bogdanovskaya V.A. et al, 1979].

The above effects do not depend on the nature of the electrode. Electrochemical measurements were conducted on electrodes made of carbon black, pyrographite, vitreous carbon or gold. Laccase was immobilised by direct adsorption onto the electrode. The potential of all the electrodes rose in the presence of oxygen and laccase. The maximum value of the potential (+1.207 V), close to the equilibrium potential of oxygen electrode, was established on carbon black electrodes which had been kept in a laccase solution (10^{-5} M) for 24 h. Adsorption of the enzyme on carbon black electrodes is practically irreversible. After immobilisation the electrode retains its catalytic properties in the absence of laccase in the solution. The enzymatic nature of electrocatalysis in this case was proved by a specific inhibition of electrocatalytic effects by fluoride and azide ions, heat inactivation, and by comparing the pH dependence of the electrocatalytic effects and the catalytic activity in the oxidation of ferricyanide ion by oxygen.

Experiments showed that the stationary potential of the electrode depends on oxygen partial pressure and the solution pH. To elucidate the nature of the stationary potential (E_{st}) established on the electrode with immobilised laccase, the effect of oxygen partial pressure and the solution pH on Est was studied. It was found that $\partial E_{st}/\partial pH = 10^{-12}$ mV and $\partial E_{st}/\partial pH = 60$ mV. These values are close to the calculated coefficients of the Nernst equation for the O_2/H_2O system. Rotating disc-electrode experiments did not detect intermediate hydrogen peroxide in the solution. The observed electrochemical process on the electrode with immobilised laccase is governed by the reaction of oxygen four- electron reduction to water:

$$O_2 + 4e^- + 4H^+ \xrightarrow{immobilized.laccase} 2H_2O \tag{2}$$

Thus, the date available demonstrate the possibility of the enzymatic electrocatalysis of oxygen reduction via a direct electron transfer along the electrode → oxygen molecule circuit and may serve as a basis of creating efficient electrodes of biocatalytic oxygen reduction. That was the first observation of DET-effect. This phenomenon was investigated experimentaly in detailes [Bogdanovskaya V.A. et al., 1979; Bogdanovskaya V.A. et al., 1980; Varfolomeev S.D. et al, 1982].

It was studied the dependence of the current on the potential under potentiostatic conditions at different concentrations of laccase immobilised on the surface of carbon and at various concentrations of hydrogen ions and partial pressures of oxygen. The rates and, correspondingly, the currents of the electrochemical reduction of oxygen in a certain range of concentrations and potentials depend linearly on the surface concentration of the immobilised enzyme. The linear relationship between the reaction rate and the concentration of the enzyme was observed in the 0.6-1.2 V range of potentials measured with respect to hydrogen electrode in the same solution. In this range the enzymatic rections is the rate-determining stage

The kinetic results can be represented in the form of the following equations:

$$\partial E/\partial \log i = \partial E/\partial \log P_{O_2} = 0.5\ \partial E/\partial pH = 26 \pm 3\ mV \tag{3}$$

or

$$i \approx [e][O_2][H^+]^2 \exp\left[\frac{E}{(26 \pm 3)mV}\right] \tag{4}$$

e - surface concentration of enzyme.

The theoretical analysis of the reaction mechanism includes a search of a process scheme whose kinetic description would give Eq. (4). A correct method of analysis seems to involve the discussion of the simplest reaction schemes fitting the experiment and a gradual complication of the mechanisms if the conclusions do not agree with the experimental results.

The following basic assumptions were made in the analysis of the mechanism of the catalytic effect of laccase in oxygen reduction:

1. The active site of the enzyme forms a complex or a chemical compound with oxygen molecule. This is implied by the fact that in the process of action of the enzyme no semireduced oxygen forms (free radicals, hydrogen peroxide) were obseved in the solution.
2. Proton addition reactions proceed under equilibrium conditions and can be characterised by equilibrium constants. This assumption is substantiated by the results of studies of the kinetics of proton tranfer reaction .
3. The reactions of electron tranfer from the electrode to the active site of the enzyme follow the equation of slow discharge and their rate is defined by the electrode potential [Skorchelletti V.V. et al,1974; Krishtalin L.I. et al, 1979]:

$$i_a = k_a(E) = k_0 \exp\left[\frac{nF}{RT}\alpha E\right] \tag{5}$$

$$i_c = k_c(E) = k_0' \exp\left[-\frac{nF}{RT}\beta E\right] \tag{6}$$

where E is the potential measured with respect to a standard electrode (e.g. hydrogen electrode): α and β are transfer coefficienta ($\alpha + \beta = 1$).

Further on, the rate constant of the electrode process of electron transfer

$$A + ne^- \xrightarrow{k(E)} A^{n^-} \tag{7}$$

will be denoted as k (E). If the electrode process is fast and reversible and can be described by the Nernst equation:

$$A + ne^- \Rightarrow A^{n^-}; E = E_0 + \frac{RT}{nF}\ln\frac{A}{A^-} \quad (8)$$

the equilibrium constant giving the ratio of the oxidised and reduced forms will be expressed as

$$\frac{A^{n^-}}{A} = K(E) = K_0 \exp\left(-\frac{nF}{RT}E\right) \quad (9)$$

The experimentally found equation of the rate of electrochemical oxygen reduction by laccase (Eq.4) shows that the rate of the process has the second order with respect to the concentration of hydrogen ions. This means that the limiting stage is preceded by two stages of proton addition. The simplest mechanism which is consistent with this fact involves the stages of addition of oxygen and two protons and the kinetic stage of reduction of the active site:

$$\begin{aligned} &e + O_2 \xrightarrow{K_S} eO_2 \\ &eO_2 + H^+ \xrightarrow{K_1} eO_2H^+ \\ &eO_2H^+ + H^+ \xrightarrow{K_2} eO_2H_2^{2+} \\ &eO_2H_2^{2+} + e^- \xrightarrow{k(E)} eO_2^-H_2^{2+} \end{aligned} \quad (10)$$

In terms of the proposed classification system, mechanism (10) will be denoted as O2ppel (oxygen-proton-proton-electron). Further on the constants describing the reaction kinetics will be presented as: K_S, the constant of dissociation of the enzyme-oxygen complex; K_1 and K_2, the constants of protolytic equilibria; k(E), the rate constant of "electrochemical discharge" (Eq.6).

For scheme (10), the equation of stationary rate of the first-order reaction with respect to oxygen is:

$$v = \frac{k_{O_2}}{K_{O_2}}[e]_0[O_2] = \frac{k(E)}{K_S K_1 K_2}[H^+]^2[e]_0[O_2] \quad (11)$$

or

$$i \sim [e]_0[O_2][H^+]^2 \exp\left(-\frac{\beta nF}{RT}E\right) \quad (12)$$

Equation (12) describes the experimentally observed dependence of the reaction rate on the concentration of the enzyme, oxygen, and hydrogen ions.

Nevertheless, this group of reaction mechanisms is highly improbable since it follows that the "elementary" stage of the enzyme is of multielectron nature. A comparison of Eqs. (12) at $\beta < 1$ with the experimental data ($\partial E/\partial \log i = 26 \pm 3$ mV) shows that $n > 2$. The transfer of 3 or 4 electrons in one elementary act is a process with a very low probability and reactions of this kind are unknown in electrochemistry. This makes us to modify the mechanism under discussion. It can be thought that the kinetic stage of electron transfer is preceded by fast equilibrium stages so that the number of electrons transferred in the rate-determining stage is substantially lower. This assumption leads us to a group of mechanisms which are in accordance with the experimental data.

Consider the mechanisms of this group. Denote as $el_{eq}^{n_1}$ the fast equilibrium stage involving the transfer of n_1 electrons with equilibrium constant K(E), as el^{n_2} the rate-determining stage involving the tranfer of n_2 electrons. The $O_2 el_{eq}^{n_1} pp\, el^{n_2}$, $O_2 p\, el_{eq}^{n_1} p\, el^{n_2}$ and $O_2 pp\, el_{eq}^{n_1} el^{n_2}$ mechanisms differ in the location of the equilibrium electrochemical stage. For these mechanisms the reaction rate is described as:

$$i \sim \frac{k_0 \exp\left(-\frac{\beta n_2 F}{RT}E\right) K_0 \exp\left(-\frac{n_1 F}{RT}E\right)}{K_s K_1 K_2}[H^+]^2 [e]_0 [O_2]_0 \qquad (13)$$

No experimental values of K_0, K_1 and K_2 are available at present; therefore, these mechanisms are indistinguishable for the time being. Fundamentally important information can be obtained from the analysis of the numerical value of $\partial E/\partial \log i$ because it gives the number of electrons transferred in each stage. It follows from Eq. (13):

$$\frac{\partial E}{\partial \log i} = \frac{\partial E}{\partial \log O_2} = \frac{2\partial E}{\partial pH} = \frac{2.3RT}{F(n_1 + \beta n_2)} \qquad (14)$$

A comparison with experimental data gives:

$$n_1 + \beta n_2 = 2 \qquad (15)$$

The values in Eq.(15) are limited by the inequalities $n_1 \geq 1$, $n_2 \geq 1$, $0 < \beta < 1$, as follows from the analysis of the physical picture of electron tranfer. Since $n_1 < 2$ (Eq.31), the only integer solution of these inequalities is $n_1 = 1$. Accordingly, $\beta n_2 = 1$; Therefore, at $\beta < 1$, $n_2 > 1$. The only nearest physically relevant integer solution of this inequality is $n_2 = 2$. The $O_2 el_{eq}^{n_1}$pp el^{n_2} mechanism is expressed as

$$\begin{aligned}
&e + O_2 \overset{K_s}{\longleftrightarrow} eO_2 \\
&eO_2 + e^- \overset{K(E)}{\longleftrightarrow} eO_2^- \\
&eO_2^- + H^+ \overset{K_1}{\longleftrightarrow} eO_2^- H^+ \\
&eO_2^- H^+ + H^+ \overset{K_2}{\longleftrightarrow} eO_2^- H_2^{2+} \\
&eO_2^- H_2^{2+} + 2e^- \xrightarrow{k(E)} eO^- + H_2O
\end{aligned} \qquad (16)$$

plus fast (non-limiting) stages

The two-electron transfer from the electrode to the active site of the enzyme in a rate-determining stage seems to be rather unusual. However, the existence of a two-electron stage in the mechanism of laccase action is in accordance with the results of structure studies which showed that the active site of the enzyme contains a two-electron acceptor formed by two copper ions. The kinetics of oxidation of a number of organic and inorganic substrates of the enzyme with oxygen in solution was examined [Malmström B.G. et al, 1974; Naqui Ali et al, 1980]. The rate of homogeneous oxidation was studied as a function of the concentration of the enzyme, substrates, and hydrogen ions.

The results of the kinetic analysis based on the data of the electrochemical kinetics of cathodic oxygen reduction and the stationary kinetics of the laccase catalysis of oxidation of organic and inorganic donors are summarised by the following scheme:

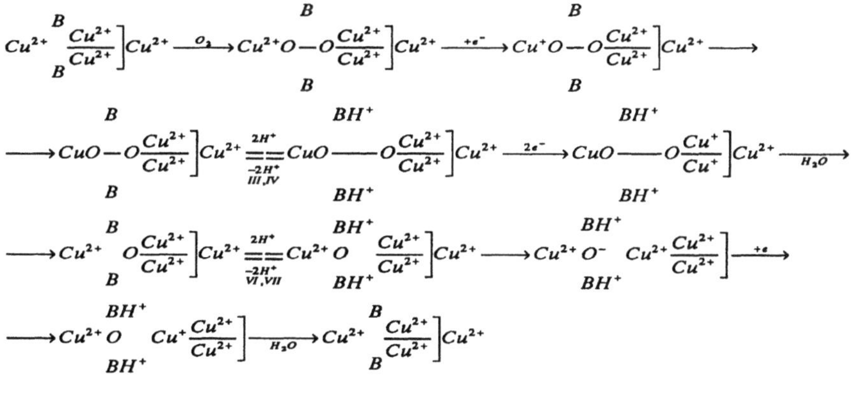

The comparing of the kinetic data and the information on the composition and structure of the active site of laccase can give a picture of the molecular mechanism of oxygen reduction by this enzyme.

It was known that the active site of laccase contains four copper ions of three types. Two copper ions of the third type are associated and can act as a two-electron donor-acceptor. Of fundamental importance is the fact that two ionogenic groups with similar (identical) pK_a values (about 5.5) were found to participate in the mechanism of catalysis. It can be believed that these ionogenic groups are two bases which are contained in the active site of the enzyme.

Stage II, being a one-electron reduction of the active site of the enzyme, seems to be required for the "strengthening" of the complex of oxygen with the active site. Stages III, IV, VI, and VII depict the protonation of bases contained in the active site of the enzyme. The two like, symmetrical bases play a fundamentally important role in the mechanism of oxidase catalysis. On the one hand, in the protonated form they facilitate the transfer of electrons from the donor (electrode) to the two- electron acceptor, on the other, they accelerate electron transport and are donors of two protons in the stage of water formation.

It follows from kinetic data that the mechanism of catalysis includes the formation of an intermediate particle of the EO^- type (stage V). This particle was detected in independent ESR studies of the enzyme [Andreasson L. et al, 1976].

Thus, the investigation of the kinetics of laccase catalysis under electrochemical conditions and at a stationary rate of oxidation of various substrates made it possible to postulate a detailed and experimentally substantiated molecular scheme of action.

HYDROGENASE

As the carbon black electrode with immobilized hydrogen use from *Thiocapsa roseopersicina* was introduced into the buffer phosphate solution saturated with hydrogen, a potential equal to 0.000±0.001 V was set at the electrode. This corresponded to the value of the equilibrium hydrogen potential [Yaropolov A.I. et al, 1983; Yaropolov A.I. et al, 1984]. In an atmosphere of argon, the stationary potential is shifted to the region of positive values. In the presence of inactivated hydrogenase the background reaction of hydrogen electrooxidation in the observed region of potentials is absent.

Addition of carbon monoxide, a hydrogenase inhibitor, to the system also leads to complete disappearance of electrochemical activity of the enzyme electrode in the H_2 atmosphere. Consequently, the observed potential is not connected with redox reactions of any groups in the protein, and it is established only in the presence of hydrogen. The presence of an active enzyme is also an indispensable condition. The observed potential (E) varies linearly with the pH of the solution and with the hydrogen concentration (pH_2). The slope values are approx. -59 and -23 mV. The partial derivatives of potential relative to pH and $\log pH_2$ can be calculated from Nernst equation for the oxidation/formation reaction of hydrogen. The values so defined (-59.2 and -29.6 mV) are in good agreement with those found experimentally.

Thus, it is shown that an equilibrium hydrogen potential is established at the enzyme electrode with immobilized hydrogenase. Logarithm of current as a function of pH and logarithm of hydrogen partial pressure at different values of overvoltage was studied. The zero order of the electrocatalytic reaction rate with respect to substrate and product concentrations was observed. It was concluded that the hydrogen molecule addition and proton ejection out of the active enzyme center take place faster than the oxidation-reduction stages (electrochemical ones in the given case) .

To find the kinetics of electron exchange between the enzyme active center and the electrode, the dependence of current on overvoltage was considered. The absence of single slope of the polarization curve in semilogarithmic coordinates attests that, firstly, there is more than one stage (two by the number of transferred electrons) during the catalytic process, and, second, both these stages are dependent on potential and occur at relatively comparable rates. Thus, the formula for the description of the polarization curve should be complicated. Application of electrochemical kinetics equations (equations type (5), (6)) to the hydrogenase kinetics results in equation [Varfolomeev S.D., 1993]:

$$i = \frac{\exp\left[2\frac{\alpha nFE}{RT}\right] - \exp\left[-2\frac{(1-\alpha)nFE}{RT}\right]}{K_1 \exp\left[\frac{\alpha nFE}{RT}\right] + K_2 \exp\left[-\frac{(1-\alpha)nFE}{RT}\right]} \tag{17}$$

where K_1 and K_2 are the constants including the surface concentration of active enzyme and definite combinations of elementary constants. At $K_1 \neq K_2$ the branches of a polarization curve in positive and negative regions will be asymmetrical about the origin of the coordinates, as distinct from the case of single stage ionization reaction.

In Eq. (17) the value of α, a transfer coefficient, is unknown. Let $\alpha = 0.5$, as for most electrochemical reactions. The optimization of the polarization curve positive branch in accordance with Eq.(17) by two parameters K_1 and K_2 leads to a good description with correlation coefficients exceeding 0.999. The transfer coefficient variation does not cause a regression improvement. Thus, Eq.(17) describes the current-potential relationship well. The positive region of the i-E curve offers an opportunity for predicting reverse reaction, that is H_2 electrochemical evolution. Consequently, the polarization curve behavior follows the suggested description, and the above assumptions relative to the catalytic reaction course hold. Alongside fast addition- dissociation stages there are two potential-dependent, one electrode redox stages. On the strengh of this it is also inferred [Karyakin A.A. et al, 1984] that electrochemical stages should follow one another, i.e. they cannot be separated by addition- dissociation stages. It is interesting to compare the mechanisms of hydrogenase action in electrocatalytic and homogeneous regimes. For homogeneous kinetics of hydrogen oxidation catalyzed by hydrogenase from *T.roseopersocona* the stages of hydrogen addition and proton elimination from enzyme active sites occured much faster than the limiting steps [Karyakin A.A. et al, 1977; Zorin N.A. et al, 1988]. The latter are stages of interaction with one electron acceptor (redox steps) which follow one another. Moreover, it was found the whole sequence of steps from homogeneous kinetics and discovered the intramolecular stage of electron exchange in the mechanism of homogeneous hydrogenase action. The only kinetic scheme was found to obey the experimental data for hydrogenase catalysis, include the stages of electrochemical or one-electron redox steps. (Hydrogen - proton - electron - transfer, - electron - proton mechanism H_2peltelp mechanism). It was concluded that the kinetic properties of hydrogenase action obtained from electrochemical and homogeneous kinetics describe the same mechanism.

It has been shown that the initial portions of polarization curves in the positive and negative regions are symmetrical about the origin of coordinates. Thus, the portion under low overvoltages is completely reversible.

However, at high negative values of overvoltage on polarization curves of hydrogen electrochemical evolution one can observed the limit in current appear. The existence of electrochemical activity dependent on hydrogen concentration and pH is implied by the fact that the reaction rate rises both in an argon atmosphere because of the product absence, and at acid pH through the increase in substrate concentration.

At high values of overvoltage in the negative region there is a fall in the enzymatic reaction rate. The hydrogen ion concentration affects only the extent of the predicted section. Note that hydrogenase inhibition is completely reversible: the i-E curves in the positive region obtained before and after keeping at negative potentials are the same. In view of the absence of any other factors in the system which have an influence on the hydrogenase activity the conclusion can be drawn that the reversible inactivation takes place at the cost of enzyme overreduction.

Thus the hydrogenase adsorbed on conductor possesses electrocatalytic activity in both the reaction of hydrogen oxidation and evolution. It is possible to describe quantitatively the observed electrocatalytic reactions. The data on the mechanism of hydrogenase action in electrocatalytic process coincide with those obtained in studying the homogeneous kinetics of this enzyme. In bioelectrocatalysis (according to the mechanism of direct electron exchange between the electrode and enzyme active site) it operated in the same way as in the solution using the electrode as one of the substrates.

PEROXIDASE

Peroxidase is a typical representative of a large group of redox enzymes, the molecule of which contains a heme as prosthetic group. Some researchers [Yaropolov A.I. et al, 1979; Gorton L. et al, 1992; Paddock R.H. et al, 1989] reported a possibility of a mediatorless catalysis of hydrogen peroxide (H_2O_2) electro- reduction on an electrode in presence of adsorbed horse radish peroxidase (HRP) according to the reaction:

$$H_2O_2 + 2H^+ + 2e^- \xrightarrow{HRP} 2H_2O$$

Similar results were obtained by use of fungal peroxidase [Kulys J.J. et al, 1990; Kulys J. et al, 1991]. Electrocatalysis of H_2O_2 reduction with peroxidase in anaerobic conditions was observed on electrodes made of various materials (gold, pyrographite, carbon black [Yaropolov A.I. et al, 1979], platinum [Wollenberger U. et al, 1990], graphite [Jonsson G. et al, 1989) and carbon paste [Gorton L. et al, 1992]) and shown to depend on the amount of adsorbed HRP. The catalytic effect on metallic electrodes with adsorbed HRP attests to the mediatorless (direct) electron exchange to occur between HRP and electrode. The maximum stationary potential at carbon black electrode in this system was 1.24 V (relative to hydrogen electrode in the same solution), which was below the equilibrium potentialof the system H_2O_2/H_2O. This is related to various concurrent electro-chemical by-reactions shifting the electrode potential to the negative region. The electrocatalytic reaction was inhibited by phenylhydrosine. The properties of immobilized HRP in H_2O_2 reduction on electrode and of o-dianisidine oxidation were compared. HRP immobilization of carbon black electrode retained about 50% HRP initial activity. The dependencies of immobilized HRP on pH and H_2O_2 concentration actually coincided in the electrochemical system and in o-dianisidine oxidation. The linear dependence of the current at the immobilized-HRP electrode on H_2O_2 concentration occured at the H_2O_2 concentration 0.5-800 μM. Deceleration of the electrochemical reaction at the H_2O_2 concentration higher than 1 mM and pH < 4.5 was due to attenuation of HRP catalitic activity. Formation of HRP-H_2O_2 inactive complex worsened the electrochemical parameters with time and upon a long-time incubation, led to almost a total inactivation. The catalytic currents at the present potential in 1 mM H_2O_2 presence became half weaker during 1 h. The stationary potential of the system fell by 100 mV during 20 h. The insufficient stability of the system with time because of HPR inactivation afforded no detailed mechanistic study of the electrochemical reaction.

GLUCOSE OXIDASE

Some publications [Scheller F. et al, 1979; Taxis du Poet P. et al, 1990] describe the effect of direct electron exchange between glucose oxidase and electrodes made of various materials. However, a possibility of mediatorless electrocatalysis of glucose electrooxidation by this enzyme is under discussion up to now.

We studied the electrochemical oxidation of glucose on carbon black electrodes with adsorptionally immobilized glucose oxidase from Penicillium vitale in anaerobic conditions. The enzyme presence shifted the stationary potential on the electrode to 0.45 V relative to the hydrogen electrode in the same buffer and increased the catalytic currents of glucose oxidation. The maximum value of the current in the system was reached at pH 6.5-7.3, which corresponds to pH of the enzyme action in glucose oxidation by oxygen.

The catalytic currents of glucose oxidation were established to decrease with time. Within 10 - 20 min., the electrode became almost totally electrochemically inactivated. Note that despite the electrochemical inactivation of electrodes, the activity of electrode-immobilized glucose oxidase in glucose oxidation by oxygen did not almost alter. A similar effect was observed when various carbon materials were used as electrode. Supposedly, the follow-up system includes not a direct but mediator-assisted electron transport where the surface groups of carbonic material of quinoid structure act as immobilized mediators. The decrease in currents (glucose oxidation) with time is explicable by irreversible redox conversion of these groups. We also obtained similar results on electrochemically pretreated electrodes in the range of potentials for performance of electrochemical glucose oxidation. After that the enzyme was adsorbrd on the electrode and electrochemical measurements taken. Thus, the electrochemical inactivation of enzyme electrode was induced by other reasons, for instance by changes in the protein region responsible for electron exchange between the electrode and glucose oxidase active site.

The essential act of elucidation of feasibility of mediatorless electron transfer involving glucose oxidases is the application of gold and platinum as electrode materials, the surface of which is lacking the groups structurally close to artificial electron acceptors of glucose oxidase. The electrodes made of these materials were lacking significantly reproducible effects of mediatorless catalysis of glucose electrooxidation at high homogeneous enzyme concentrations.

Thus, the feasibility of mediatorless glucose electrooxidation involving glucose oxidase remains undecided to date.

THE TUNNELLING OF ELECTRONS, A POSSIBLE WAY FOR THE EXPLANATION OF DET-EFFECTS

Active sites of enzymes acting as catalysts of electrode processes are localised inside the protein globule. It is evident that under catalytic conditions electrons are transferred to sufficiently long distance by the DET mechanism.

The main conclusion of the analysis given below is that the concept of the tunnelling mechanism is adequate for the description of the DET effect The acceleration of electrochemical reactions by enzymes in accordance with the mechanism of direct electron tranfer between the active site of an enzyme and the electrode is due to the involvement of at least several reaction stages. The simplest scheme contains a stage of electron transfer from the electrode to the active site of an enzyme and a reaction of a "catalytic" enzymatic transformation of an electron (or a vacancy) into the final product of the enzymatic reaction:

$$e \xrightarrow{k_1} e^{\#} \xrightarrow{k_{cat}} E + P \tag{18}$$

where e and $e^{\#}$ are the oxidised and the reduced forms of the active site, k_1 is frequency or the rate constant of electron transfer between the electrode and the active site, k_{cat} is the catalytic rate constant of the enzymatic reaction characterising the limiting stage of the catalytic process. In a stationary state with respect to $e^{\#}$, the intermediate form, with the material balance equation taken into account, the specific current from the electrode is expressed as

$$i = -\frac{nFk_{cat}k_1}{k_1 + k_{cat}}[e]_0 \tag{19}$$

where $[e]_0$ is the surface concentration of the enzyme. The most difficult problem as regards this equation is the evaluation of k_1 since k_{cat} values for enzymatic catalysis can be determined quite easily in the experiment. Also, kcat values for enzymatic reactions do not vary much and are distributed in a rather narrow zone arround the average figure of $10^2\ s^{-1}$. Consider the governing laws of bioelectrocatalysis on condition that the process of electron transfer from the electrode to the active site of an enzyme is a subbarrier tunnelling transition. The mechanism of electron tunnelling in chemical and biochemical systems is a subject of a detailed analysis in the literature [Moelwin-Hughes E.A., 1961; Kharkyanen V.N. et al, 1978]. The rate constant of electron tunnelling depends on the height and the width of the barrier:

$$k_1(r) = \nu \exp(-r/A) \tag{20}$$

where ν is the frequency factor, r is the width of the barries, A is a parameter whose value depends on the height of the kinetic barrier. Several quantum mechanical models of electron transfer allowing the values of the frequency factor and the parameter A to be calculated have been developed by now. The solution of the problem of tunnelling across restangular, triangular, and parabolic barriers with a continuous energy spectrum of electrons has been discussed in the literature [Moelwin-Hughes E.A., 1961]. K.I.Zamaraev in his works has taken into account the discrete character of the spectrum and motion of nuclei in molecules in electron tunnelling between two molecules [Zamaraev K.I. et al, 1978]. A theory of electron transfer in protein systems making it possible to calculate transfer rate constants on the basis of structure properties of the medium has been discussed by V.N.Kharkyanen et al. [Kharkyanen V.N. et al, 1978]. It is to be emphasised that in all models of electron tunnelling transition mentioned here the dependence of transition rate constant on distance is expresses by Eq. (20). The physical sence and numerical values of the parameters ν and A can vary depending on the model.

For our purposes, i.e. for the explanation of DET effect, it seems sufficient to use a simple, very crude model giving an idea about the magnitude of possible effects.

According to the model described in [Moelwin-Hughes E.A., 1961], the frequency factor ν has a physical sence of the frequency of collisions of an electron with the barrier (the velocity of an electron in a molecule or a crystal) and equals $\sim 10^{16}\ s^{-1}$. The value of the parameter A having the dimension of distance is defined by the difference of electron energies and the height of the barrier. It is important to note that the magnitude of A only slightly depends on the shape of the barrier, the main factor determining its numerical value being the energetics of the process.

It is of interest to evaluate the distances within which the rate of electron transfer to the active site of an enzyme is greatly exceeding the rate of catalysis. In this case the enzymatic reaction is the rate-determining stage. For carbon electrodes, it was found that for a parabolic barrier A = 0.594 Å. It was shown that for $k_{cat} = 10^2\ s^{-1}$, $r_{cr} = 19.1$ Å. This means that if an electron-accepting particle in the active site of an enzyme is located at a distance from the electrode surface shorter than 19 Å, the rate of donation of electrons to the active site will surpass the rate of the catalytic stage and the catalytic enzymatic reaction will be the rate-determining stage.

The model of electron transfer in bioelectrocatalysis can be verified by studying the dependence of the measured bioelectrocatalytic currents on the distance between the electrode and the electron-accepting point in the active site of an enzyme. Such an attempt was made by A.I.Yaropolov et al. [Yaropolov A.I. et al, 1981]. Since the range of distance under discussion is of molecular size, the only means of varying the distances seems to be the use of monolayers of substances of different dimensions placed between the electrode surface and the active site of an enzyme. With this aim in view, a comparative study of DET-effects was conducted on electrodes obtained by enzyme adsorption directly on carbon and on carbon coated with a lipid monolayer.

According to studies performed by O.M.Poltorak, cholesterol adsorbs on carbon black in two orientations. At low concentrations cholesterol is adsorbed in a "flat" orientation, the thickness of the lipid layer being 6 Å. At high cholesterol concentrations the molecules of the lipid are arranged vertically" and the thickness of the dielectric layer is 17.8 Å [Behko E.M. et al, 1979]. These results were obtained on the basis of analysis of cholesterol adsorption isotherms featuring two plateaus at different concentrations of the lipid [Behko E.M. et al, 1979]. These results were checked and quantitatively confirmed [Yaropolov A.I. et al, 1981].

Electrodes without lipid monolayer and also with a monolayer of cholesterol in two orientations were prepared. Laccase was adsorbed on the surface of the lipid and electrocatalytic properties of the obtained systems were studied in the reaction of electrochemical reduction of molecular oxygen. Electrodes with a monolayer of adsorbed lecitin were also obtained and studied.

Experiments on adsorption of the enzyme on the surface of cholesterol monolayers were performed under the conditions of a complete saturation of the surface with the lipid. It was demonstrated that protein adsorption does not lead to the desorption of cholesterol or any noticeable migration of the lipid to the solution. Laccase under the conditions of adsorption and catalysis processes was shown to be adsorbed on the lipid surface practically irreversibly. Enzymatic activity was detected only on the electrode and was not found in the solution.

A quantitative analysis of the relative efficiency of electron transfer gives the following results is slightly dependent on the potential. The flows of electrons determined from the value of current from the electrodes of the same surface area without cholesterol and with cholesterol in a flat orientation virtually coincide. The stationary flow of electrons strongly diminishes if electrodes with cholesterol adsorbed in a vertical orientation are used. Thus, the efficiency of electron transfer in the system with vertically oriented cholectrol is about 1/5 as against the previous case. An independent series of experiments conducted on electrodes with a preadsorbed lecitin layer showed that the efficiency of electron transfer becomes about 1/270 as compared to the first case.

Thus, the concept of the tunnelling of electrons accounts for the observed DET-effects. If the site accepting electrons from the electrode is located at a distance shorter than r_{cr} (r_{cr} 20-30 Å; the radii of protein molecules are 10-30 Å), the rate of its "occupation" with electrons will be much higher than the rate of the enzymatic reaction and the experiment will reveal the effects of a drastic acceleration of the electrochemical reaction by a protein catalyst.

DET-EFFECT BIOSENSORS

At present, the solution of a wide range of problems related to testing biologically active compounds needs the development of high-sensitivity assays, simple, readily available and apt to being automated. Functioning of a biosensor supposes a combination of biochemical and physico-chemical interactions. The interaction of a biologically sensitive material with a test substance alters the system parameters converted by an inverter into relevant physical changes. The DET-effect can find its application in biosensors of a few types, the functional principles of which are described below.

Laccase-based potentiometric immunosensor. The capacity of laccase to catalyze the mediatorless electroreduction of oxygen and the feasibility of laccase-assisted immunoassay underlied the elaboration of a new type of biosensor [Ghindilis A.L. et al, 1991; Ghindilis A.L. et al, 1992]. The test antigen was covalently immobilized on an electrode made of carbonic material. Addition of laccase-antigen conjugate into the reaction unit led to the conjugate-antigen binding on the electrode surface. This rapidly changed the electrode potential due to the laccase-assisted catalysis of electroreduction. The preliminary addition of the test antigen into the reaction unit decelerated the growth of electrode potential due to the competition for binding the free and immobilized antigens. From the measurements of the rates of electrode potential at various antigen concentrations, the calibration curve can be plotted. The feasibility of insulin and human immunoglobulin quantitation by this method was demonstrated. The assay sensitivity was 10^{-7} M. The assay time was 20 min. Despite the evident advantages of this method (the second substrate of the enzymatic reaction should not be added), its sensitivity to the test compound is not high. It was developed another method also based on laccase-assisted DET-effect catalysis but with amperometric recording using signal accumulation. This afforded to essentially sensitize the assay, up to $1O^{-9}$ M. As example, consider the scheme of human IgG assay. Antibodies against human IgG were covalently immobilized on the surface of the electrode of carbon black modified by amfifilic polymer based on polyethyleneimine. The electrode was placed into the solution containing the test antigen and laccase/antigen conjugate. The formed complex of laccase/antigen conjugate with antibody, immobilized on the electrode stimulated the rise of the electrode potential due to the catalytic reduction of oxygen. The charge accumulated on the electrode was measured by the current arising from the forced potential +0.075 V.

Reagentless amperometric sensors based on coimmobilized oxidases and peroxidase. The mediatorless electrocatalysis by peroxidase was used for creation of bienzymic electrons to assay various metabolites [Gorton L. et al, 1992; Kulys J.J. et al, 1990; Gorton L. et al, 1991; Kacaniklic V. et al, 1993]. The functioning principle of two-enzyme electrode was the following. The electrode made of various materials (usually carbonic) carried the coimmobilized oxidase, catalyzing the oxidation of a definite metabolite, and HPR. The oxidase- catalyzed reaction yielded H_2O_2 reduced on the HRP-electrode . The cathode current of H_2O_2 reduction was proportional to the test metabolite concentration.
In the work [Jonsson G., 1989] the linear calibration curves for hydrogen peroxide were obtened from 0.1 to 500 μM on the electrode with immobilized HRP for a flow-injection system. The biosensors to assay of glucose, alcohol, L- and D-amino acids were developed [Kulys J.J. et al, 1990; 57]. The biosensors were based on relative oxidases coimmobilized with HRP on the electrodes made of graphite and carbon paste. The enzymes were stabilized by various methods of covalent crosslinkage of the enzymes with the proteins and charged polymers.

REFERENCES

Andreasson L.-E., Reinhammar B., (1976) Biochim.Biophys. Acta 445, 179-597.
Beh'ko E.M., Kamyshnyi A.L., Chukhrai E.S, Poltorak O.M. (1979) Kolloidnyi Zhurnal 41, 534-539.
Beh'ko E.M., Kamyshnyi A.L., Chukhrai E.S., Poltorak O.M. (1979) Kolloidnyi Zhurnal 41, 211-217.
Berezin I.V., Bogdanovskaya V.A., Varfolomeev S.D., Tarasevich M.R., Yaropolov A.I. (1978) Dokl. Acad. Nauk USSR, 240, 615-619.
Berezin I.V., Varfolomeev S.D. (1980) Enzym.Eng.5, 95-100.
Berezin I.V., Varfolomeev S.D., Yaropolov A.I., Bogdanovskaya V.A., Tarasevich M.R. (1975) Dokl.Acad.Nauk USSR, 225, 105-108.
Bogdanovskaya V.A., Varfolomeev S.D., Tarasevich M.R., Yaropolov A.I. (1979) Electroanal.Chem.Bioelectrochem. Bioenergetics 104, 393-403.
Bogdanovskaya V.A., Varfolomeev S.D., Tarasevich M.R., Yaropolov A.I. (1980) Elektrokhimiya 16, 763-768.
Chernavskaya N.M., Chernavskii D.S. The Tunnelling Transport of Electrons
Csoregi E., Jonsson-Pettersson G., Gorton L. (1993) J. Biotechnol. (in press).
Eddowes M.J., Hill H.A.O. (1981) Biosci. Rep.1, 521-532.
Ghindilis A.L., Skorobogat'ko O.V., Yaropolov A.I. (1991) Biomed.Sci.2, 520-522.
Ghindilis A.L., Skorobogat'ko O.V., Gavrilova V.P., Yaropolov A.I. (1992) Biomed.Sci., 7, 301-304.
Gorton L., Bremle G., Csoregi E., Persson B., Jonsson-Pettersson G. (1991) Anal.Chem.Acts. 249, 43-49.
Gorton L., Jonsson-Pettersson G., Csoregi E., Johansson K., Dominguez E., Marko-Varga G. (1992) Analyst., 117, 1235-1241.
Hill H.A.O., Higgens I.J. (1981) Philos. Trans.R.Soc. London A 302, 267-273.
Hopfield J.J., (1974) Proc.Natl.Acad.Sci.US, 71, 3640- 3644.
Ianniello R.M., Lindsay T.Y., Yacynych A.M. (1982) Analyt.Chem., 54, 1098-1106.
Johansson G., Hinduston Antibiot Bull. (1978) 20, 117-126.
Jonsson G., Gorton L. (1989) Eletroanalysis, 1, 465-468.
Jonsson-Pettersson G. (1991) Electroanalysis, 3, 741-745. 54.
Kacaniklic V., Johansson K., Marko-Varga G., Gorton L., Jonsson-Pettersson G., Csoregi E. (1993) Biosensors and Bioelectronics (in press).
Karyakin A.A., Varfolomeev S.D. (1977) Uspekhi Knimii 55, 1524-1549.
Karyakin A.A., Varfolomeev S.D., Berezin I.V. (1984) Dokl.Acad.Nauk USSR 275, 110-114.
Kharkyanen V.N., Petrov E.G., Ukraiuskii I.I. (1978) J. Theor.Biol. 73, 29-50.
Krishtalin L.I. (1979) Electrode Reactions. The Mechanism of Elementrary Act., Nauka, Moscow, p.244.
Kulys J.J., Bilitewski U., Schmid R.D. (1991) Sensors and Actuators (B), 3, 227-232.
Kulys J.J., Schmid R.D. (1990) Bioelectrochemistry and Bioenergetics, 24, 305-309.
Kulis J.J., Svirmickas G.-J.S. (1979) Dokl.Acad.Nauk USSR 245, 137-140.
Malmstrom B.G., Andreasson L.-E., Reinhammar B., (1974) in: The Enzymes, ed.Boyer, P.D., Academic Press 12, 507-579.
Miyawaki O., Wingard L.B. (1984) Biotechnol.Bioeng., 26, 1364-1370.
Moelwin-Hughes E.A., Physical Chemistry, v.1 Pergamon Press, London-New York-Paris, 1961.
Naqui Ali, Varfolomeev S.D. (1980) FEBS Letter 113, 157-160.
Paddock R.H., Bowden E.F. (1989) J.Electroanal.Chemistry, 260, 487-492.
Potasek M.J., Hopfield (1977) Proc.Natl.Acad.Sci.SU, 74, 3817-3821.
Scheller F., Struad G., Neumann B., Kuhu H., Ostrowski W. (1979) Bioelectrochem.Bioenergetic, 6, 117-122.
Skorchelletti V.V. (1974) Theoretical Electrochemistry, Khimia, Leningrad, p.567.
Taxis du Poet P., Miyamoto S., Murakami T., Kimurta J., Karube I. (1990) Anal.Cjim., 235, 255-261.
Varfolomeev S.D., Methods in Enzymology (1988) 137, 430-440.
Varfolomeev S.D., Bachurin S.O., Aliev K.V., (1978) Dokl. Acad.Nauk USSR 239, 100-103.
Varfolomeev S.D., Bachurin S.O., Ali Naqui, (1980) J. Mol. Cat. 9, 223-226.
Varfolomeev S.D., Berezin I.V. (1981) Dokl.Akad.Nauk SSSR 260, 1192-1195.
Varfolomeev S.D., Berezin I.V. (1982) Adv.Physical. Chemistry (ed.Ya.M.Kolotyrkin), Mir Publishers, Moscow, 60-95.
Varfolomeev S.D., Yaropolov A.I., Bachurin S.O., Ch.D. Toaj (1973) in Physico-Chemical Aspects of Electron Transfer Processes in Enzyme Systems, IFIAS, Stockholm, 144-151.
Varfolomeev S.D., Yaropolov A.I., Berezin I.V., Tarasevich M.R., Bogdanovskaya V.A. (1977) Bioelectrochem. Bioenergetics 4, 314-326.
Varfolomeev S.D., Yaropolov A.I., Karyakin A.A. (1993) J. Biotechnology 27, 331-339.
Wollenberger U., Bogdanovskaya V., Bobrin S., Scheller F., Tarasevich M.R. (1990) Analyt.Letters, 23, 1795-1800.
Yaropolov A.I., Halovik V., Varfolomeev S.D., Berezin I.V. (1979) Dokl. Acad.Nauk USSR, 249, 1399-1402.

Yaropolov A.I., Karyakin A.A., Gogotov I.N., Zorin N.A., Varfolomeev S.D., Berezin I.V. (1984) Dokl. Akad.Nauk USSR 274, 1434-1437.
Yaropolov A.I., Karyakin A.A., Varfolomeev S.D. (1983) Vestn.Mosk. Univ.Khim. 24, 523-528.
Yaropolov A.I., Karyakin A.A., Varfolomeev S.D., Berezin I.V. (1984) Bioelectrochem.Bioenergetics 12, 267-277.
Yaropolov A.I., Tarasevich M.R., Varfolomeev S.D. (1978) H.R.P.B.B. 5, 18-24.
Zamaraev K.I., Khairutdinov P.F. (1978) Uspekhi Khimii 47, 992-1017.
Zorin N.A., Karyakin A.A., Gogotov I.N., Varfolomeev S.D. (1988) Biokhimia 53, 728-734.

FROM MOLECULAR CHARACTERIZATION TO MOLECULAR MANUFACTURING AND MOLECULAR ELECTRONICS

Claudio Nicolini

Institute of Biophysics, University of Genova
and Polo Nazionale Bioelettronica, Portoferraio (LI), Italy

INTRODUCTION

Advances on the preparation techniques and on the fundamental properties of LB films allow organic materials to be transferred one after another from a water surface into solid substrates to form monolayers. Recent advances in designing supermoleculecular aggregates by manipulating biomolecules and monolayer packages has proven the feasibility to utilize LB in devices, justifying the attempt to apply the films of this type within present silicon technology for the production of bioelectronics devices. Recently, a step forward in real-time instrumentations was also realized. The atomic surface becomes visible through the Scanning Tunneling Microscope (STM), while femto-second laser spectroscopy can follow the time evolution of chemical reaction kinetics. A major distinction of molecular nanotechnology from bulk semiconductor technology is the bottom-up approach of building devices from the atomic scale up. Steps towards molecular manufacturing and molecular electronic devices are presently being taken in several laboratories starting from the characterization via tunneling/atomic force microscopy and other molecular probes of both Langmuir-Blodgett and self-organizing monolayers containing properly engineered 2-D materials either organic or biological of desired semiconducting properties. The object of this presentation is to summarize our most recent developments in the characterization of polymers down to atomic resolution and to outline the molecular manufacturing and molecular electronics carried out in my laboratories.

MOLECULAR CHARACTERIZATION

Nanogravimetric balance

It is well-known that for biomolecular compounds, especially for proteins, the area per molecule cannot be calcutated from the π - A isotherm obtained by Langmuir-Blodgett film technique because of the difficulty in estimating the actual surface concentration of spread molecules. Moreover, the area at the air/water interface can differ strongly from that at the solid substrate. We have then constructed a device named nanogravimetric balance, which (Fig.1) allows to determine the mass of deposited monolayer, the area per molecule on the solid substrate and to estimate the packing degree in the film [Facci et al., 1993] (Fig. 1). The method uses quartz resonators, changing the oscillation frequency after depositing additional mass in accordance with Sauerbry equation:

$$\Delta f/f_0 = \Delta m/Arl \qquad (1)$$

From Neural Networks and Biomolecular Engineering to Bioelectronics
Edited by C. Nicolini, Plenum Press, New York, 1995

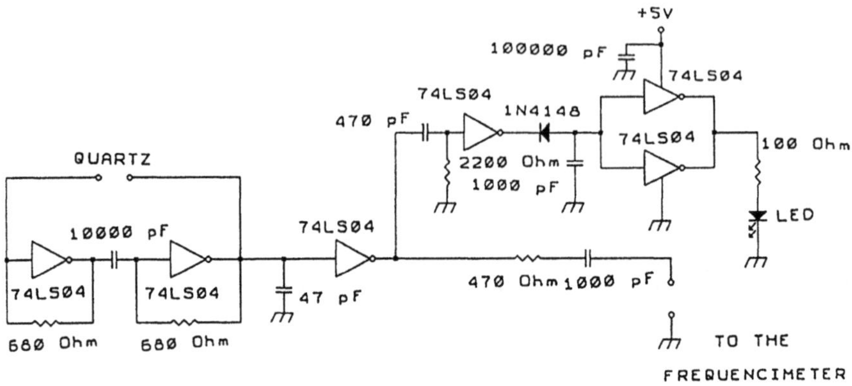

Figure 1 Electronic scheme of the nanogravimetric balance driving circuit.

where Δf is a frequency shift, f_0 is the initial resonator frequency, Δm is deposited mass, A is a covered area, r is the quartz density and l is the resonator thickness.

The balance was calibrated by depositing cadmium arachidate LB films (Fig.2). After the calibration the method was applied to measurements of the thin films later described.

Atomic microscopy

Scanning tunnelling microscope (STM) and Atomic Force Microscopy (AFM) have been used to study biological objects and these studies showed atomic resolution microscopy to be a useful tool for obtaining not only the high resolution structure but also the electronic properties of such objects.

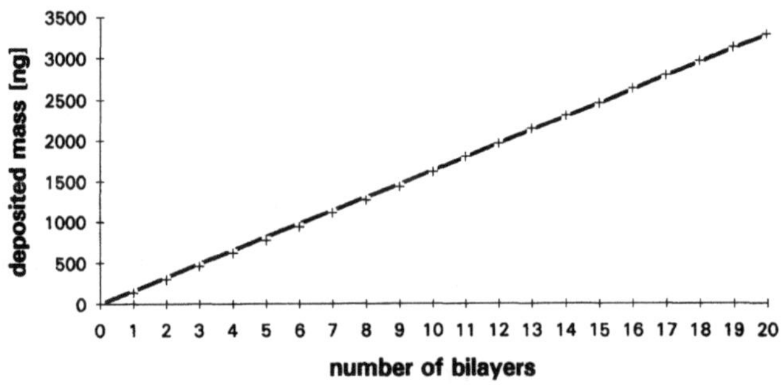

Figure 2 Calibration curve of the nanobalance performed by depositing 20 layers of cadmium arachidate.

Surface optimization and calibration. A traditional substrate for STM experiments is highly oriented pyrolitic graphite (HOPG) easy to prepare and having areas of atomically flat surface with dimensions of thousands angströms. Unfortunately, biological objects show very high mobility on the graphite surface due to its hydrophobicity. Another bad property of graphite is its ability to show DNA-like features. With both problems we came in touch during our experiments with proteins and DNA. One of the possible solutions is to use another substrate such as gold. However, it is not very easy to prepare good gold surface for the STM experiments. Furthermore, taking into account that the surface will be still metallic we are forced to use chemical treatment to fix the biological molecules on the surface. We thereby explore some metal/metal-organic interface/biological object (M-MOI-

BO) structure, such as silane monolayer [Nevernov et al., 1994]. Different silanizing agents are used for the modification of the quartz, metal and other surfaces for the covalent immobilization of proteins [Dubrovsky et al.,1993]. However, silanization by traditional methods does not ensure uniform distribution of the silane derivative at the support surface. For fabricating more regular polyorganosiloxane films of graphite we used a method developed by M.Malmquist and Olofsson [1989], whereby the heated vapour of silane derivative adsorb under vacuum onto the support surface. The ethoxyl (methoxyl) groups are hydrolyzed to silanol groups and can react with the available hydroxyl groups on the surface and with other silane monomers. Epoxy groups exposed on the surface of polymerized siloxane film can be used for the protein immobilization. These groups react with the lysine primary amino groups present at the protein surface. Langmuir-Blodgett films of antibodies have been immobilized as monolayers onto quartz supports prepared using this method and their surface density onto silanized surfaces at different surface pressures was determined from fluorescence measurements [Radicchi et al., in preparation]. Images obtained by STM on silanized graphite samples show very stable tunnelling conditions within a wide range of parameters and display absolutely flat areas over more than 1000 Å. Incidentally, artefactual DNA-like features do not appear with silanized graphite preparations, contrary to what has been observed with pure graphite. To prove the existence of the silane layer over the graphite surface we obtained the atomic resolution images of silanized graphite, showing atomic scale structures different from the one obtained from the freshly cleaved graphite (Fig.3). This structure is visible by eyes as well as it is possible to see additional spots on Fourier transform images (Fig.3).

We can conclude that after the high-temperature silanization of graphite surface the fabricated polyorganosiloxane film is monomolecular and two-dimensionally ordered. The specific adsorption of biological objects like proteins onto the conductive surface using the described M-MOI-BO structure appears useful in the field of scanning probe microscopy studies of biopolymers, as well as for molecular electronics and biosensor technology [Nevernov et al., 1994a]. An atomic structure of higly oriented pyrolitic graphite (HOPG) is typically used for the calibration of lateral scanning X and Y piezomovers of STM. A specimen of HOPG was prepared for calibration procedure by making a fresh dissection of graphite surface with adhesive tape. This sample was placed on sample holder of STM-20-92 (MDT/Asse-Z; Moscow/Padua) and measured in constant-height (CH) and constant-current (CC) modes in air. Mechanically prepared platinum tip was used in this measurement. For each parameter typical measurement falls in a range determined by the following values: bias voltage of tunneling gap (Vt) - -1 +1 V, tunneling current (It) - 0.5 - 1.2 nA, scanning velocity (Vs) - 600-10000 Å/s [Nevernov et al., 1994b]. The results of this measurement show that it is possible to obtain atomic resolution images of graphite surface within a wide range of working parameters in CH mode, as well as in CC mode. We used Fast Fourier Transform (FFT) for these images to get the parameters of graphite surface atomic structure in the current coordinate system of STM-20-92. With the help of these parameters, and with a knowledge of real parameters of graphite surface, avaible from a number of publications, we can write a system of linear equations:

$$\frac{K_X}{K_Y}=\frac{\frac{\sqrt{3}}{2}C\sin\alpha_c}{\frac{1}{2}C\cos\alpha_c-b\cos\alpha_\beta}\qquad\qquad \frac{K_Y}{K_X}=\frac{\frac{\sqrt{3}}{2}b\sin\alpha_\beta}{\frac{1}{2}b\cos\alpha_\beta-a\cos\alpha_\alpha}\tag{2}$$

where Kx and Ky are multiplication factors for the current calibration coefficients respectively for X and Y piezomovers; a, b, C and α_i are parameters obtained from measurements and from the literature. The solution of these equation gives Kx = 2.785, Ky = 2.12. Utilizing these factors a new set of measurements was done for 7 graphite surface images in CH mode with the same starting I_t and different V_t in the range from -1 to 1 V with a step of about 0.2 V. The process has testified the quality of our horizontal calibration. Vertical calibration along the axis of the tunneling current perpendicular to the surface was not done on graphite, since its surface corrugation is too small to supply the precision needed.

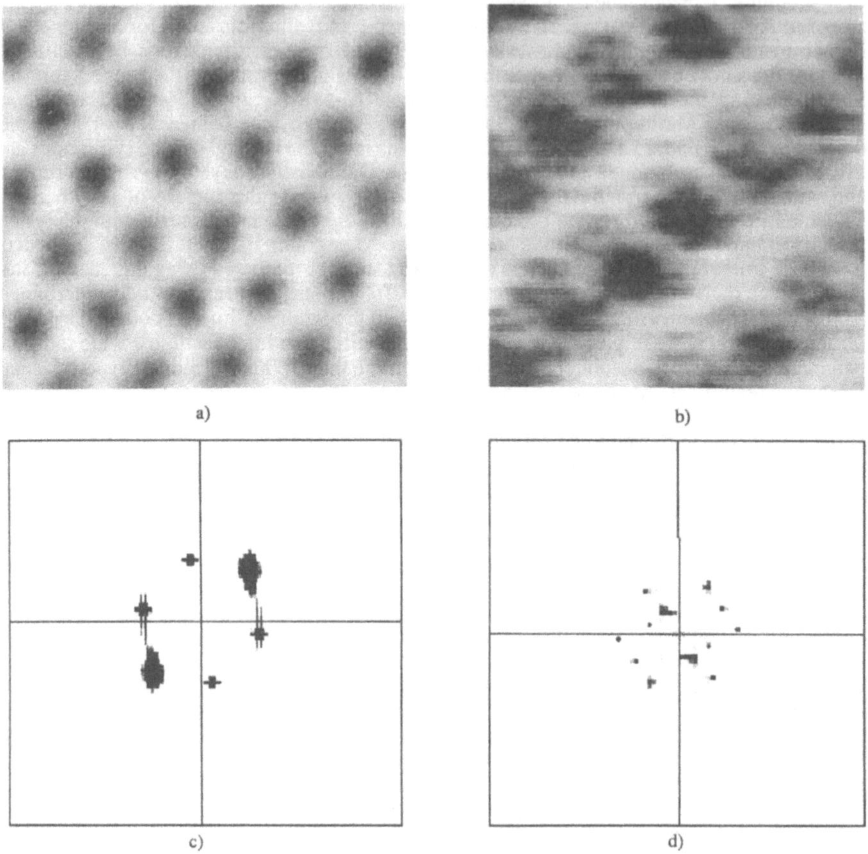

Figure 3 Grey scale top-view experimental images of graphite (a) and silanated graphite (b) and their corresponding two-dimensional FFT (c and d).

Biopolymers preparation and characterization. As a substrate for the STM visualization of biopolymers we used highly ordered pyrolitic graphite (HOPG, MDT Corp). Despite the high hydrophobicity of graphite, the mobility of these biopolymers is indeed low due to the high molecular weight of chromatin; furthermore the weak bonding between the horizontal layers makes it easy and fast to obtain the atomically flat surface using adhesive tape. Chromatin was extracted from calf thymus with two different procedures; while with the first procedure, called the "cold water" procedure, we obtain undigested native chromatin known to mantain to some extent the higher order structure [Nicolini et al., 1989], the second procedure, called the "nuclease digestion" method, utilizes a mild micrococcal nuclease digestion causing a significant size reduction of the chromatin which does not preserve the quaternary DNA structure [Nicolini et al., 1989]. The chromatin extracted with both procedures have been dialyzed overnight against TE buffer (Tris 10 mM pH 8, EDTA 1 mM pH 8) at 4°C. All the samples have been diluted in TE to a final concentration of 50 mg/ml; this concentration has been evaluated in SDS 0.2% (ε=21000 cm^{-1} M^{-1}) and corresponds to an optical absorbance of 1.3 OD in a 1 cm path length cuvette. With this dilution multimolecular aggregates of the samples are avoided. A drop of 10 µl of the diluted solution has been deposited on a graphite; the TE buffer image on graphite has been acquired as a reference to evaluate the background and any artefactual aggregation. DNAs extracted from the various samples with the usual procedure (phenol, chloroform and isoamylic alcohol), have been separated on 0.8 % agarose gel stained with 0.5 mg/ml ethidium bromide. After running the gel has been visualized on a UV light transilluminator and a

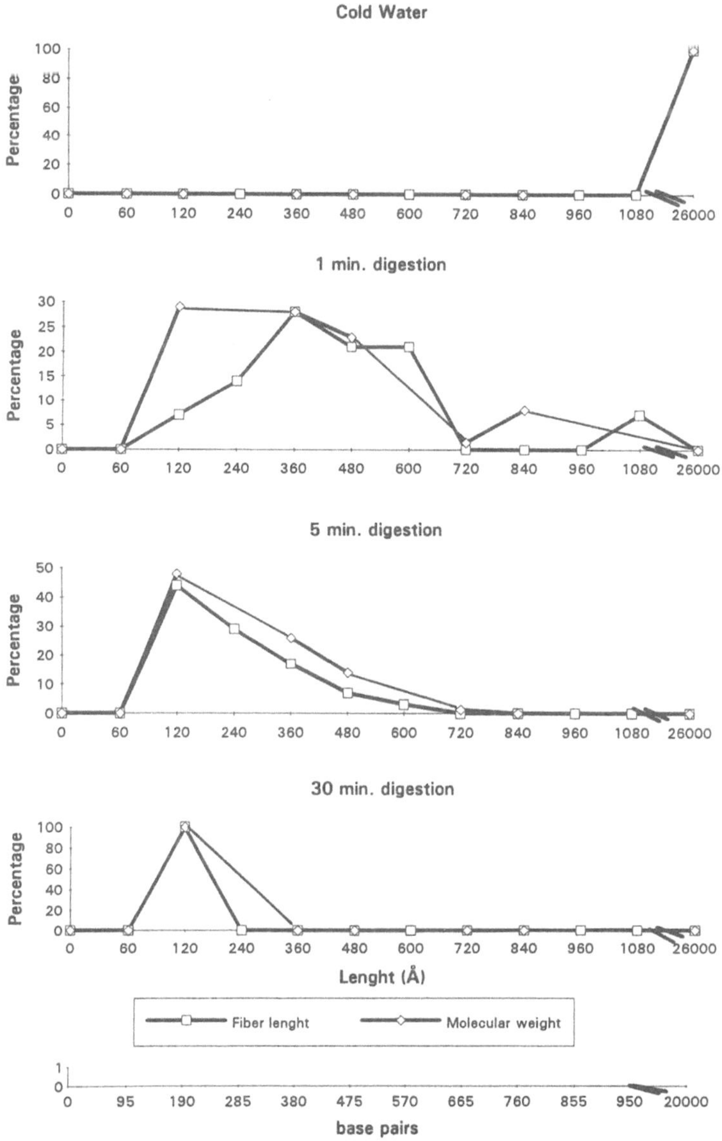

Figure 4. Chromatin-DNA fiber length. Fiber length and molecular weight of chromatin isolated by the"cold water" procedure and by "limited nuclease digestion", for 1, 5 and 30 minutes.

picture has been taken; the electrophoretic pattern has been acquired via a Grundig telecamera interfaced with a personal computer. The two different procedures gave rise to quite different electrophoretic profiles (Fig. 4); namely DNA fragments extracted from the cold water chromatin were longer than 25 Kbp (1.65×10^7 daltons), while the DNA corresponding to the nuclease digested chromatin presents a defined nucleosomal ladder. In particular, in the sample digested for 1 minute are evident fragments the size of which ranges from the mononucleosome (about 190 bp, 1.2×10^5 daltons) to polynucleosomes (up to seven nucleosomes) yielding nucleofilament 100 Å wide and up to 700 Å long (as it appears indeed by STM). In the sample digested for 5 minutes, DNA corresponding to mono- and dinucleosomes is prevailing, but it is possible to see also fragments from tri- and tetranucleosomes (therefore yielding fragments mostly 100-200 Å long). Circular dichroism spectra between 200 and 300 mm have been acquired to confirm the two distinct structural features typical of the two chromatin preparations [Nicolini et al., 1977; 1989].

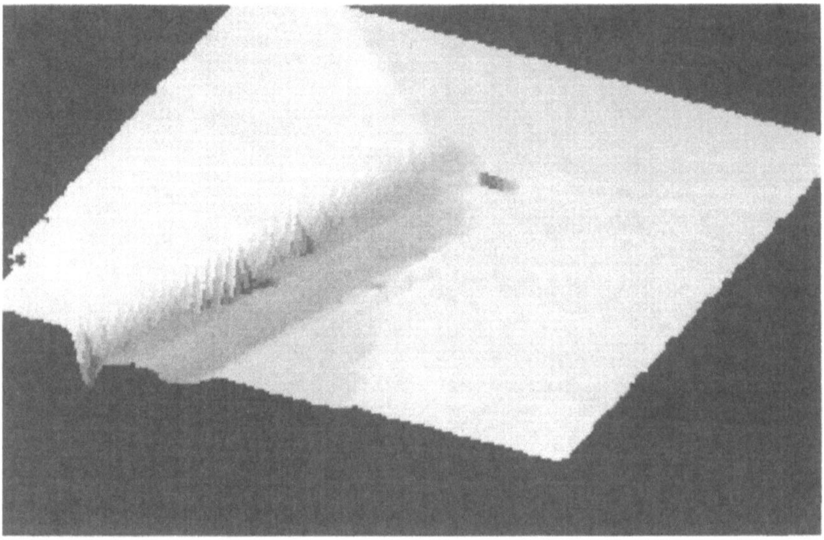

Figure 5 Three-dimensional view of an STM scan of the "cold water" chromatin sample.

The liquid samples were dyalized overnight before the deposition against TE buffer. This has been done to minimize the presence of salts on the surface after the deposition. A large amount of salts leads to unstable tunnelling conditions because of the high resistance of the salt layers. Each graphite specimen was cleaved before the deposition of 10 μl drop chromatin solution on its surface. The sample was left undisturbed on air in freezer under the temperature of 4°C. After 1.5 hours of drying the sample was placed on the STM sample holder of commercially available STM (Asse-Z/MDT Corp.; Padua/Moscow) and was dried at room temperature for approximately 30 minutes. The tunneling was initiated immediately after the evaporation of the last amount of free water. This has been done to avoid further dehydration of the biopolymers [Nicolini et al., 1993a]. All the STM images were obtained using the same mechanically cut tip consisting of 90% Pt, 10% Ir alloy (Goodfellow, England). The diameter of Pt/Ir wire was 0.5 nm. Before the measurements the tip was tested with pure graphite surface and displayed the required atomic resolution (see earlier). Each graphite specimen was prepared with a delay of two hours. During these two hours the solutions were kept at 11°C. The preservation of native chromatin structure was controlled by circular dichroism analysis before and after each experiment. The images were obtained in constant-current STM mode. In this mode the tunneling current is continously monitored and the tip sample distance rapidly adjusted by the STM electronic feedback, whereby the tip sample distance is kept constant and the feedback signal follows closely the contours of the conductive surface. Despite the fact that chromatin, histone proteins and DNA are insulators, a number of their STM observations have been published. Possibly because other physical phenomena are responsible for the formation of such images (see later). For each specimen were taken images of 30 Å-scale to prove the quality of the tip, and of 128 Å-scale to check the flatness of the surface and the stability of the tunneling conditions (level of noise, absence of oscillations in the feedback system). Then for each sample a set of images of dimension 1000-2000 Å was taken to cover the entire scan area. When the chromatin fibers were found, we repeated the scan of the same area or used backward scanning technique to prove the reproducibility of the image. On the acquired large scale scans we found several objects which we analyzed in detail for each of the four samples previously described (see table 1 and Fig. 4). Fig. 5 shows a scan which was taken on the cold water sample. The objects which are seen on the image could be described as a combination of two parallel strongly interlinked fibres. The dimension of these two combined fibres was estimated to be over 20,000 Å in length and about 800 Å in width.

Table 1. Percentage of chromatin fiber with given width (Å) as obtained by STM for the four different chromatin preparations.

Sample	0-30	65±35	135±35	205±35	275±35	345±35	415±35	800±20
Cold Water	0	0	0	0	0	0	0	100
1 min. nuclease digestion	0	46	6	12	21	9	6	0
5 min. nuclease digestion	10	64	13	8	1	2	0	0
30 min. nuclease digestion	0	10	0	0	0	0	0	0

In the short nuclease digestion (1 min.) sample, we could observe a number of objects all of which have long narrow shape with recurrent "bumpy" structures. Estimation of the dimension of these objects gave lengths which vary from 100 to 1050 Å and widths from 50 to 380 Å. From the lenght distribution shown in Fig. 4 it can be clearly seen that the lengths are around 400-450 Å with only a fiber at 1050 Å. The width distribution for the same sample, however, shows a bimodal distribution around 90 Å and 250 Å. In a typical scan for the long nuclease digestion (5 min.) sample we find instead fibers which seem to be arranged like beads on a string (Fig. 4); completely missing are the long and narrowly shaped objects which were observed on the short nuclease digestion sample. Their length distribution curve in this case shows high frequency in the range of 100 to 280 Å with a maximum at 100-140 Å and a narrow tail stretching up to 560 Å (Fig. 4), in perfect agreement with gel electrophoresis; their width distribution apppears centered around 100 Å with only a tail up to 320Å. For this sample we made also width distribution histogramms classified by the length of the objects, showing that there is no significant shift in the width frequency distribution as function of fiber length. STM image of the fourth sample (Fig. 7) provides unique details at the angström resolution on the geometric feature of single mononucleosomes, an oblate-like shape quite compatible with independent earlier X-ray and neutron crystallography studies [Travers and Klug, 1987]. Striking is the clear appearance of the two turns of DNA fiber 20 Å-wide, superfolding around the octamer core 70 Å wide, 110 Å long and 70 Å high; the nucleosomal DNA superhelix appears about 4 times wider than the DNA double-helix itself. For comparison, in similar STM images linear double-stranded l-DNA looks like long rods 20 Å wide with a period of 30-40 Å which can be attributed to the helical pitches of DNA molecule. Comfortingly, Atomic Force Microscopy of 19 images of cold water chromatin obtained with Park AFM (Park Scientific Intstruments Co., USA) evidenced fibers having width, height and length summarized in table 2. With the exception of one fiber of about 110 Å and five fibers at about 700 Å ± 35 Å the average width is 363.4 ± 40 Å, the average height is 80.2 ± 22 Å and the length ranges between 2,000 and 11,000 Å.

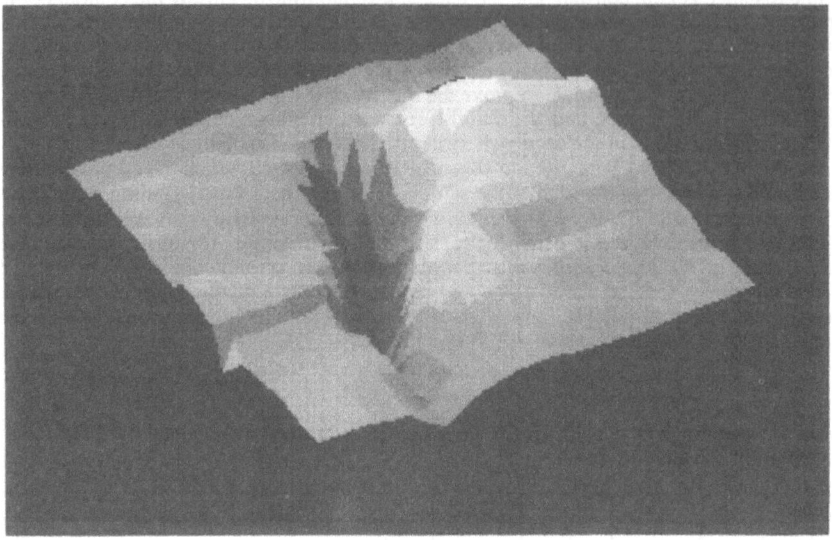

Figure 6 Three-dimensional (from above and from below) view of an STM image of the chromatin fiber, 5 minute nuclease digestion.

Table 2. Cold water chromatin fibers imaged with AFM.

Image number	Width [Å]	Height [Å]	Lenght [Å]
1	460	150	11000
2	395	30	3940
3	314	77	2244
4	466	116	9564
5	407	60	9065
6	367	40	3626
7	345	56	4630
8	311	60	5468
9	453	145	2168
10	116	68	4559
11	487	111	4651
12	363	64	4968
13	300	65	4117
14	763	98	7381
15	639	40	5133
16	366	42	3059
17	687	106	3219
18	672	86	2442
19	720	60	3600

Ab initio simulation. TEOSTM is a simulation program implemented "in house" [Kurnikov et al.], able to make "synthetical" images of biological structures as seen by STM microscope. To produce an image ab initio we carried out four basic operations:
- mathematic description of the molecule under study;
- quantum mechanical calculation ;
- simulation of STM image;
- proper display of the image.

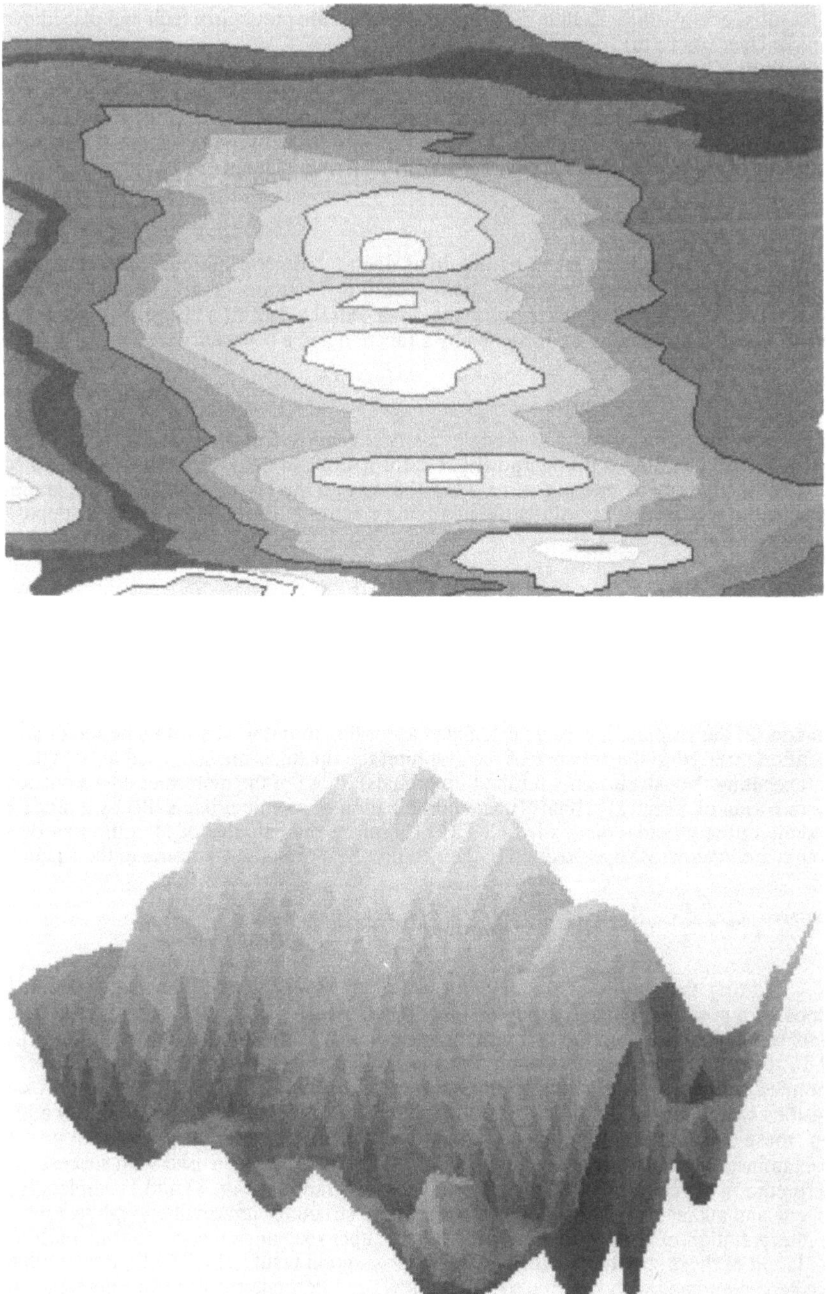

Figure 7 Two-dimensional (above) and three-dimensional view (below) of the STM scan of a mononucleosome.

Four programs function together with TEOSTM:
- a molecular modelling software (Insight II-Builder);
- a potential energy minimization software of the molecule previously built and placed over a graphite layer (min.exe)
- a program of quantum calculations (cndo.exe)
- a program of visualization of simulated images (molpos.exe, modified version of mi.exe).

By using these four programs together we reach the purpose of having the simulated STM image of the molecules of interest. The tunneling current between the graphite of the layer and the tip of the microscope is calculated according to the relation:

$$I \infty v\ (D\alpha(v)M^2{}_{\alpha t} + D_\beta M^2{}_{\beta t}) \qquad (3)$$

where we write the current taken by the tip as due to two contributions: one relative to an electron extracted from a "layer α", corresponding to atoms of graphite which have a correspondent atom on the lower plain, the other extracted from a "layer β" which do not have it. These contributions are mediated by a function and a constant:

$D\alpha(v)$, D_β

which takes into consideration that while layers α heavily feel the presence of an external electrical field (V), layers β undergo slight deformations in the Fermi state. $M^2{}_{\alpha t}$ and $M^2{}_{\beta t}$ represent the energy of interaction between the tip and the layers α and β of the graphite, calculated to second order perturbation involving electronic states of the molecule deposited on the graphite. Thereby,

$$M_{\alpha/\beta} = <\Psi_t|H|\Psi_{\alpha/\beta}> + \sum_j \frac{<\Psi_t|H|\Psi_j><\Psi_j|H|\Psi_{\alpha/\beta}>}{E_j E_F} \qquad (4)$$

derived from the theory called "super-exchange" which explains the passage of current from the graphite to the tip as mediated by electronic states (states j) of molecule under study. In equation [2] the graphite layers are calculated according to flat waves with a period of spatial repetition referred to the network of the graphite; the tip states are supposed to be type s, p, d, f (executing the calculus only for the f tip orbitals); those of the molecules are calculated as superposition of atomical orbitals (linear combination of atomic orbitals Ψl) by a procedure of semiempiric calculations called CNDO. Therefore the calculus of Hamiltonian occurs through the integral of superposition of the two involved orbitals, according to the equation:

$$<\Psi_t|H|\Psi_j> = \frac{E_j E_F}{2} + \int <\Psi_t(x,y,z)\Psi_j(x,y,z)dxdydz> \qquad (5).$$

Adding the results of the calculus provided by equation [3] in the equation [2], supposing a possible course for the two functions D which mediate the behaviour of carbon atoms α and β of the graphite and moving the results of equation [2] and equation [1], we obtain the tunneling current measured at the tip. Therefore by scanning the tip over the examined sample, we can calculate the current in different positions of the tip (different positions which reflect the spatial coordinates x, y, z of the Schrödinger function) and "to map" these currents in a 2-D representation: this map will represent the simulated image of the examined biological structure. The software TEOSTM has been used with success in the interpretation scheme of STM images produced by graphite (Fig. 8) and by molecules of benzene and alkanes on top of it (not shown). These structures are small enough to adhere to the interpretation of conduction proposed by the super-exchange theory: for this reason the simulation of these structures with TEOSTM yields good results (Fig. 8). Chains of alkanes have been simulated with success, so that it allows an interpretation of experimental images: the chain of alkanes appears placed parallel rather than perpendicular to the plain of graphite. Proteins, nucleosomes, DNA chains and LB films of lipids are too big (more than 20 Å) for to be descibed by simple super-exchange process.

The current revealed by the STM microscope is constituted by electrons which, by tunnel effect, move from the examined biological structure to the tip of the microscope; however, to keep the system neutral it is necessary that the graphite "gives again" electrons to the biological sample: we need a passage of electrons between the graphite and the tip through the absorbed molecule. However, this apparent success of STM technique to visualize biological structures of several tens or even several hundreds of angstroms thick gives rise to a confusion among theoreticians concerning a possible physical mechanism of STM image formation of such structures.

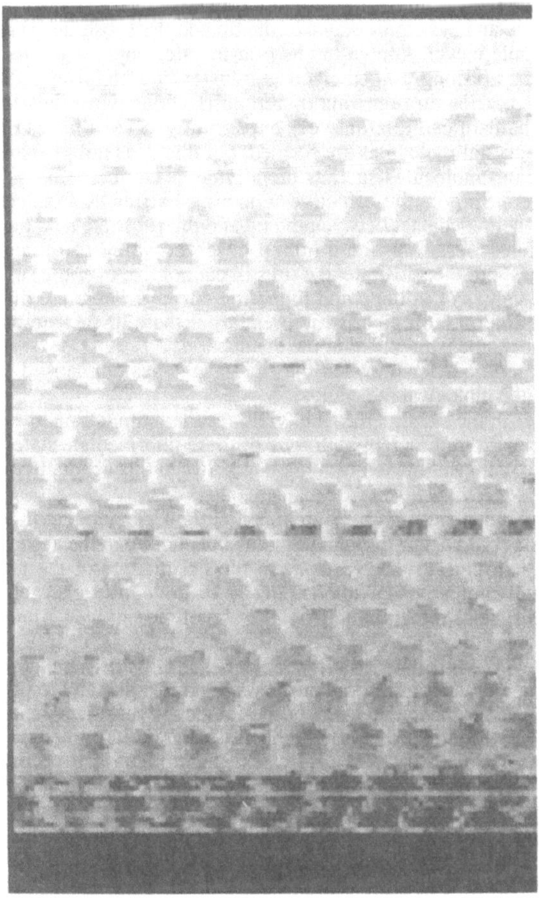

Figure 8 Theoretical STM images of graphite.

Biological materials are generally non-conductive in the usual sense of the word. If put aside the special case of metal-containing proteins, biomolecules do not contain electron or hole states near Fermi-energies of the materials usually used for tip and substrate in STM experiments. Thus, an electron or a hole cannot "jump" to some intermediate states on DNA or proteins and the only physical mechanism for the electrical current through these molecules during STM operation seems tunneling. Simple theoretical considerations show that the tunneling current should exponentially depend upon the distance between the tip and the substrate. The logarithm of this dependence is proportional to the square root of the height of the effective tunneling barrier between the tip and the substrate which in the absence of the adsorbate is about the Fermi energy of the tip and substrate materials, so:

$$I(z)\infty V\exp(-\frac{2\sqrt{2}mE_F}{h})\infty V\exp(-2z) \qquad (6)$$

we assumed here that z is expressed in Å units and E_F took a value of 4 eV. Accurate experimental measurements in vacuum confirm this rapid decay of the tunneling current with tip-substrate distance and evaluated tip - substrate distance to be 1.4 Å before Van-der-Waals contact in normal experimental conditions. Unconductive materials including biopolymeres can influence the value of the tunneling current through the interaction between the electronic shells of the molecule and electronic states of the tip and the substrate. This mediation effect of a bridging molecule is well-known in electron-transfer physics and recent experimental and theoretical studies of long-ranged electron transfer in proteins and DNA show that the effective tunneling barrier through this materials is about two times lower than that in vacuum. This mechanism can qualitatively explain how STM images of monolayeres of alkales or of liquid crystal molecules are formed, but this does not give an explanation how huge (on atomic scale) molecules such as biopolymeres can be seen by STM. Even if we propose 4-time decreasing of the effective tunneling barrier in an adsorbate and 10^5 W resistence of the tunneling contact between the tip and the substrate near their direct Van-der-Waals contact which is less than the minimal observed value of tip-substrate resistence before the tip crushes into the substrate, the tip still cannot rise over the substrate higher than 15-20 Å at normal values of voltage and tunneling current. There are different theories to explain the conduction on biological structures larger than 20 Å, theories which consider different physical phenomena involved in the conduction:

- a fall of the energetic barrier between the tip and the system caused by the interaction between the tip and the biological molecule;
- atoms or molecules in the substructure that yield resonant tunnelling: the electron provided by the graphite jumps from the graphite to one of these intermediate states of the biological molecule, then a second phenomenon of tunneling between this conductive orbital and the tip happens;
- every big molecular structure on air contains a certain percentage of water: conduction through the molecule could occur through the ionic conduction in the absorbed water;
- thermic energies can give to the electrons in the Fermi state of the graphite and of the examined biological molecule sufficient energy to cover the energetic amount which separates the Fermi state of the graphite from the one of the tip: we will not have any tunnel effect.

These aspects of the STM imaging are still not clear even considering that the above different physical phenomena might act in a cooperative way and it is difficult to predict which one is the dominant effect. We have good reasons to consider the theory super-exchange not sufficient to explain the conduction through the system graphite-molecule absorbed -air-tip; in this case we need to identify which physical phenomena can be predominant in the conduction of these systems. We have hereby taken a semiempirical approach by introducing in our original version of TEOSTM the above new physical parameters to best fit the experimental conduction of the various biological complex structures under study with STM.

Previously reported biological adsorbates looked like depressions relative to constant current level of uncovered substrate surface. If the only mechanism of the adsorbate imaging is a modulation of the interactions between the tip and substrate electronic states this would mean that the tip over the uncovered substrate is more than 30 Å apart as the tip move toward the substrate over an adsorbed biomacromolecule. This situation seems to be very unprobable in view of reasons given earlier; so it seems reasonable to try to introduce an alternative physical mechanism of formation of "negative" STM images of large adsorbates [Nevernov et al, 1994 b]. Similar depression of 20-30 Å depth is frequently apparent in free DNA samples, in mononucleosome samples (100 Å long and 70 Å wide) and in polynucleosomes of different length which appeared as depressions 20-30 Å depth [Nevernov et al, 1994 b]. As a rule "the biopolymers" were associated to steps on the graphite surface while to the left and right of the objects can be seen depressions smaller in sizes. The exact similarity in the orientation of the smaller depression relative to the main objects points that they are "reflections" of the same object due to multi-tip effect with the distance about 200 Å between alternative protrusions of the tip. In partially digested chromatin multiple objects of the width of about 100 Å and the length from 200 to 1000 Å looked also like depressions of the depth of 30 - 50 Å. These objects correspond to polynucleosomes of different length due to the random character of the process of digestion.

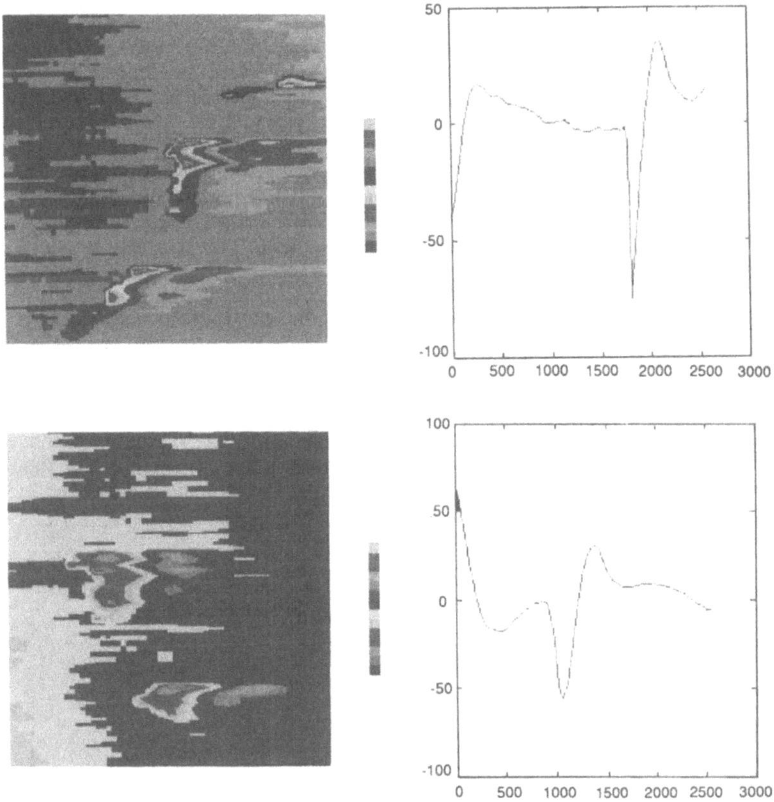

Figure 9 STM image of a single polynucleosomal fiber, obtained on constant current mode, with the tip moving from left to right (above) and from right to left (below).

Fig. 9 shows images of a single "polynucleosome rod" obtained during one experiment, whereby above it is shown a constant current profile obtained when the tip was moving from the left to the right and below the profile recorded while the tip returned from the right to the left. On both images the "polynucleosome rod" appears as a depression accompanied by a protrusion, probably due to multiple-tip effect. It is interesting to note that the order of the "negative" and the "positive" images of the object is different on the direct scan and on the backward scan. Such a great difference between the images obtained simultaneously in time but in opposite scanning directions indicates the important role of the friction in the formation of these images. From joint AFM and STM investigations it is known that normal conditions in STM experiments the forces which act between the tip and the substrate can be huge on the microscopic scale that is 10^{-9} - 10^{-6} N. If we recall that the characteristic force value which is needed to break a covalent bond is 1eV/Å=1.6 x10^{-9} N we can imagine that such forces can deform substantially the substrate during the scan. To the effect of these deformations have been attributed for example anomalous measurements of barrier-height. In order to explore a possible appearance of large unconductive adsorbates as depressions on constant-current STM images we have analyzed by the same well-known non-conductive adsorbates like dendrimers with the width of 107 Å (Fig. 10).

A water solution of $4 \cdot 10^{11}$ particles/ml was prepared. The substrate, Highly Oriented Pyrolytic Graphite (HOPG), was immersed for 5 min. in 1% Glutaraldehyde solution, then rinsed in pure water and dried with nitrogen flow. This action was carried out in order to facilitate the adhesion of the dendrimers, whose surface is very reach in amino groups, whith the graphite substrate. A drop of 10 µl of sample solution was spread on the HOPG plate and dried in air for 1 h. Tungsten tips were used during the experiments. Dendrimer mages were obtained using typical tunnelling voltages of 0.2-0.3 V and currents of 1-3 nA. Images were recorded both during forward and backward scans. Fig.10 (above) shows the results of such measurements. In the forward scan it is possible to see similar features in both the images. In the lower-left corner, for instance, it is present a corrugated depression with an average diameter of about 120 Å It is clear that the tip interacts rather strongly with the sample in all the region corresponding to the feature at issue.Moreover, the negative feature shows a deeper grove in its right part in both the scanning directions, as it is clearly evident from the average cross-sections of the images measured on the area of the features, fig. 10 (below). These results, therefore, seem to point out the possibility of imaging large adsorbate on HOPG by following the negative corrugation which is leaved by the adsorbate itself when staying on the substrate coated by buffer or GA (as in the present case). The images would appear because the tip, during its motion, removes the large non conductive adsorbate from its place and follows the residual corrugation. The fact that the corrugation results to be deeper and sharper in the right side of the feature in both scanning directions, points once more out the interaction between tip and sample which takes place during the scanning from left to right, pushing the particle, and which cannot take place any more during the right to left motion of the tip, as the adsorbate is already displaced and, maybe, destroyed by the tip action itself.

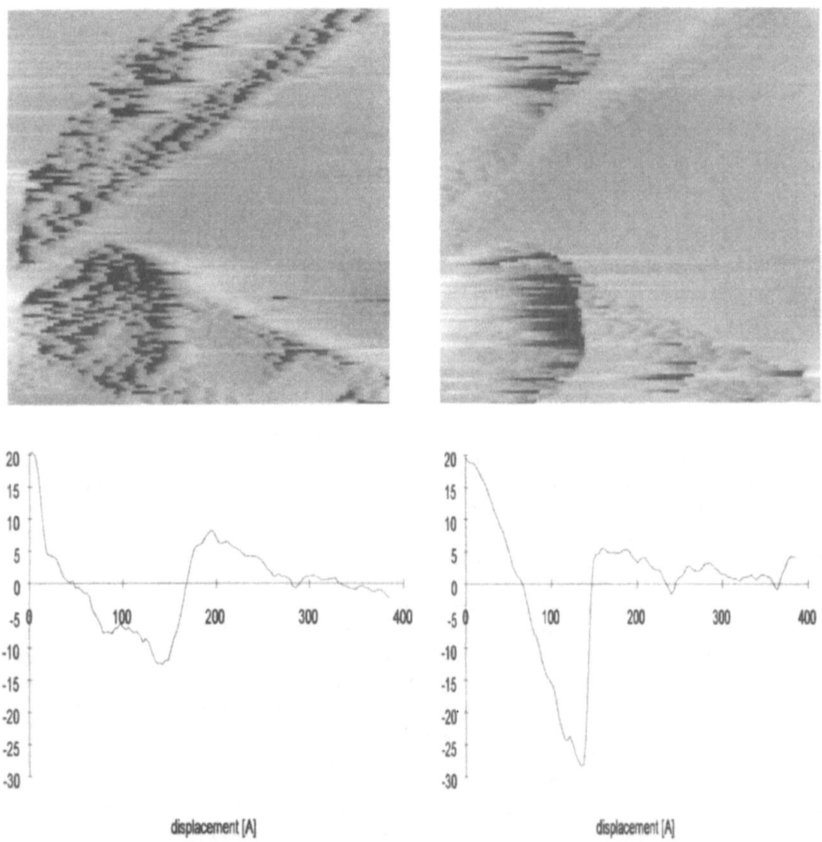

Figure 10 As Figure 9, but for a dendrimer of 107 Å diameter.

To make estimations for characteristic deformations of the substrate surface under the action of the force from the tip, it is sufficient to evoke an elastic behaviour; the expression for vertical displacements of points at the boundary surface of the elastic half-space takes then a form:

$$\delta z = \frac{F(m-1)}{2\pi G m R_0} \tag{7}$$

where G and m are elastic constants of the halfspace, F the value of the applied normal force and $R_0 = \sqrt{x^2 + y^2}$ is the distance in the horizontal plane to the point of the application of the force. From equation (5) the characteristic size of the depression of the substrate under the mechanical action of the tip (designated as R_h on the figure 10) will be:

$$R_h = \sqrt{\frac{F(m-1)}{2\pi G m}} = \sqrt{F} \times 10^5 \text{ Å} \tag{8}$$

From equation (6) we see that at normal conditions of the STM experiment when the repulsive force acting between the substrate and the tip is of the order 10^{-10} -10^{-6} N, the characteristic size of the protrusion of the graphite surface is about 1~100 Å. Thus we see that the deformation of the substrate during scanning can really be of the size of large biological structure.

X-ray Low Angle Scattering

Three silicon plates covered by IgG multilayers were used for X-ray analysis. The modified position-sensitive detector after calibration at the Institute of nuclear physics, Russian Academy of Sciences (Novosibirsk, Russia). and testing in the Institute of Crystallography of Russian Academy of Sciences, (Moscow), was installed in the Institute of Biophysics (Genova) at the X-ray small-angle scaterring diffractometer AMUR-K. Detector was connected with electronic control device and PC. New enhanced version of software was installated at the PC. While the scattering of LB-films on the solid supports here reported is measured with monochromatic beam, the solution investigations of biopolymers require high intesity of primary beam and are carried out without monochromator. Therefore the adjusting of difractometer AMUR-K was carried out for two kinds of measurements - with and without monochromator. The specimen holder was modified and the measurement can be done up to Θ =12.5°. Key parameters of our difractometer AMUR-K can be summarized as follow:

- Radiation type	X-ray CuKα, λ=1.542 Å MoKα FeKα
- Linear resolution of position-sensitive detector	0.15 mm
- Detector window	10 x 100 mm
- Step of measurements	0.07 mm
- Maximal frequency count	25 000 Hz
- Sample - detector distance	670 - 720 mm
- Angle range of the measurements	0.03 - 8.00°
moving detector	up to 25°

Highly-oriented polymer with period 210 Å was used as standart for calibrating of the difractometer. Fig. 11 shows the X-ray diffraction curves obtained by a film of lipids from archeobacteria. The position of the Bragg reflection peak clearly visible in curve 1, appears reproducible. According to the Bragg law:

$$n\lambda = 2d\sin\theta \tag{9}$$

we can get indeed, even in the case of one peak, an indicative measurement of thickness for each layer, that for the shown curve is 38.6.

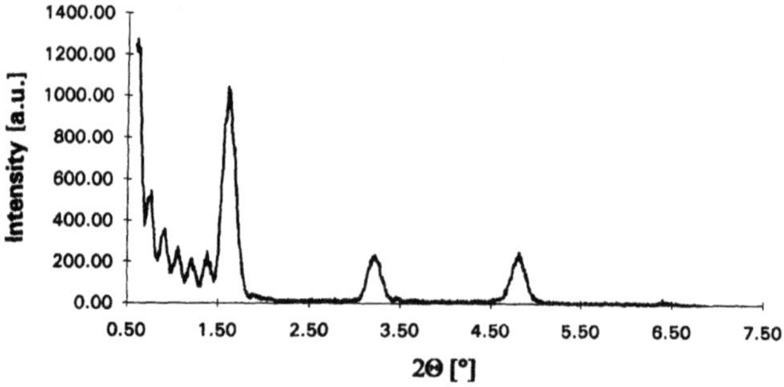

Figure 11 Scattering pattern from 10 bilayers of Cd-arachidate.

Ellipsometry

The measurements were carried out with PCSA null ellipsometer having He-Ne (633 nm) laser as a light source. The data was treated according to the two-layer model, whereby the first lower layer accounts for the imperfections which always exist on the surface of silicon substrate in the form of traces of polishing, intrusions of the polisher, thin oxide layer etc. In order to prove the applicability of the measurements at model multi-angle of incidence were carried out. Substrate and one- to five-monolayer films were measured at the angle of incidence, from 50° up to 80° with 5° step. The measurements of the bare substrate were treated according to the one-layer model, and the resulted parameters of the surface layer were used in the treatment of the measurements of the films according to the two-layer model. To compare the applicability of the models the data on the films were treated by one-layer model as well. The mean values of the measurements at different angles are given for the thickness and the index of refraction, the discrepancy of the values obtained under different angles is given as an error. The errors of the one-layer treatment are too large, achieving for the refraction index the value of 0.5. The values of the refraction index of one- and five-monolayer films differ very much: 2.8 and 1.9 respectively. In general, the refraction index of the film can vary due to some changes occurring in the film as the number of monolayers grows, but neither the obtained value of it (2.8), nor the amplitude of change (0.9) can be attributed to an organic matter [Tronin et al., 1993]. On the contrary, the two-layer treatment gives rather self-contained figures. The discrepancies are reasonable and the refraction index of the film remains more or less the same. The mean values almost coincided with that obtained only from the measurement at angle of incidence of 70°, which then was used in the experiments. The thickness of the bound layer was determined as the difference between the total thickness and that of the RAM monolayer measured before the reaction.

Infrared spectroscopy

The infrared region of the electromagnetic spectrum extends form the red end of the visible spectrum to the microwave region. The region includes radiation at wavelengths between 0.7 and 500 μm, or in wavenumbers, between 14,000 and 20 cm^{-1}. The spectral range used most is the mid-infrared region, which covers frequencies from 4000 to 200^{-1} (2.5 to 50 μm). Infrared spectrometry involves examination of the twisting, bending, rotating, and vibrational motions of atoms in a molecule. Upon interaction with infrared radiation, portions of the incident radiation are absorbed at specific wavelengths. The multiplicity of vibrations occurring simultaneously produces a highly complex absorption spectrum that is uniquely characteristic of the functional groups that make up the moleucle and of the overall configuration of the molecule as well. The infrared spectrum of a compound is essentially the super position of absorption bands of specific functional groups, yet subtle interactions with the surrounding atoms of the molecule impose the stamp of individuality on the spectrum of each compound. For qualitative analysis, one of the best features of an infrared spectrum is that the absorption or the lack of absorption in specific frequency regions can be correlated with specific stretching and bending motions and, in some cases, with the relationship of these groups to the rest of the molecule. Thus, when, interpreting the spectrum, it is possible to state that certain functional groups are present in the material and certain others are absent. With this one datum, the possibilities for the unknown sometimes can be narrowed so sharply that comparison with a library of spectra of pure compounds permits identification. Many useful correlations have been found in the mid-infrared region. This region is divided into the "group frequency" region, 4000-1300 cm^{-1} (2.50-7.69 μm), and the fingerprint region, 1300-650 cm^{-1} (7.69-15.38 μm). In the group frequency region the principal absorption bands are assigned to vibration units consisiting of only two atoms of a molecule - that is, units that are more or less dependent on only the functional group that gives the absorption and not on the complete molecular structure. Structural influences do reveal themselves, however, as significant shifts from one compound to another.

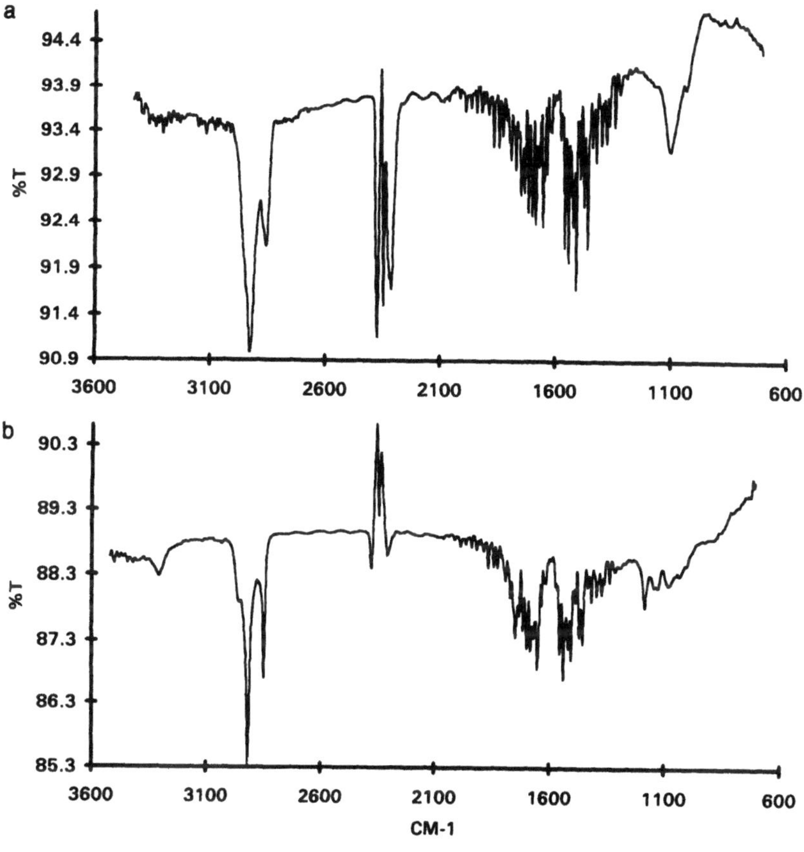

Figure 12 Infrared spectrum of thin film of lipids from archaebacteria with (above) and without (below) the insertion of valinomycin.

In the derivation of information from an infrared spectrum, prominent bands in this region are noted and assigned first. In the interval from 4000 to 2500 cm^{-1} (2.50-4.00 μm), the absorption is characteristic of hydrogen stretching vibrations with elements of mass 19 or less. The C—H stretching frequencies are especially helpful in establishing the type of compound present; for example, C==C—H occurs around 3300 cm^{-1} (3.03 μm), aromatic and unsaturated compounds around 3000-3100 cm^{-1} (3.33-3.23 μm), and aliphatic compounds at 3000-2800 cm^{-1} (3.33-3.57 μm). When coupled with heavier masses, the hydrogen stretching frequencies overlap the triple-bond region. The intermediate frequency range, 2500-1540 cm^{-1} (4.00-6.49 μm), is often called the unsaturated region. Triple bonds, and very little else, appear from 2500 to 2000 cm^{-1} (4.00-5.00 μm). Double-bond frequencies fall in the region form 2000 to 1540 cm^{-1} (5.00-6.49 μm). By judicious application of accumulated empirical data, it is possible to distinguish among C==O, C==C, C==N, N==O, and S==O bands. The major factors in the spectrum between 1300 and 650 cm^{-1} (7.69 and 15.38 μm) are single-bond stretching frequencies and bonding vibrations (skeletal frequencies) of polyatomic systems that involve motions of bonds linking a substituent group to the remainder of the molecule. This is the fingerprint region. Multiplicity is too great for assured individual identification of the bands, but collectively the absorption bands aid in identifying the material (Fig. 12).

Surface potential gauge

Surface potential measurements on molecular monolayers such as those achievable by means of the Langmuir-Blodgett (LB) technique is a well established and common technique for investigating the monolayer at the liquid-gas interface and it is often affected by strong shielding disturbances form the polar molecules of the liquid phase (in most of the cases water solution). Moreover, it is well known that during and after the deposition, LB films can change their structure with respect to that at the liquid-gas interface, vanishing the results obtained by surface potential measurements at the liquid-gas interface. Therefore, we have developed a technique for the investigation of mono and multilayers on the solid substrates [Erokhin et al., 1994]. The instrument (fig. 13) based on the capacity modulation of a capacitor allows to measure the potential arising even from one monolayer and generated by the charge distribution of the molecules of the film deposited on a fixed electrode, thereby monitoring the anisotropy (i.e. preferential orientation) of molecular films; in certain cases we can assay also the functional properties of molecules (like the light-induced charge separation in metalloproteins) arranged in LB films.

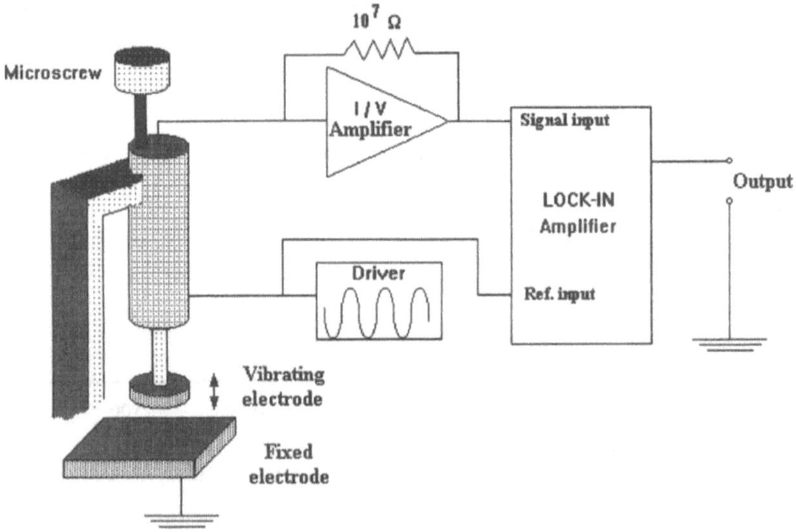

Figure 13 Scheme of the surface potential gauge.

MOLECULAR MANUFACTURING

The potential of the Scanning Tunneling Microscope (STM) as an ultrahigh-resolution instrument for surface modification was recognized almost immediately after atomically resolved imaging was demonstrated a decade ago. Since then, controlled manipulation of single atoms on surface has been achieved under highly ideal conditions.We must emphasize that other Scanning Probe Microscopy like Atomic Force Microscopy (AFM) were developped, thus extending the applications of the new nanotechnologies. Supramolecular engineering represents the tool being presently developed to fabricate nanostructures.

Fatty-acids and their surface modification by scanning tunnelling microscope

Investigations of the structure and properties of molecules, adsorbed on the surface of solids gives important information about the basic principles of intermolecular and interfacial interactions. There was proposed a number of applications of such systems in the future electronic device technology. Ones of the most interesting objects of this kind are mono- and multilayers of amphiphilic molecules fabricated with Langmuir-Blodgett (LB) technique.

In this paragraph we report on the local modification of the internal structure of LB film with STM tip voltage pulses, as can be seen by the difference in constant current images before and after such treatment [Nevernov et al.,1994c]. From the number of molecules which may be used for the fabrication of LB films we selected icosanoic (arachidic) acid [$CH_3(CH_2)_{18}COOH$]. The deposition technique for this substance is well developed and LB films of it were studied with a number of experimental methods. Structure of LB films of cadmium fatty acid salts was also studied by scanning tunnelling microscopy (STM). Icosanoic acid was dissolved in 1 ml of chloroform with the concentration 2 mg/ml and spread on the surface of water and compressed to a surface pressure of 30 mN/m. The water contained 10^{-4} M $CdCl_2$. This layer was deposited on the surface of freshly cleaved highly oriented pyloric graphite. This sample was studied on air with a commercially available STM (Asse-Z/MM-MDT Corp.; Padua/Moscow) with the possibility of lithography (voltage pulse) operation. We used mechanically prepared Pt/Ir tips as a probe. All the measurements were done in the constant-current imaging mode.

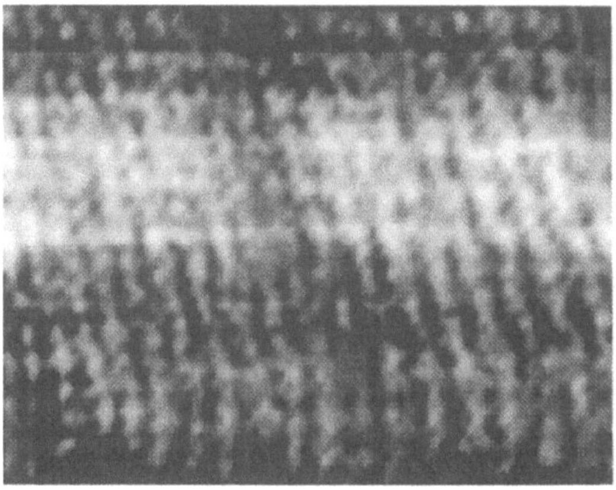

Figure 14 Top-view STM image of Cd-arachidate, 8,2x7,3 nm.

We obtained STM images which showed surface corrugation of about 0.2 nm. On Fig. 14 the molecular resolution STM image is presented. Measurement of the unit vectors of Cd arachidate surface packaging gives the values of 0.52 nm and 0.36 nm which gives the surface area per one molecule of about 0.18 nm^2. This value is in a good agreement with the area per molecule during the deposition and with other published reports. Three pulses voltage were applied in different places of the scan area. The parameters of the pulses were the following: the voltage was 20 V, while the tip was moved by 2 nm away from the normal tunnelling position, and the pulse duration was 10 msec. Spots of about 50 nm in diameter can be easily seen on the top-view image presented on the fig. 15. The depth of these

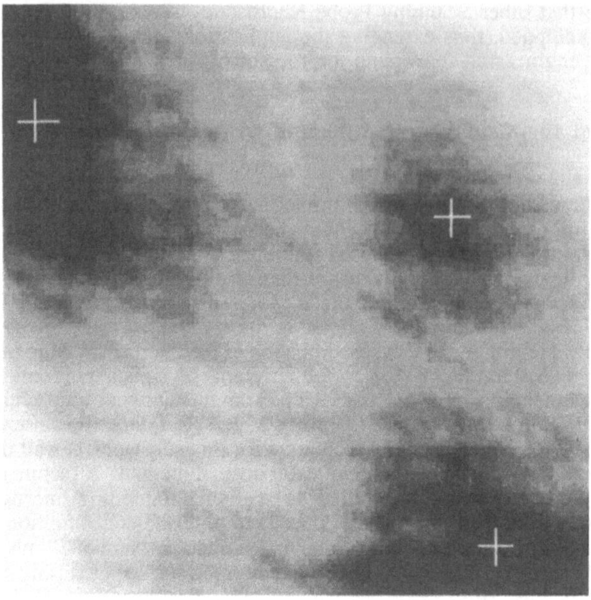

Figure 15 Top-view STM image of Cd-arachidate bilayer after the application of three 20V pulses, 191x191 nm.

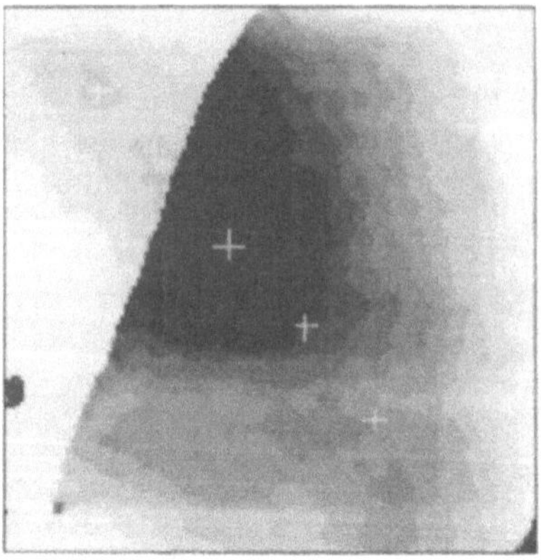

Figure 16 - 510x510 nm top-view STM image of Cd-arachidate bilayer after five 20 V pulses.

features was about 0.2 nm. Places where the pulses were applied are marked with white crosses, as apparent by the cross-section along the X direction.

In fig. 16 the top-view STM image of Cd arachidate bilayer shows a step, which we consider to be graphite step under the bilayer. The result of application of five 20 V pulses using the same technique is also presented in this figure. Pulses were applied starting with the right bottom corner of the scanning area and moving to left top corner. The result of this manipulation is a short line approximately 200 nm long and 80 nm wide. Its depth is about 0,6nm ± 30%. On the same image we can see that our manipulations made no effect on the edge. When the same experiments were made using pure freshly cleaved graphite, no changes in graphite structure were evident. The results of our experiments show that after the voltage pulse application spot-like depressions appear on the constant current STM images of Cd arachidate bilayer. In constant current mode we measure the signal of the feedback system, which manipulates the tip-sample distance in order to keep the tunnelling current constant. This means that in the centre of the spot tunnelling current is lower than on the edge. The drop in the tunnelling current according to our opinion can be due to the drop in the conductivity of a single molecule and/or to reorganisation of bilayer structure due to the destruction of chemical bounds.

If we consider the features appeared after the influence of voltage pulses to be the result of conductivity changes in the bilayer of Cd arachidate, the possible reason for these changes can be the following chemical modification:

$$CH_3(CH_2)_{18}COO^- \xrightarrow{HOH,\,e} CH_3\text{-}R_1\text{-}CH_2OH + CH_3\text{-}R_2\text{-}COO^-,$$

where R_1: $-(CH_2)_n-$, R_2: $-(CH_2)_m-$, $n+m=16$

or:

$$(CH_3(CH_2)COO)_2\,Cd \xrightarrow{HOH;\,HCl;\,e} 2\,CH_3(CH_2)COOH + CdCl_2,$$

In this case we can conclude that the charge transfer process is concentrated in saturated hydrocarbon chain and do not depend on the specific properties of fatty acid molecules. Here we suggested that there were no packing modification of the layer and its thickness was the same. In the case of chemical bound destruction accompanied with the removement of some quantity of substance from the layer the process is similar to the electron beam resist treatment. It is very interesting that in this case the results of this treatment may be seen immediately, with the same STM technique without additional chemical treatment. The results on STM lithography can be interesting for developing future data storage devices. There are a lot of attempts to use STM for the fabrication of nanometre scale erasable features. For example recently the fabrication of erasable marks over the surface of composite material was reported. These materials show phase transitions under the application of voltage pulses accompanied by electric conductivity changes. In our work we also observed changes in electrical conductivity of Cd arachidate bilayer. So this kind of material can be also used for these purposes and it appears that the parameters and characteristics of the material could be optimised to produce the required chemical modification of internal structure of LB Cd arachidate film.

Thermophilic lipid monolayer and valinomycin insertion

Studies based on the analysis of the rRNA 16S partial sequence homologies made in the late 70's by Woese, revealed the existence of a consistent group of prokaryots, named archeobacteria, scarcely related to all the others (eubacteria). These studies have been confirmed by the comparative analyses of the physiological, morphological and biochemical characteristics of significative representatives of the new group. The typical feature of archeobacteria is their capacity to grow in extreme environmental conditions like high temperature, salt or methane concentrations. It was also found that an enormous pH difference of up to 4 units exists across the membrane, the internal pH being around 6 and the external between 1.5 and 2.5 . The characteristics just described are probably due to the unusual structural features of their membrane lipids, which differ from the ones of eubacteria and eukariotes in the fact that the former are all based on the condensation of glycerol or more complex polyols with two isoprenoid C_{20}, C_{25}, C_{40} chains and have also an sn-2,3-diacylglycerol stereoconfiguration (whereas for naturally-occurring glycero-phosphatides and diacylglycerols an sn-1,2 stereo-configuration is normal). Previous works have shown that the dimensions and bipolar structure of these tetraether lipids yield to a different type of membrane, which is now organized in amphiphilic monolayers lacking a midplane region. The membrane of Sulfolobus solfataricus, a thermophilic archaebacterium, is formed by lipids of varying complexity, which after chemical treatement (acid hydrolisis) yield to two main types of components: a symmetric lipid in which the hydrophilic parts correspond to two glycerols (glycerol-dialkyl-glycerol tetraether, GDGT) and an asymmetric one in which one of the glycerols is substituted by a single branched-chain nonitol (glycerol-dialkyl-nonitol tetraether, GDNT). In the membrane, the nonitolic heads of GDNT are located outside the cell. In fact, a fundamental aspect of the behaviour of GDNT relies on the slow motion of its external portion: the steric constraints imposed by the hydrogen bonds between the adiacent nonitols induce a much stronger immobilization of this part with respect to GDGT and also to other monopolar lipids. The strong cooperative interactions of the nonitol groups, can play, in addition to a mechanical role, also a physiological one, by creating an electrical barrier due to dipole-dipole interactions that could explain the cell capacity of mantaining the very high pH-gradient. In the membrane, a variety of different polar groups is linked to the free hydroxyl group of glycerol and to the one of nonitol. In recent papers [Accossato et al., 1994; Troitsky et al., 1994] excellent results were obtained from the study of thermophilic lipids properties in terms of stability and preferential orientation in respect to their capacity to form monolayers. Langmuir-Blodgett technique is used for deposition of thin films of the lipids from archaebacteria. Nine different compounds are studied. Highly uniform multilayers are prepared from some of them. Due to ad hoc chemical modifications the films are made stable in water solutions (Fig. 17). Study of the monolayers at the air-water interface and deposited films proves U-shape of the molecules with close packing of isopranyl chains in the monolayer. Thin uniform layers of surfactants were produced on solid substrates of pratically any thickness proportional to the thickness of the monomolecular layer. It is possible to create systems of complex strucutre where monolayers of different types of molecules possessing various properties are arranged in the required order across the thickness of the film, i.e. to accomplish a "construction" at the molecualr level. One can embed also different biological molecules into the monolayers or adsorb them on the latters to produce biologically active systems. The mixed films of two lipids with valinomycin are prepared and studied to check the possibility of dissolution of some functional molecules in these lipid matrixes. To reach uniform distribution of valinomycin in the multilayer with high concentration the films consisting of alternating layers of the lipid molecules of one type and valinomycin were deposited (Fig. 18). Archae 9 films display absolute stability in water solutions caused by salt formation For this reason, the Archae 9 was used also as a matrix for dissolution of valinomycin. Surface pressure-area isotherms in Fig. 17 shows that distribution of valinomycin molecules is much more uniform with respect to the case of Archae 5 matrix because the curve for the mixture is shifted considerably to the region of higher area values.

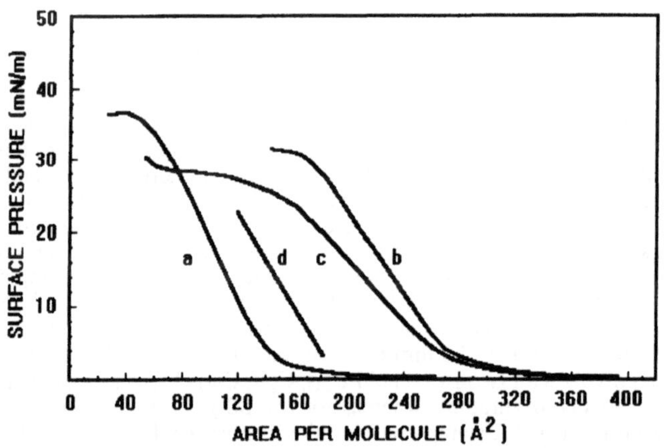

Figure 17 Surface pressure-area isotherms of the monolayers of Archae 9 (a), valinomycin (b), and mixture of Archae 9 with valinomycin in molar ration of 1:0.5 (c). Subphase is distilled water. Solid line (d) shows the calculated position of the isotherm of mixed monolayer. Areas per molecule are equal to 1.47, 2.48 and 2.87 nm^2 for the curves a, b and c accordingly.

Finally, to avoid precipitation of valinomycin in the Archae 9 film under the condition of its uniform distribution with high concentration in the matrix technique of depositing the alternating monolayers was worked out (Fig. 18).

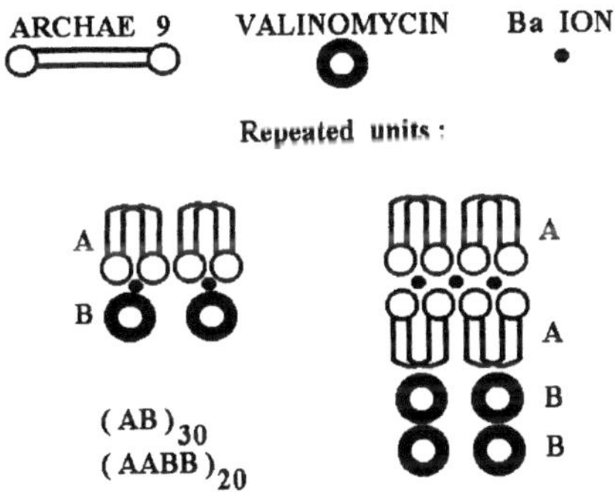

Figure 18 Schemes of the structures of alternating monolayers created on the basis of the Archae 9 barium salt and valinomycin.

Transfer ratios for both sequences of alternation were practically equal to unity and no precipitation of bulk crystallites was observed with optical microscopy. Film of the ABABAB... type deposited onto sapphire substrate were recrystallized in several days probably because of weak adhesion of the film to hydrophylic surface. Indeed, the films of both types are stable in water solutions. Structural studies of such "superlattices" are in progress. However, it is obvious that the technique used gives us the possibility to distribuite uniformly considerable amount of the functional molecules in the lipid matrix. To sum up, the application of the LB technique appeared to be successful for depositing the films of the lipids from archaebacteria in spite of the molecules used possess the structure unusual for the compounds, which form highly uniform LB films. This possibility seems to be provided by U-shape of the molecules at the air-water interface with close packing of isopranyl chains at the same time, so that behaviour of the molecules at the surface and their interaction with water are similar to those for usual surfactant molecules [Troitsky et al., 1994]. Instability of the films of native compounds as well as some of their derivatives in aqueous solutions is caused by the presence of hydrophilic groups, which interact with water stronger than with each other. This interaction can/not be changed pratically when varying the composition of water subphase. Introduction of carboxyl groups into hydrophilic parts of the molecules changes the situation in cardinal manner. Interaction of the monolayers with each other becomes very strong due to salt formation with metal ions dissolved in water subphase. The films become stable in aqueous solutions and quality of deposition improves considerably.

Synthetic Lipid monolayer

Fullerene. Fullerenes are compounds of extreme interest for their potential utilization as superconducting materials when their films are properly dopped or imbedded in lipid membrany. Depositions at the air-water interface of uniform C_{60} films with the use of LB technique. This has been recently achieved by mixing monolayer of C_{60} with specially selected surfactant compound [Berzina et al., 1994]. The surfactant compound used (surfactant) forms itself amorphous uniform LB films and C_{60} molecules distribute homogenously in this matrix. Another property of the surfactant compound is the solubility in hexane whereas C_{60} does not disolve in the latter. One-component C_{60} film is obtained after treatment of one sample with hexane. To achieve this result mixtures of C_{60} with several surfactant compounds were studied (benzene $C_{16}H_{33}$ - BEDT-TTF, $C_{17}H_{35}$ - DC-TCNAQ, hexane, surfactant) and finally the one selected was surfactant. Surface pressure area isotherms were recorded to evaluate the degree of C_{60} dissolution in matrix monolayers. Deposited LB films were studied by optical microscopy, electron diffraction and trasmission electron microscopy techniques to record changes of the structure and morphology under variation of the conditions and selection of the compounds. One component C_{60} monolayers at the air-water interface as well as C_{60} films transferred with horizontal lift technique were investigated also for interpretation of the data on the mixed multilayers.

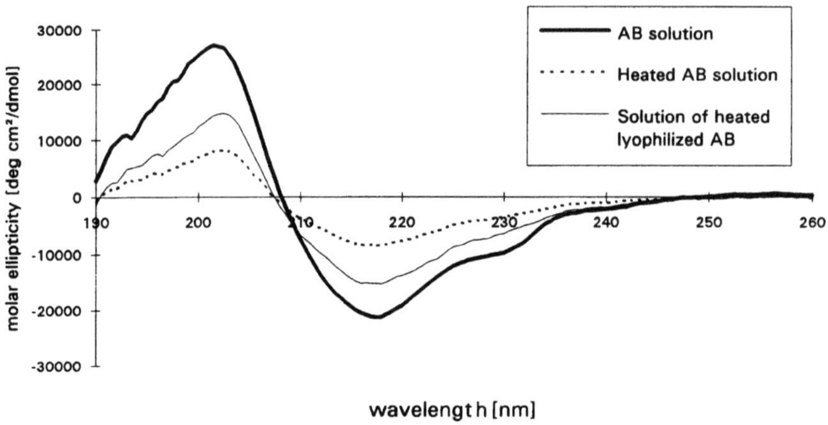

Figure 19 CD spectra of antibody solutions. This picture shows the loss of secondary structure upon heating. Dehydration provides only partial stabilizing effect.

Conductive LB films of heptadecylcarboxymethylene BEDT-TTF. Surfactant compounds containing bis-ethylenedithio-tetrathiofulvalene (BEDT-TTF) donor group appeared to be very promising for production of highly conductive LB films. New donor compound, i.e. heptadecylcarboxymethylene BEDT-TT contains strong hydrophilic group which does not prevent close packing of BEDT-TT were deposited from the water subphase containing $FeCl_3$. Conductivity of the films reaches 4 $Ohm^{-1}cm^{-1}$. Some structures parameters are determined by electron and small-angle X-ray diffraction studies. From these data possible structure models are proposed and peculiarities of the structure have been discussed [Troitsky et al., 1994].

Antibody LB films and effect of glutaraldehyde fixation

It was found, that the secondary structure of antibodies is preserved in LB films after heating till 150° C, while in the solution it is denatured at 60° C (Fig. 19). It was also shown, that dehydration performs only small stabilizing effect with respect to molecular ordering of the film.

Functional properties of the layer were improved after the heating, namely the saturation level was increased for 25% and the reaction rate 6-fold (Fig. 20). These improvements are likely to be due not to modification of each molecule, but to increased regularity of total molecular system. Activity is preserved in LB films after heating until 150°C.

LB films and protein LB films, are rather flexible formations. Molecules in films can be rearranged during and after deposition onto solid substrates in order to provide thermodinamically stable organization. The structure of the immobilized Langmuir-Blodgett films of IgG and their immunological activity were studied by means of ellipsometry. The dependence of the film thickness on the surface pressure of deposition gives evidence of the tilting of molecules with an increase of the pressure. Below pressure 30 mN/m the thickness of the film is approximately 4 nm which coincides with the smallest dimension of the IgG molecules. Between 30 and 40 mN/m the thickness increases sharply achieving the value of about 10 nm which is equal to the largest molecular dimension. The further increase of pressure does not come out in the growth of the thickness. This means that the films are transferred from the water-air interface in the form of 2-D ordered monomolecular layer. The dependence of the immunological activity on the pressure of deposition was shown to have a descending pattern. We propose different mechanisms which explain the decrease of the immunological activity at the IgG molecules in the film with the increase of the surface density such as the blocking of the active sites and the decrease of the conformation mobility of the Fab fragments [Dubrovsky et al., 1993]. In another work [Facci et al., 1993] we tried to fix the preferential anisotropic organisation of monolayer at the water surface by cross-linking protein molecules by glutaraldehyde. Nanogravimetric (Fig. 21) and surface potential measurements revealed, that providing cross-lincking of antibodies before each deposition, it was possible to make the film anisotropic. Surface potential measurements have shown increased potential through the film, which was even able to quench the reaction with glutaraldehyde.

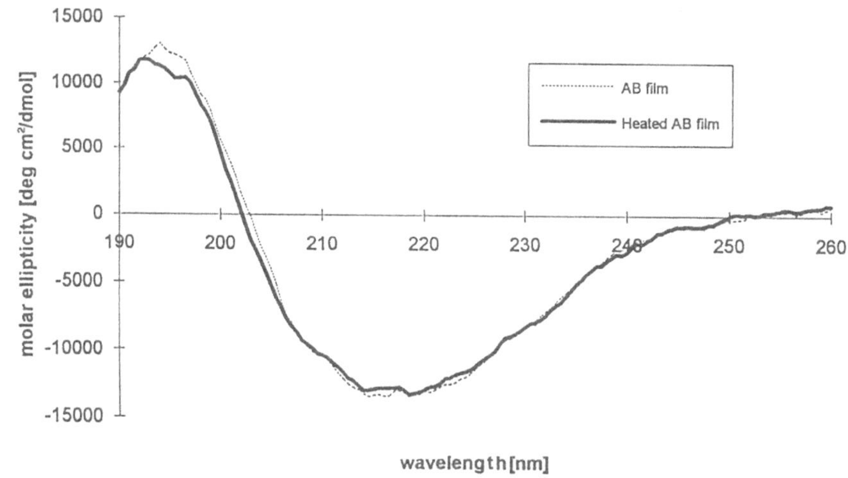

Figure 20 CD spectra of antibody LB films (20 monolayers). This picture shows the secondary structure preservation after heating (150° C, 30 minutes)

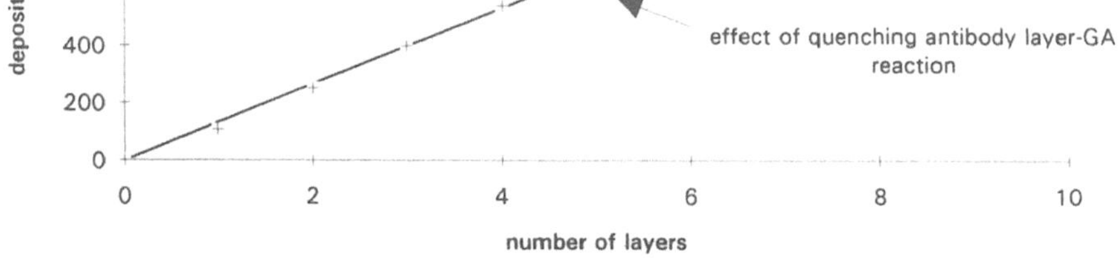

Figure 21 Deposited glutaraldehyde mass after each antibody layer deposition showing that the reaction with GA is quenched after 4 antibody layers and that it restarts increasing the reaction time.

MOLECULAR ELECTRONICS

In the field of the molecular electronics the actual growth of interest for the Langmuir-Blodgett (LB) films of organic and inorganic molecules is due to the possibility of making biolelectronic devices at a molecular level. Exploiting the properties of the molecules and of the films formed by them, it is possible to build sensors, new materials, memories, rectifiers and optical devices. Worthy of notice are recent accomplishements in the realization of "bona-fide" nanoelectronic devices based on biological materials capable to work at room temperature [Erokhin et al.,1994; Nicolini, 1993].

Monoelectronic devices

Ultrasmall CdS particles were formed by exposing deposited cadmium arachidate Langmuir-Blodgett bilayers to atmosphere of H_2S. STM images of the resultant films reveals the presence of particles with sizes of about 40 - 60 Å. Also voltage-current characteristics were measured by STM on the structure "tip - tunnelling gap - CdS particle - tunnelling gap - graphite substrate". Steps in voltage-current characteristics indicate the appearance of single electron process (Coulomb blockade) at room temperature (Fig. 22).

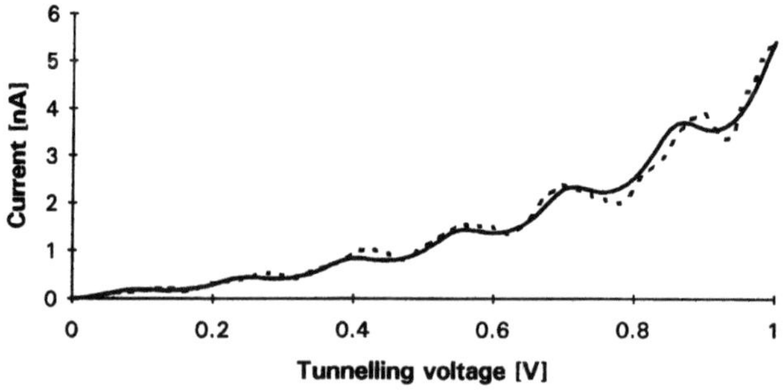

Figure 22 Theoretical (• • •) and experimental (——) voltage-current characteristichs of 50 Å CdS particles, at room temperature.

Unusual behaviour of Voltage-Current (V/I) characteristics in systems containing two electrodes and ultra small conductive particle between them, separated by two tunnelling junctions, is very interesting and has no analogies in other systems. Theory explains the phenomenon through the appearance of "Coulomb blockade" due to passing of electrons through the intermediate granule. The presence of an electron in the granule changes strongly the energy of the particle (due to its small capacity). This energy has to be overcome by external voltage in order to provide the possibility for one more electron to be tunnelled to the granule. Increasing the voltage one can observe discrete increase in the current in correspondence to those voltage values which overcome the Coulomb blockade potential. The increase in current takes place every time the voltage increases of:

$$\Delta V = e/C \tag{10}$$

where e is the electron charge and C is the capacity of the granule.

The process is called monoelectronic because it is possible to distinguish current steps in V/I curve due to unitary increase in the number of electrons in the granule. Step-like behaviour of the current was observed in several works at low temperature. The value of the temperature is very important for observing such phenomena and the following equation must held true to allow the monitoring of steps in V/I characteristics:

$$e^2/2C > kT \tag{11}$$

Thus, the temperature at which monoelectron phenomena can be observed is limited by the capacity of the granule and therefore by its dimensions. Rough estimations, assuming spherical shape, give 90 Å as limiting value of the granule radius: for bigger radii Coulomb blockade cannot take place at room temperature. Technologically, the fabrication of structures suitable for observation of monoelectron phenomena, such as "electrode - tunnelling gap - nanometric granule - tunnelling gap - electrode" is very difficult. Using STM for realising such structures can be very fruitful, because, when the particle is formed, it can be found with STM and the tip, placed just above the granule, acts as the upper electrode. Very few works, which perform these measurements, are reported in literature. A method, resulting in the formation of CdS inside arachidic acid film by exposing initial cadmium arachidate Langmuir-Blodgett (LB) film to H_2S atmosphere, was recently implemented [Erokhin et al., 1993]. During such treatment, protons of H_2S become bond to the acid head groups and sulphur atoms, bond to cadmium, form CdS phase. The formation of CdS was checked by optical absorbance and electron diffraction. Sizes of the formed particles were estimated to be about 100 Å if the initial film contained 20 - 30 monolayers. It is interesting to mention, that the spacing of the LB film itself was decreased from 49.0 Å to 39.0 Å, as was measured by X-ray diffraction. The aim of our recent work was to investigate by STM (Fig. 22) a bilayer of cadmium arachidate after exposing it to H_2S, to find granules of nanometer sizes and to measure with the STM tip local V/I characteristics on "graphite - tunnelling gap - CdS particle - tunnelling gap - tip" structure. A bilayer of cadmium arachidate was transferred onto the graphite surface according to standard procedure. For providing the chemical reaction the sample was placed into a chamber, containing H_2S, for 30 minutes.

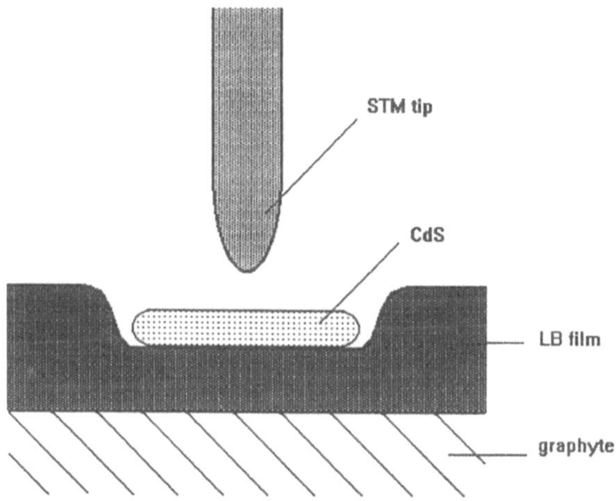

Figure 23 Experimental configuration of "graphite-CdS particles-STM tip" structure

STM measurements were performed using a device (Asse-Z/MM-MDT), which allows to measure local V/I characteristics. For the measurements of V/I characteristics STM tip was placed over the desired point (CdS particle, identified on previously obtained image) in constant current mode. When the tip was above this point, feedback was switched off and the tip - substrate voltage was swept from -0.5 to 0.5 V. CdS particles are well distinguishable in the picture. Sizes and shapes of the particles are not equal eachother, but, in general, sizes are in the range 40 - 60 Å (nevertheless it is possible to find also particles with sizes outside the range). The surface of the particles is rather flat. This fact becomes understandable if we suppose that CdS particles are small monocrystals. The hypothesis is also in agreement with light absorption data, showing the existence of the original CdS band structure in the particles after the reaction with H_2S, and with electron diffraction data, demonstrating that the lattice spacing value of the particles is the same as in bulk crystal. The characteristic was obtained by placing the tip above the CdS particle, the position of which was determined from previously acquired image. Despite some noise, steps in V/I characteristics are well distinguishable. Steps in the characteristics are equidistant and correspond to the value of voltage of about 0.2 V. Taking into account that the particles have in-plane dimensions of 40 - 60 Å and their surface is flat, we can conclude, that the most probable shape of them is disk-like one and the thickness of the disk is a couple of lattice unit cells of CdS. As a conclusion, nanometer scale CdS particles were formed by exposing cadmium arachidate LB film to H_2S atmosphere. Their sizes measured by STM were found to be small enough to allow monoelectron phenomena. Thus, such treatment of the Cadmium Arachidate films results in the creation of a new material, where nanometer scale monocrystal semiconductor particles are embedded into insulating LB matrix. This material displays new kinds of phenomena, particularly, but probably not only, monoelectron ones, which allow to study fundamental properties of systems with decreased number of dimensions and from technological point of view can permit the construction of new types of devices, such as monoelectron transistors.

Electronic Materials

Several biofilms previously reported are sensitive to the effect of electron beam and loose conductivity under local irradiation. Thus, desirable insulating patttern can be created in the conducting material, which is important for technological lithographic applications. In a separate line of investigation, however, in order to determine their electrical propérties, few produced thin biofilm have been deposited over a suitable interdigitated micro-electrode, which is made by sapphire support on which aluminum conductive tracks are fabricated by a photolitography process. Because of the dimensions of these conductive tracks (six to eight magnitude higher) relatively to the dimensions of the polymers used, it was necessary to deposit more layers (ten). In this way, we ensured a good regularity of the films. The experimental arrangement allows us to measure the current that flows in the sample, versus amplitude and frequency of the incoming wave. The data obtained wtih four distinct class of thin biopolymers films, a conductive lipid and three insulators, two proteins and one fatty-acid, are summarized in Fig. 24 and Fig. 25, respectively in DC and AC current.

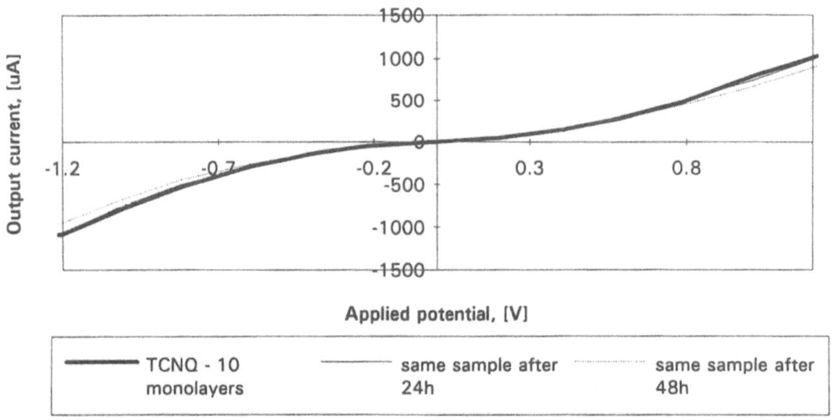

Figure 24 Voltage-current characteristics of conductive lipids.

Ion- and immuno-sensors

The methods of high-temperature support silanization and fixation of the antibody monolayer by glutaraldehyde provides the ordered multilayer films immobilized to the solid surface. Antibody monolayers immobilized on the polyorganosiloxane film can be used as sensing elements in immunosensors based on the effect of total internal reflection of excited fluorescence (TIRF) and of surface plasmon resonance (SPR). Surface plasmons generated on the presence of silane film and immobilized antibodies has been also studied with STM. This gives important information about the local surface characteristics of specific antibody-antigen interactions on the surface, which is being used to build effective immunosensors. Worthy of notice are recent developments in progress leading to the construction of nanogravimetric sensors and ion-biosensors, both highly stable and highly sensitive and both based on the previously described thin film technology of exceptionally thermostable lipids and antibodies.

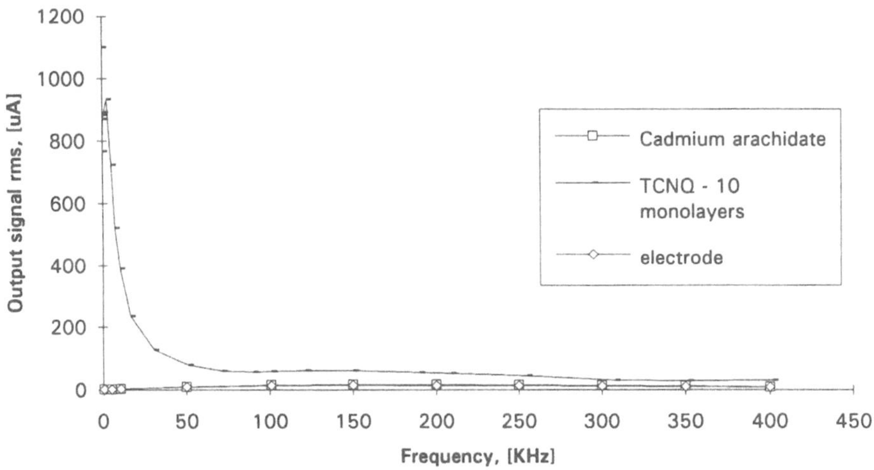

Figure 25 Frequency response of lipids and proteins. b) is the zoomed representation of the curves in a) relative to proteins and non-conductive lipids.

CONCLUSIONS

This overview was intended to give an updated status of the research being presently carried out at molecular level, with particular emphasis on the activities of my laboratories at Marciana and Genova, as reported at the 1993 Elba Forum within the framework of the Italian National Project "Technologies for Bioelectronics". Progress in this area proceeds at high rate world-wide and a complete review of the related fields was outside the scope of this chapter, being the subject of other contributions in this and successive volumes of the Elba Forum series now established by Plenum Publishing Corporation.

ACKNOWLEDGEMENT

The work here summarized was partly supported by Technobiochip, industrial research company among leading multinational and national industries located in Marciana (Island of Elba). The author would like to thank Mr. Fabrizio Nozza for the technical assistance in preparing this manuscript.

REFERENCES

Allinger N.L., 1977, *Amer.Chem.Soc.*, 99:8127

Amrein M., R. Drr,.Stasiak A., Gross H, Travaglini H, 1989, *Science* 243:1708

Andrade J.D., Vanwagenen R.A., Gregonis D.E., Newby K. & Lin, J.N., 1985,. *IEEE Trans.*, ED 32:1175-1179

Arscott P.G, Bloomfield H., 1992, *Methods in Enzimology*, 221:490

Attridge J.W., Daniels P.B., Deacon J.K., Robinson G.A. & Davidson, G.P., 1991, *Biosens. Bioelectron.*, 6:201-214
Averin D.V. and Likharev K.K., 1986, *Zh. Eksp. Teor. Fiz.*, 90:733
Averin D.V, and Likharev K.K., 1991, (eds Altshuler, B.L., Lee, P.A., and Webb, R.A.) *Elsevier*, Amsterdam.
Beebe T.P. Jr., Wilson H, Ogletree D.F., Katz J.E., Balhorn R,.Salmeron M.B, Siekhous W.J., 1989, *Science* , 251:370-372
Bein T., 1993, *Nature*, 361:207-208
Berzina T.S., Troitsky V.I., Neilands O.Ya., Sudmale I.V., Nicolini C., 1994, *Thin Solid Films*, in press
Binnig G., Rohrer H., Gerber C., and Weibel E., 1982, *Phys.Rev.Lett.*, 49:57
Binnig G., Rohrer H., 1982, *Helv.Phys.Acta* , 55:726
Blackford B.L., Jericho M.H, Malhern P.J., 1991, *Scan.Microsc.*, 5:907-918
Bottomlay L.A et al., 1992, *J.Vac.Sci.&Tech.*, A 10:591-595
Chen C., Hamers R., 1991, *J.Vac.Sci.Technol..*, 9:503
Chiang S , 1992, in:"Scanning Tunneling Microscopy", *Springer Verlag*, Berlin Heidelberg, Ch.7
Clemmer C.R., Beebe T.P., 1991, *Science*, 251:640
Cricenti A et al , 1992, *Polymer Preprints*, 33:741
Cricenti A., Scarselli M. et al., 1992, *Polymer Preprints*, 33:741
Cricenti A., Selci S., Felici A.C, Generosi R., Gori E., Djaczenko W., Chiarotti G., 1989, *Science*, 245:1226
Cricenti A., Selci S., Scarselli M., Generosi R., Amaldi R., Chiarotti G.
Cricenti A., Selci S. et al., 1991, *J.Vac.Sci.&Tecnol.*, B9(2):1285
Cricenti A., Selci S.,.Felici A.C., et al., 1989, *Helvetica Physica Acta*, 62:701
Doveret M.H, Esteve D., and Urbina C., 1992, *Nature*, 360:547-553
Driscoll R,.Youngquist M.G. and Baldeschvieler J.D, 1990, *Nature* , 346:294-296
Dubrovsky T.B, Demcheva M.V., Savitsky A.P., Mantrova E.Yu., Yaropolov A.I., Savransky V.V, Belovolova L.V., 1993, *Biosensors & Bioelectronics*
Dubrovsky T.B., Erokhin V.V., Kayushina R.L., 1992, *Biolog. Membr.*, 6(1):130-137, (Harwood Acad. Publ. GmbH)
Dunlap D.D., Bustamante C., 1990, *Nature* , 342:204
Dunlap D.D., Bustamante C., 1990, *Nature* , 344:641-644
Durig U, Gimzewski J.K. and Pohl D.W., 1986, *Phys.Rev.Lett.*, 57:2403
Durig U, Zuger O. and Pohl D.W., 1988, *J.Microsc.*, 152:259
Edström R.D., Meinke M.H., et al., 1990, *Biophys.J.;* 58:1437
Engel A, 1991, *Ann.Rev.Biphys.Chem.*, 20:79-108
Erokhin V., Feigin L., Ivakin G., Klechkovskaya V., Lvov Yu., and Stiopina N., 1991, *J. Makromol. Chem., Makromol. Symp.*, 46:359-363
Erokhin V., Facci P., and Nicolini C., 1993, *Biosensors and Bioelectronics*, in press
Erokhin V., Facci P., and Nicolini C., 1994, *Rev. Scient. Instr.*, in press
Frommer J., 1992, *Angew.Chem.I nt. Ed. Engl*l., 31:1298-1328
Fulton T.A, and Dolan G.J., 1987, *J. Phys. Rev. Lett.*, 59:109
Gourdon A, 1992, *New Journal of Chemistry*, 16(10):953-957
Guckenberger G., Hacker B. et al., 1991, *J.Vac.Sci.&Technol.*, B9(2):1227
Guckenberger R., Hartmann T., Wiegrabe W., Baumeister W., 1992, Chap.3 in "Scanning Tunneling Microscopy" (Eds.: R.Wiesendanger, H. Guntherodt), *Springer*, Berlin
Haussling L., Michel B., Ringsdorf H., Rohrer H., 1991, *Angew. Chem. Int. Ed. Engl.*, 30(5):569
Hentschke R., Schurmann B.L., Rabe J.P., 1992, *J.Chem.Phys.*, 96:6213
Janzen A.F and Seibert M., 1980, *Nature*, 286:584-585
Keller D., Bustamante G., and Keller R.W., 1989, *Proc. Natl. Acad. Sci. USA*, 86:5356
Kurnikov I.V., Sivozhelezov V.S., Redchenko V.V and Gritsenko O.V., 1992, *Molecular Engineering* , 1:53
Lee I., Atkins E., Miles M., 1991, *J.Vac.Sci.and Technol.*, B9:1107
Lukashev E.P., Kononenko A.A., Noks P.P., Gaiduk V.I., Tseitlin B.M., Rubin A.B. and Betskii O.V., 1989, *Proc. Acad.Sci. USSR* , 304-306
Lyn M., 1975, (ed. Weetal, H.H.), *Marcel Dekker*, New York, ch. 1, 13
Malmqvist M., Olofsson, G., 1989, *U.S. Patent* 4,833,093
McMaster T., Carr H., Miles M.J., Cairnes P., Morris V., 1991, *Macromolecules*, 24:1428
Miles M.J., McMaster T. et al., 1990, *J.Vac.Sci.&Technol.*, A8:698
Mullen K., Ben-Jacob E., Jaklevic R.C., and Schuss Z., 1988, *Phys. Rev.*, B37:98-105
Nawaz Z., Pethica J.B., 1992, *Surf.Sci.*, 264:261-270; 1994 a, *Surface Science*, in press
Nevernov I., Kurnikov I., Facci P., and Nicolini C., 1994 b, *J. Appl. Physics*, in press
Nevernov I., Facci P., Dubrovskaya S., Erokhin V. and Nicolini C., 1994 c, *Langmuir*, in press
Nicolini C., Staub R., Gussoni A. and Nevernov I., 1994, *J. Mol. Biol..*, in press
Nicolini C. and Kendall F., 1977, *Physiological Chem. Phys.*, 9:265-283
Nicolini C., Vergani L., Diaspro A.,Scelza P., 1988, *Biochem. Biophys. Res. Com.*, 155:1396-1403
Nicolini C., Erokhin V., Catasti P., Antolini F. and Facci P., 1993, *Biochem. Biophys. Acta*, 1158:273-278

Nicolini C., 1994, *Biosens. and Bioelec.*, in press
Peterson I.P., and Russell G.J., 1985, *Thin Solid Films*, 134:143-152
Rabe J.P., Buchholz. S , 1991, *Science*, 253:424
Sabo Y, Kononenko A.A, Zakharova N.I., Chamorovski S.K. and Rubin A.B., 1991, *Proc. Acad.Sci. USSR*, 316-318
Schön G., and Zaikin A.D., 1990, *Phys. Rep.*, 198:237
Selci S., Cricenti A., 1991, *Physica Scripta*, T35:107-110
Smotkin E.S., Lee C., Bard A.J., Campion A., Fox, M.A., Mallouk, T.E., Webber, S.E., and White, J. M., 1988, *Chem. Phys. Lett.*, 152:265-268
Specht M., Ohnesorge F., Heckl W., 1991, *Surf. Sci. Lett.*, 257:L653
Tachibana H., Azumi R., Nakamura T., Matsumoto M. and Kawabata Y., 1992,*Chemistry Letters*, 173-176
Tieke B., 1990, *Advanced Material* s, 2(5):222-231
Troitsky V.I., Berzina T.S., Katsen Ya. Ya., Neilands O. Ya and Nicolini C., 1994, *Synt. Met.*, submitted.
Tronin A., Dubrovsky T., DeNitti C., Gussoni A., Erokhin V., Nicolini C., 1994, *Thin Solid Film* , 238:
Wilkins R., Ben-Jacob E., and Jaklevic R. D., 1989, *Phys. Rev. Lett.*, 63:801
Woese G.R., Fox G.E., 1977, *Proc. Natl. Acad. Sci. USA* , 74:5088-5090

QUATERNARY CHROMATIN-DNA STRUCTURE IMAGED BY SCANNING NEAR-FIELD OPTICAL MICROSCOPE: A COMPUTER SIMULATION

Paolo Facci and Claudio Nicolini

Institute of Biophysics, University of Genova, Via Giotto 2, 16153 - Genova, Italy

ABSTRACT

Scanning tunnelling optical microscope, the last kind of scanning probe microscope, seems to be very useful for biophysical imaging science. Our purpose is to understand the behavior of such a device, considered as a particular kind of "near field" microscope, when used to obtain images of chromatin fibers or biopolymers in general. Chromatin fibers of 30 nm in diameter were taken as a sample. They were shaped like a helical array of dielectric oblate ellipsoides, the nucleosomes, and their interaction with a laser beam was calculated in the near field range, as detected by a sharp dielectric scanning tip. Theoretic images of this sample were obtained as a function of several parameters of the simulation model.

Moreover, an insight into the theoretic resolution achievable with such a microscopy is presented as a function of the tip size. The results and the weak invasiveness of the technique itself point out this kind of microscopy as one of the most suitable for imaging biopolymers and biostructures up to nanometric resolution.

INTRODUCTION

The recent discovery of the scanning tunneling optical microscopy [Pohll et al., 1984; Courjon et al., 1989; Reddick et al., 1989; Betzig et al., 1986; Betzig et al., 1991] has suggested that this technique can be used in the field of the biological imaging because of its high resolution [Allegrini et al., 1971] and little invasiveness.

In this kind of microscopy the sample is placed on a surface where a total internal reflection takes place and modulates the arising "evanescent wave". This interaction is probed at subwavelenght distances by a nanometric dielectric tip which performes scannings over the surface [Pohll et al., 1984; Courjon et al., 1989; Reddick et al., 1989; Betzig et al., 1986; Betzig et al., 1991]. This technique allows to overcome the resolution limit of optics due to the diffraction, reaching nanometric resolution even with visible wavelenghtes [Betzig et al., 1986; Betzig et al., 1991; Allegrini et al., 1971].

In this study, we try to understand the potentialities of this kind of microscopy in the biophysical imaging of complex biopolymers such as Dna superstructures, treating the evanescent microscopy as a particular case of the more general "near field" microscopy [Düring et al. 1986; Labani et al., 1990; Girard and Courjon, 1990; Girard and Bouju, 1992; Cites et al., 1992]. A suitable model for 30 nm chromatin fibers was found in literature [Nicolini, 1986; Zeitz et al., 1983; Belmont et al., 1985; Diaspro and Nicolini, 1987] and *ad hoc* modified to match the requirements of the study. Therefore, chromatin superstructure

From Neural Networks and Biomolecular Engineering to Bioelectronics
Edited by C. Nicolini, Plenum Press, New York, 1995

was shaped like an helical array of oblate ellipsoidal dipoles, the nucleosomes, considered as the main scattering centres in our sample [Belmont et al., 1985].

A dielectric tip, modeled as a nanometric spherical dipole [Labani et al., 1990], was numerically scanned over a chromatin fiber and the total intensity scattered by the helical array was detected by the tip in the near field range [Labani et al., 1990].

In this way, theoretic images of chromatin fibers where obtained as a function of the various parameters of the sample, mainly: the radius and the pitch of the solenoid, the number of nucleosomes per helix turn, their dimentions and shape, their orientation with respect to the helix axis, their dielectric constant, the tip size, the polarization state and wavelenght of the incident laser beam. These data show that this kind of microscopy can provide highly resolved images of chromatin fibers and that it is even sensitive to fine structural changes in the geometrical parameters of the model.

Moreover, with the help of a simple but useful sample, two dielectric spheres of 5 nm in radius, some interesting results about the resolution achievable with this microscopy were obtained.

THE MODEL

Quaternary chromatin-Dna structure was shaped, according to literature [Nicolini, 1986; Belmont et al., 1985; Diaspro and Nicolini, 1987], like a helical array of oblate ellipsoids, in number of 6 per helix turn. The ellipsoids played the role of nucleosomes. They were assumed to be the main scattering centres [Nicolini, 1986; Belmont et al., 1985] and their behavior was analyzed within the Born approximation [Belmont et al., 1985]. Each ellipsoid had its principal axis of 11, 11 and 5.5 nm and could be tilted with respect to the helix frame. We choose as nucleosome dielectric constant a value of 1.68^2 [Belmont et al., 1985].

We considered the fiber laying on a surface on which total internal reflection takes place and interacting with the incoming light beam.

Therefore, each nucleosome, impinged on light, contributed to the total scattered light according to its position, orientation, shape, dimentions and dielectric properties. For each scattering centre, the polarizability tensor was computed in the principal axes frame and then in the laboratory one.

The light, scattered by each nucleosome, was collected by a spherical dipole of subwavelenght dimensions [Labani et al., 1990] at each scanning point and was used to form the theoretic image.

THE SIMULATION

Theoretic images of chromatin quaternary structure were obtained computing, at each scanning point, the amount of the light scattered by the nucleosome in the near field approximation.

In fact, the electric field scattered by a radiating dipole of polarizability **P** is given by [Bohren and Huffmann, 1983]:

$$\mathbf{E} = k[\mathbf{P} - (\mathbf{P}\cdot\mathbf{n})\mathbf{n}]\frac{\exp(ikr)}{r} + [3\mathbf{n}(\mathbf{n}\cdot\mathbf{P}) - \mathbf{P}](\frac{1}{r^2} - \frac{ik}{r})\frac{\exp(ikr)}{r} \qquad (1)$$

where k is the wave number of the radiation (we supposed elastic scattering, so k does not change because of scattering) and **n** is the unit vector in the direction of the observer.

In the near field range, that is to say considering only the short range interaction, the previous formula becomes [Labani et al., 1990]:

$$\mathbf{E} = [3\mathbf{n}(\mathbf{n}\cdot\mathbf{P}) - \mathbf{P}]\frac{\exp(ikr)}{r^3}. \qquad (2)$$

For each scanning point $\mathbf{E}_{tot}$ is given by:

$$\mathbf{E}_{tot} = \sum_{i=1}^{N} \mathbf{E}_i , \tag{3}$$

where N is the number of nucleosomes and $\mathbf{E}_i$ is the field scattered by the i-th nucleosome.

Thus, the field scattered by the dipole-like probe is estimated and its intensity is computed in an arbitrary point inside it [Labani et al., 1990]. Calculating this value at each scanning point, a matrix is filled with the intensities of scattered light which gives the image of the scanned sample in constant height mode. Images were obtained also as a function of the state of polarization of the incident beam, mainly linear polarization and randomly polarized radiation. In this last case, images were calculated making an average of the collected intensities corresponding to various polarization states of the incident laser beam.

RESULTS AND DISCUSSION

In this section we present the results obtained by our sample on chromatin fibers of different geometry. These images were obtained as a function of various physical parameters involved in the sample.

Fig.1 A & B show a 5 helical turns chromatin fragment imaged with linear polarized (A) and unpolarized (B) light.

It is clear that a strong dependence upon the polarization state of the incoming light occurs. In fact, in fig.1 A, the incident beam is polarized so that the electric component of the electromagnetic radiation oscillates in the plane defined by the helix axis and the perpendicular to the surface. This causes a higher resolution in the helix axis direction, namely, the helix pitch is easily detectable, but little information about the fiber diameter and the helical structure is available, showing a pearl-like arrangement in the fiber.

Fig.1 B is obtained by using unpolarized light. In this case, the detected signal is the result of an inchoerent superposition of "all" the polarization states of the incident beam. In this case, information about all the directions is here available and the helical nature of the scattering array is more evident.

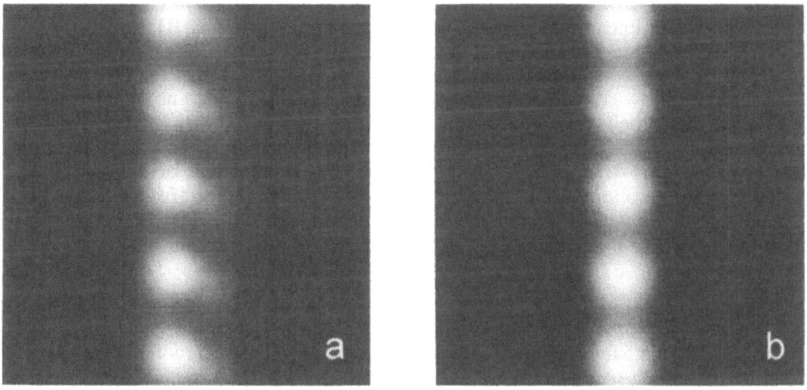

Figure 1 Chromatin fiber segment of 5 superhelical turns; area 55 X 55 nm^2, the Y axis is the vertical one, the X axis is the orizontal one: (A) R=10nm, P=11nm, β=0, incident electric field $\mathbf{E}_0(0, \sqrt{1/2}, \sqrt{1/2})$, probe radius 1nm, probe tip quote 31.5nm; (B) R=10nm, P=11nm, β=0, unpolarized incident electric field, probe radius 1nm, probe tip quote 31.5nm.

Fig.2 A & B show the images obtained on a different chromatin sample. In the imaged fragment, the nucleosomes have 90° tilting with respect to the helix axis (see Appendix), i.e. it is possible to detect a fine structural change in the fiber organization.

This change is not so evident in the case of linear polarized light fig.2 A, resulting only in a weak broadening of the pearl-like structure in the fiber axis direction. However, in

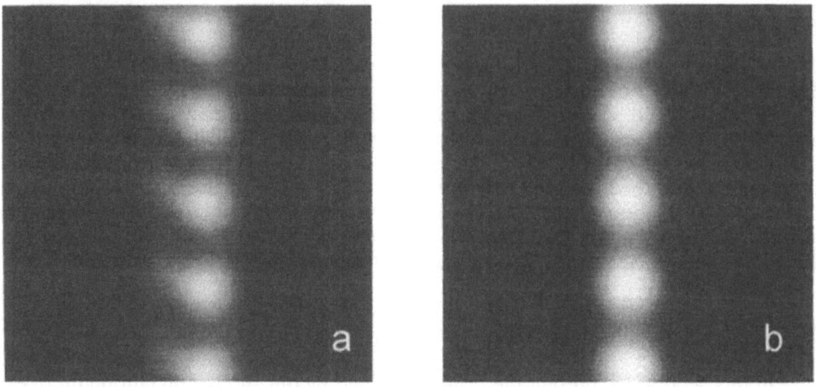

Figure 2 Chromatin fiber segment of 5 superhelical turns; area 55 X 55 nm^2, the *Y* axis is the vertical one, the *X* axis is the orizontal one: (A) R=10nm, P=11nm, β= /2, incident electric field $\mathbf{E}_0(0, \sqrt{1/2}, \sqrt{1/2})$, probe tip radius 1nm, probe tip quote 31.5nm; (B) R=10nm, P=11nm, β= /2, unpolarized incident electric field, probe tip radius 1nm, probe tip quote 31.5nm.

the unpolarized images the results of this change is drastic. In fact, the helical structure is still evident but it is strongly modified with respect to fig.1 B.

These results point out the high sensitivity of this microscopy even on a rather complicated sample as the one in question, suggesting that fine structural modifications are detectable even if the single nucleosome is not resolved. We carried out an analysis by means of a simple but useful sample, two dielectric spheres of 5 nm in radius placed on the scanning plane and acting as scattering probes, in order to quantify the resolution achievable on the scanning plane. With this model we obtained the dependence of the corrugation parameter (difference between maximum and minimum intensity in the image) as a function of the tip radius and height for linear polarized and unpolarized light (fig.3 A & B).

The dependence here depicted is very interesting because of two main features. In fact, it shows that the corrugation has a maximum for a finite value of the tip radius, i.e. too small tips do not allow enough contrast in the image because they do not catch enough light due to their dipolar behavior. Moreover, the corrugation parameter for polarized light is higher than for unpolarized one. These two results should be considered when we project an experiment with this kind of microscopes. The same model was also used to find the dependence of the resolution limit upon the radius of the scanning tip (fig.4 A & B), resulting in a linear one. This behaviour let us write the dependence of the resolution limit y as a function of the sample-tip distance Q_{tot}. In fact, from fig.4 we can write:

$$y = aR_o + b, \tag{4}$$

where a and b are the parameters of the line of fig.4 and R_o is the tip radius. One expects that for $Q_{tot} = 0$ y vanishes, so that:

$$y = d' Q_{tot}. \tag{5}$$

But as

$$Q_{tot} = R + Q + R_o \tag{6}$$

where R is the sample radius and Q is the spacing between the sample and the tip, we have:

$$aR_o + b = d' R_o + d' (R + Q), \tag{7}$$

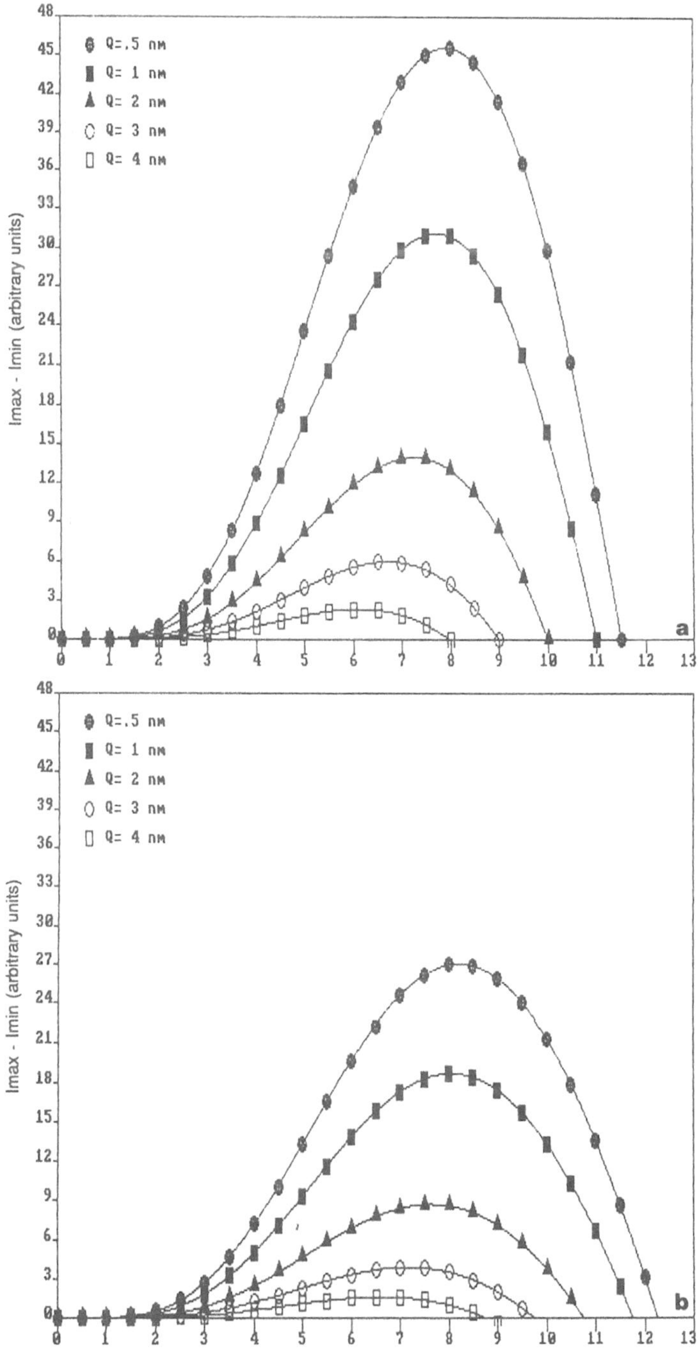

Fig.3 Corrugation parameter I_{max}-I_{min} as a function of the probe tip radius for various Q values: (A) incident electric field $\mathbf{E}_0(0, \sqrt{1/2}, \sqrt{1/2})$; (B) unpolarized case.

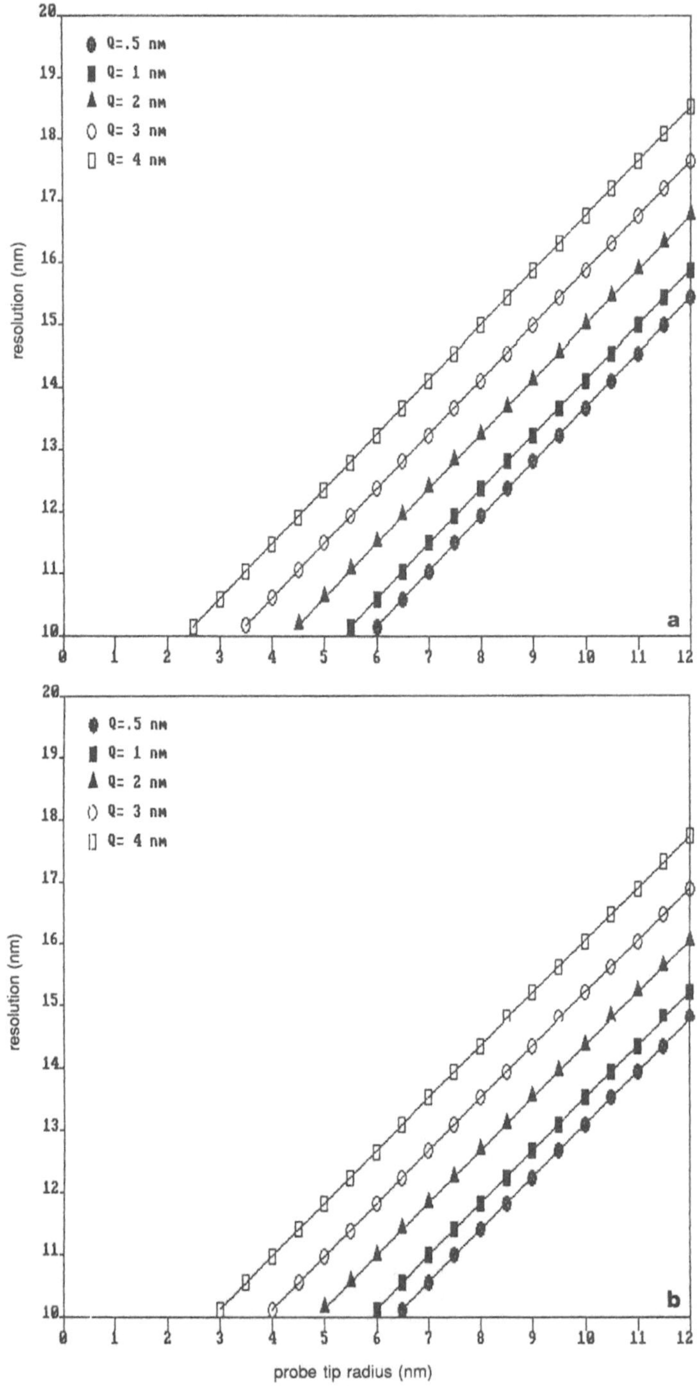

Fig.4 Planar resolution limit as a function of the probe tip radius for varius Q values: (A) incident electric field $\mathbf{E}_0(0, \sqrt{1/2}, \sqrt{1/2})$; (B) unpolarized case.

which is true for each Ro only if $a = a'$ and $b/a = R + Q$. This implies:

$$y = aQ_{tot}. \tag{8}$$

The value of a obtained from least square method is 0.8. This formula facilitates the extimation of the resolution limit achievable with this technique, providing a useful parameter in projecting the experimental work.

CONCLUSIONS

At the conclusion of this work, we can note that this kind of microscopy can provide very good results in biophysical imaging science. In fact, we showed that it has enough sensitivity, even when used on a rather complicate sample like the one we used for chromatin, to detect fine structural changes taking place inside the sample. Moreover, a simple consideration about the theoretic resolution allows to believe that this microscopic technique matches the requirements of many of the biophysical imaging problems in the range from microns to nanometers.

Ultimately, even if this microscopy cannot reach the resolution achievable with other scanning probe microscopies (like Scanning Tunnelling Microscopy or Atomic Force Microscopy), it has very important advantages, mainly the lower invasiveness (due to the fact that it performs scanning less close to the sample) and because it requires neither conductive samples, contrary to STM, nor rather hard one, contrary to AFM.

APPENDIX

Assuming the ellipsoids as composed of an isotropic dielectric with scalar dielectric constant then, given an incident electric field $\mathbf{E}_0$, the induced ellipsoid dipole momentum, $\mathbf{P}$, is given by:

$$\mathbf{P} = \begin{bmatrix} \alpha_x & 0 & 0 \\ 0 & \alpha_y & 0 \\ 0 & 0 & \alpha_z \end{bmatrix} \mathbf{E}_o \tag{A1}$$

in the orthogonal coordinate system with axes x, y and z oriented parallel to the ellipsoid axes. If we assign to the ellipsoid principal axes radii the following values: $x=b$, $y=b$, $z=a$ then:

$$\alpha_x = \alpha_y = \frac{V}{4\pi}\left[\frac{\varepsilon - 1}{\left(\frac{(1-L_3)(n^2-1)}{2}\right)+1}\right]; \qquad \alpha_z = \frac{V}{4\pi}\frac{\varepsilon - 1}{L_3(n^2-1)+1}; \tag{A2}$$

with $V = 4b^2a/3$ ellipsoid volume, $n^2 = \varepsilon/\varepsilon_0$, ε and ε_0 respectively the dielectric constants of the ellipsoid and of the surrounding medium,

$$L_3 = \left(\frac{(1+f^2)}{f^2}\right)\left(1 - \frac{[\operatorname{arctg}(f)]}{f}\right) \tag{A3}$$

where $f = [(b/a)^2 - 1]$.

In terms of a different orthogonal coordinate system, if [C] is the transformation matrix connecting old and new systems we will have:

$$\mathbf{P}=[\mathrm{C}]\begin{bmatrix}\alpha_x & 0 & 0\\ 0 & \alpha_y & 0\\ 0 & 0 & \alpha_z\end{bmatrix}[\mathrm{C}]^t\mathrm{E}_o \tag{A4}$$

If $\mathbf{n}_x$, $\mathbf{n}_y$, $\mathbf{n}_z$ are the three orthogonal unit vectors parallel to ellipsoid principal axes and $\mathbf{n}_1$, $\mathbf{n}_2$, $\mathbf{n}_3$ are the ones of the lab frame, then we can choose $\mathbf{n}_z$ as the unit normal vector to the nucleosome disk face and consider a helical arrangement for the ellipsoids such that their centers fall on the helix:

$$\mathbf{h}=(R\cos\varphi)\mathbf{n}_1+(R sin\varphi)\mathbf{n}_2+\frac{P\varphi}{2\pi}\mathbf{n}_3 \tag{A5}$$

where R is its radius, P its pitch and φ the angle about the helix axis. With this stating the unit tangent vector, **t**, and the two unit normal vectors to the helix, **n** and **l**, are given by:

$$\begin{aligned}\mathbf{t}&=\frac{\left[(-R sin\varphi)\mathbf{n}_1+(R\cos\varphi)\mathbf{n}_2+\frac{P}{2\pi}\mathbf{n}_3\right]}{C_1}\\ \mathbf{n}&=(\cos\varphi)\mathbf{n}_1+(sin\varphi)\mathbf{n}_2\\ \mathbf{l}&=\frac{P}{2\pi C_1}\left[(sin\varphi)\mathbf{n}_1-(\cos\varphi)\mathbf{n}_2\right]+\frac{R}{C_1}\mathbf{n}_3\end{aligned} \tag{A6}$$

where $C_1=[R^2+(P/2\ )^2]^{1/2}$.

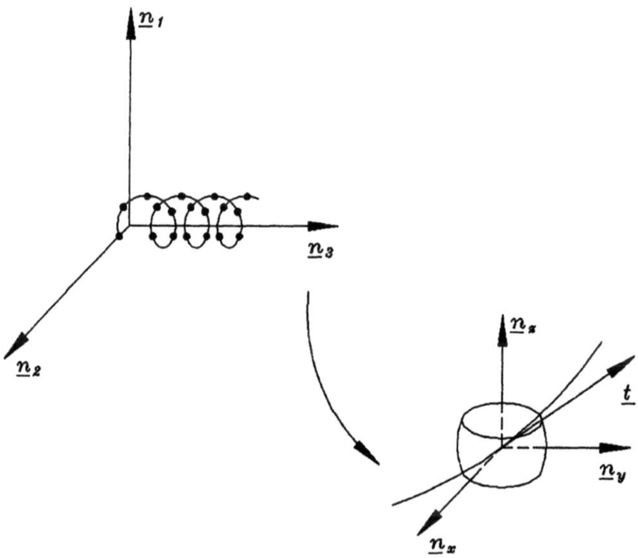

Fig.5 The nucleosome and the lab frame.

We now assume a constant orientation of each ellipsoid with respect to the helix such that, for each of them,

$$\mathbf{n}_x = \mathbf{n}$$
$$\mathbf{n}_y = -(\cos\beta)\mathbf{l} + (sin\beta)\mathbf{t}$$
$$\mathbf{n}_z = (sin\beta)\mathbf{l} + (\cos\beta)\mathbf{t} \tag{A7}$$

where β is the angle between **t** and the unit normal vector to the ellipsoid face, $\mathbf{n}_Z$, fig.5.

Then, by the above equations, we can now determine the induced dipole moment of any ellipsoid in the lab frame, given an incident field $\mathbf{E}_0$.

ACKNOWLEDGEMENTS

This work has been supported by Technobiochip.

REFERENCES

Allegrini, M., Ascoli, C., and Gozzini A., 1971, *Opt. Commun.* 2 (9):435.
Belmont, A., Zeitz S., and Nicolini, C., 1985, *Biopolymers* 24:1301.
Betzig, E., Lewis, A., Harootunian, A., Isaacson, M., and Kratschmer, E., 1986, *Biophys. J.* 49:269.
Betzig, E., Trautman, J.K., Harris, T.D., Weiner, J.S., and Kostelak, R.L., 1991, *Science* 251:1468.
Bohren C.F., and Huffmann, D.R., 1983, "*Absorption and Scattering of Light by Small Particles*", Wiley, New York.
Cites, J., Sanghadasa, M.F.M., and Sung, C.C., 1992, *J. Appl. Phys.* 71(1):7.
Courjon, D., Sarayeddine, K., and Spajer, M., 1989, *Opt. Commun.* 71:23.
Diaspro, A., and Nicolini, C., 1987, *Cell Biophys.* 10:45.
Dürig, U., Pohl, D.W., and Rohner, F., 1986 , *J. Appl. Phys.* 59(10):3318.
Girard, C., Courjon, D., 1990, *Physical Review B* 42(15):9340.
Girard, C., Bouju, X., 1992, *J. Opt. Soc. Am. B* 9(2):298.
Labani, B., Girard, C., Courjon D., and Van Labeke, D., 1990, *J. Opt. Soc. Am. B* 7:936.
Nicolini, C., 1986, "*Biophysics and Cancer*", Plenum Publ. New York.
Pohl, D., Denk, W., and Lanz, M., 1984, *Appl. Phys. Lett.* 44:651.
Reddick, R., Warmack, R., and Ferrel, T., 1989, *Phys. Rev. B* 39:767.
Zeitz, S., Belmont A., and Nicolini, C., 1983, *Cell Biophys.* 5:163.

BUILDING UP SUPRAMOLECULAR EDIFICES DEDICATED TO A SPECIFIC FUNCTION

André Barraud

Commissariat à l'Energie Atomique, France

INTRODUCTION

For the last ten years, the concept of molecular electronics, remarkably well exemplified by the human brain, has prompted physical chemists to build up artificial molecular edifices tailored to carry out simple functions, such as molecular recognition or intermolecular cooperativity. This new branch of physical chemistry, namely molecular architecture, is presently explored in two main directions : i. the synthesis of supermolecules in which the functional groups are held together by chemical (covalent) bonds, and ii. the building up of functional supra-molecular edifices, made up of several molecules held together only by physical forces, each of the partners bearing a functional group. These two strategies differ mainly by : i. the amount of chemical synthetic work, which is much lower in the second method, and ii. the stability of the resulting molecular machinery, which is higher in the first strategy. Only the construction of supramolecular edifices will be concerned in this article.

THE COHESION FORCES

Physical cohesion forces are much weaker than covalent bonds. They arise from several mechanisms, ordered in decreasing strength :

- electrostatic attraction forces between two oppositely charged moities,
- complexation forces, which express the energy stabilisation of a couple of molecules sharing an electron pair available from one of them.
- H bonds, which do the same upon hydrogen sharing.
- interactions between permanent electrical dipoles, which induce orientation at long range and attraction at short distance.
- interactions between induced resonant vibrating dipoles (called Van der Waals forces), which are attractive at very short range.
- etc.

THE BUILDING UP TECHNIQUES

Although big efforts are spent in this direction, scissors and tweezers do not operate yet at the molecular level : molecules cannot be ordered at an individual level and the physical chemist is only provided with collective methods of molecular organisation, which arrange billions of molecules simultaneously and (fortunately) create enough molecular edifices to reach the sensitivity of the characterisation equipment.

From Neural Networks and Biomolecular Engineering to Bioelectronics
Edited by C. Nicolini, Plenum Press, New York, 1995

One of the most simple and efficient techniques to order molecules collectively is the Langmuir-Blodgett method [I. Langmuir, 1920; G. Gaines Jr., 1966] and its physical and chemical variants. This method uses amphiphilic* molecules. A one molecule thick film is made at the water surface and then transferred onto a solid substrate, eventually on top of other previously deposited layers. If the amphiphilic molecule has been well designed, the resulting lamellar film (called a Langmuir-Blodgett or LB film) is well organised and all the molecules have the same orientations and positions relative to one another and to the substrate.

This historical, one molecule method, has been the starting point to several variants [A. Barraud, et al, 1987], which allow the building up of LB films with more than one molecule : the technique of alternate LB layers, the use of self organising mixtures and the semi-amphiphilic technique. These techniques are complemented by a specific solid state chemistry performed in situ in the LB films to give the supramolecular assembly its final properties.

Alternate LB films can be obtained on a double-compartmented Langmuir trough in which two independent films are formed [A. Barraud, 1987]. By programming the path of the substrate successively through each of the films, polar heads of two **different** molecules can be positioned face-to-face in **adjacent** layers. According to the path program, three different structures can be obtained with different symmetries and different chemical properties [A. Barraud, 1987] (fig 1).

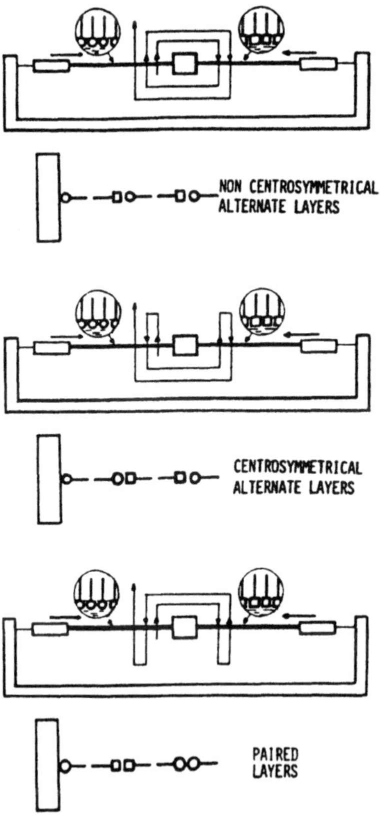

Figure 1 The technique of the LB alternate layers.

---

* an amphiphilic molecule possesses both a hydrophilic part (often named the polar head) and a hydrophobic tail (often an aliphatic chain).

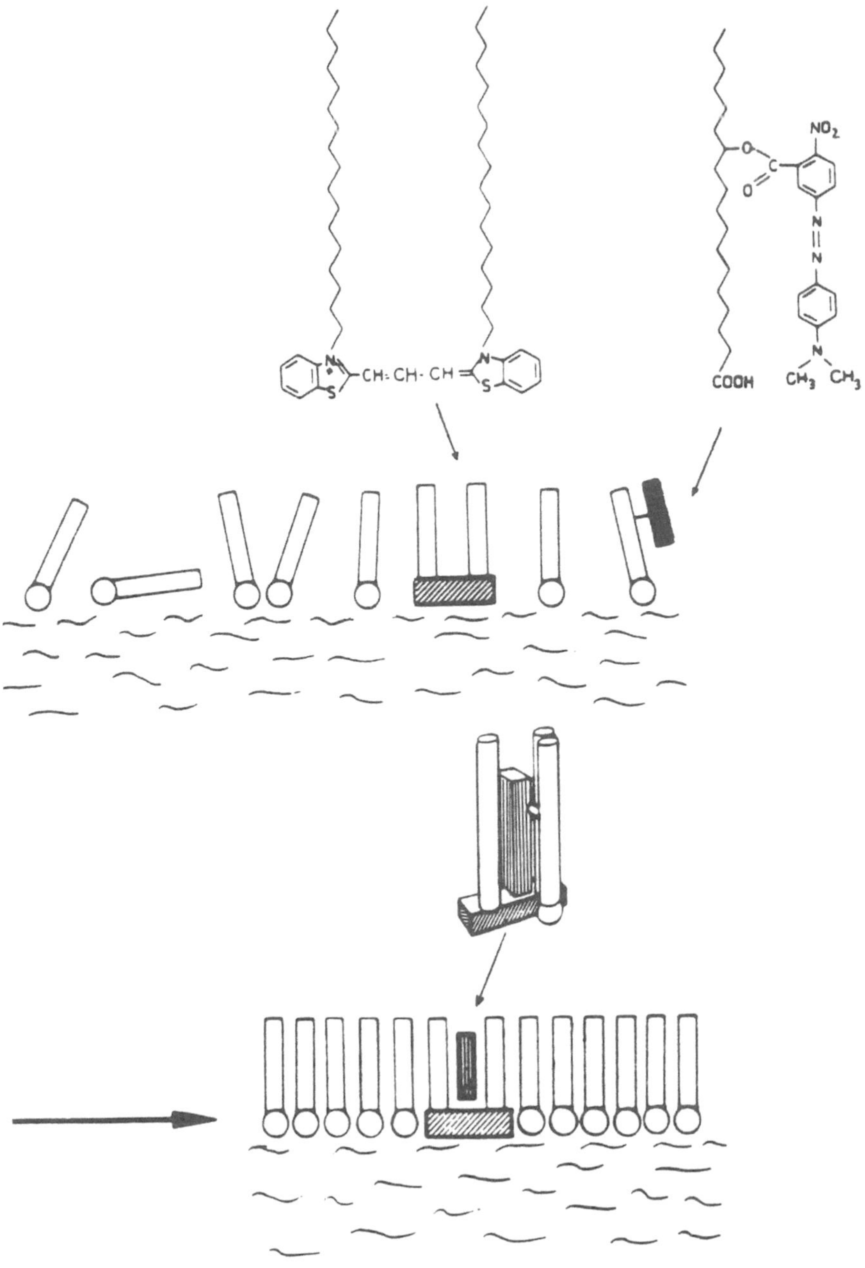

Figure 2 Self-organising mixture of a cyanine molecule (left) and a molecular conjugated "wire" (right). Upon film compression, the two molecules find each other and interlock (after E. Polymeropoulos, 1980).

The technique of self-organising mixtures takes advantage of couple formation to assemble different molecules **in the same** monolayer, contrary to the technique of alternate layers which groups two molecules from different layers. One of the above mentioned physical forces is used to "marry" molecules [E. Polymeropoulos et al, 1980]. Couples are formed upon repeated attempts : only couples of marriageable molecules are long lasting, the others break along and try other partners, till everyone is married (fig 2). Such an organisation is automatic and requires only an on purpose design of the molecules. Combined with the alternate technique, it gives a simple access to molecular assemblies involving three, or even four different molecules positioned in a controlled way.

One more degree of freedom is provided by the **semi-amphiphilic** technique [A. Barraud, 1987]. Here also the stability of a semi-amphiphilic compound** is provided by physical forces, most of the times electrostatic attraction between two molecular ions. A straightforward advantage of this technique is the simplification of the synthetic chemistry associated with the grafting of aliphatic chains on the functional heads of molecules: a fragile molecule can be taken as it is, and associated with an easily Langmuir-Blodgettable partner (fig 3). The whole semi-amphiphilic compound behaves as a normal amphiphilic molecule and, if properly designed, can give rise to high quality LB films.

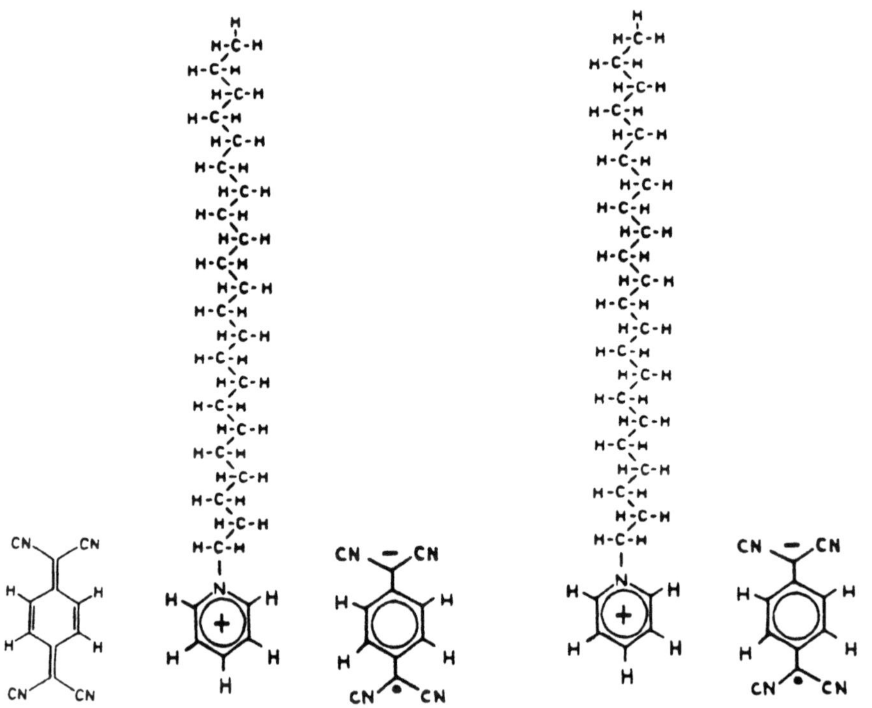

Figure 3 Semi-amphiphilic compounds: Left: a mixed valence docosylpyridinium (DCP^+) tetracyanoquinodimethane ($TCNQ^-$, $TCNQ^\circ$) semi-amphiphilic complex; Right: a DCP^+TCNQ^- semi-amphiphilic salt. These semi-amphiphilic compounds, which behave like an amphiphilic molecule, allow the fabrication of monolayers of TCNQ, although TCNQ bears no aliphatic chain.

Another crucial advantage of the semi-amphiphilic method arises when it is used in conjunction with chemical manipulations of the films, generally used as a final step to adjust the properties of the molecular assemblies when necessary [A. Barraud, 1987]. If molecules

** a semi-amphiphilic compound is made of several partners, some being amphiphilic, some being ordinary, non-amphiphilic (water soluble for instance).

have been properly designed, LB films are two-dimensional solids and chemistry which takes place in them is **solid state** chemistry, i.e. the reactions are lattice controlled. In this context, semi-amphiphilic compounds give the designer an extra degree of freedom because the reaction control is provided by the unperturbed hydrophobic lattice of the amphiphilic partner (which does not participate in the reaction), while the polar partner, which undergoes the reaction, can freely (no hydrophobic tail) reorganise to the post-reactional minimum energy configuration aimed at by the designer, without endangering the lamellar structure [A. Barraud, 1987].

Another peculiar feature makes this chemistry very attractive : because LB films are very thin, small molecules can easily diffuse from outside, which is not the case in classical, bulk solid state chemistry. Hence chemistry in LB films combines the high degree of control of solid state chemistry and the high versatility and richness of solution chemistry. For this reason it is a very useful tool to complement the physical techniques used in supramolecular architecture.

THE GOAL OF SUPRAMOLECULAR ARCHITECTURE

Would the properties of a supramolecular edifice just be the sum of the properties of the component molecules, the system would be very unattractive : no one would give it the name of "intelligent" or "smart" material. The goal of supramolecular architecture is to design the molecular edifice in such a way that an **extra property** is promoted: this property belongs to none of the partners in particular, but arises from a clever molecular assembling which changes the environment of the molecules and gives them an unexpected chemical or physical behaviour. This is how the biological molecular machineries work. The challenge here is not to copy nature, which has adjusted sophisticated systems for millions years to perform a specific function in the very specific conditions of life, but to design and build up simple, totally artificial structures which exhibit non trivial properties even in rough environmental conditions, which are representative of industrial requirements. Examples of these properties are :

- enhancement of the chemical reactivity of normally inert molecules.
- electrical conduction and magnetism, or, speaking in a more general way, cooperative phenomena.
- molecular recognition.

This sort of properties, which were so far specific to biology, are now being understood and transferred to very simple chemical edifices. Three examples will be taken to illustrate the concepts and strategies followed: the forced solid state solution, the di-oxygen trap and the LB conducting films.

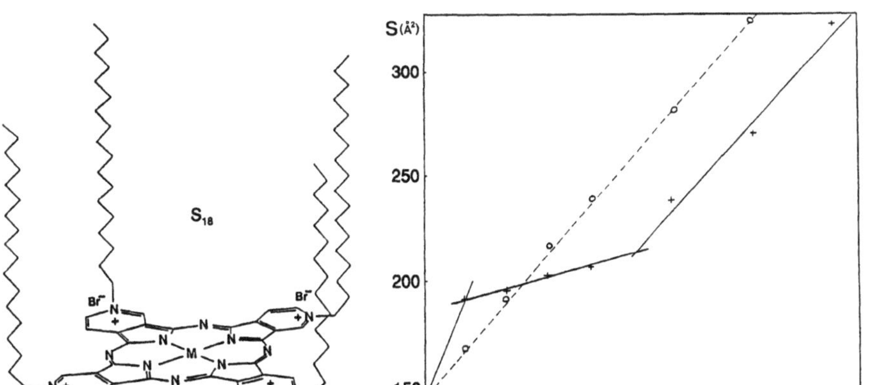

Figure 4 Mixed monolayers of Vanadyl S_{18} (left) and a diluting molecule (stearic alcohol ······, or stearic acid ——). The area of the group of molecules[S_{18} + n diluting molecules] is plotted versus n, the number of diluting molecules per S_{18}.

THE FORCED SOLID STATE SOLUTION

When two dissimilar molecules are put together in a monolayer, they generally do not mix, phase segregation occurs and the film is a collection of patches of each molecule. The consequence is a simple additivity of the two molecular areas at the surface of water. As an example, attempts to dilute the amphiphilic Vanadyl phtalocyanine of fig.4 (called S_{18} for short) by stearic alchool leads to no dilution effect as shown from the VO ESR spectra [S. Palacin et al., 1988] and the molecular areas simply add up as shown from the dashed curve of fig.4 (22 $Å^2$ increase per stearic alcohol).

However this mutual repulsion can be overcome by designing an interface between S_{18} and the diluting molecule which be liked by both. Upon adding one molecule of stearic acid per S_{18} (solid curve in fig.4), the area increase at the water surface is tremendously high (40 $Å^2$), which denotes a strong interaction between the two species. The IR absorption spectrum confirms it: stearic acid converts spontaneously into stearate to make a salt with one of the four positive nitrogens of S_{18}. To find room to do this, the stearic acid molecule opens the "flower-bud" (fig.5), which explains the 40 $Å^2$ increase. Further addition of stearic acid induces almost no area increase (5 $Å^2$ per molecule) because the flower-bud is already opened. This situation holds till five stearic acid molecules have been added, four of them making a salt with the nitrogens and one with the central ion of S_{18}. The four peripheral stearates serve as an interface to attract similar further stearic acid molecules: indeed further addition of stearic acid (which remains unionized) actually dilutes S_{18} (as pointed out by the ESR spectra) giving rise to a true solid state solution. This could not be obtained with stearic alcohol, which does not act as an interface, because uncharged.

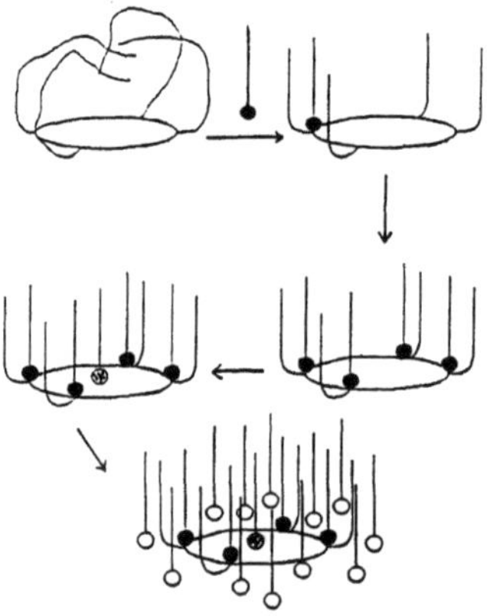

Figure 5 The mechanism for the formation of the stearate (●) interface between S_{18} and the diluting stearic acid (○) molecules.

THE DIOXYGEN TRAP

This molecular machinery applies the same working principle as hemoglobin, but is totaly artificial, without protein or water, so as to work in hostile conditions.

The active center of hemoglobin is a Fe^{II} porphyrin. Fe^{II} is six-coordinated: its four equatorial bonds keep it fixed at the center of the porphyrin macrocycle, while its two axial bonds, which like to work **simultaneously**, are normally little reactive and unable to bind dioxygen. Hemoglobin is activated when the two Fe^{II} axial bonds are made to work **unsymmetrically** : when the surrounding protein offers a histidine to one of the bonds, the other one is activated and complexes dioxygen.

For practical reasons, we adopted a very simple molecular edifice the heart of which was the amphiphilic porphyrin shown in fig.6; this porphyrin macrocycle is designed to anchor flat on water or on a hydrophilic substrate owing to its four hydrophilic COOH groups. This well defined structure made us rid of the organizing protein. Histidine was also replaced by an amphiphilic imidazole, which still possesses a complexable nitrogen. Fe^{II} was replaced by Co^{II}, less oxidizable and still six-coordinated. Two symmetrical and two asymmetrical supramolecular assemblies (fig.6) were built up using the above components to check the validity of the concept transfer [C. Lecomte et al., 1985].

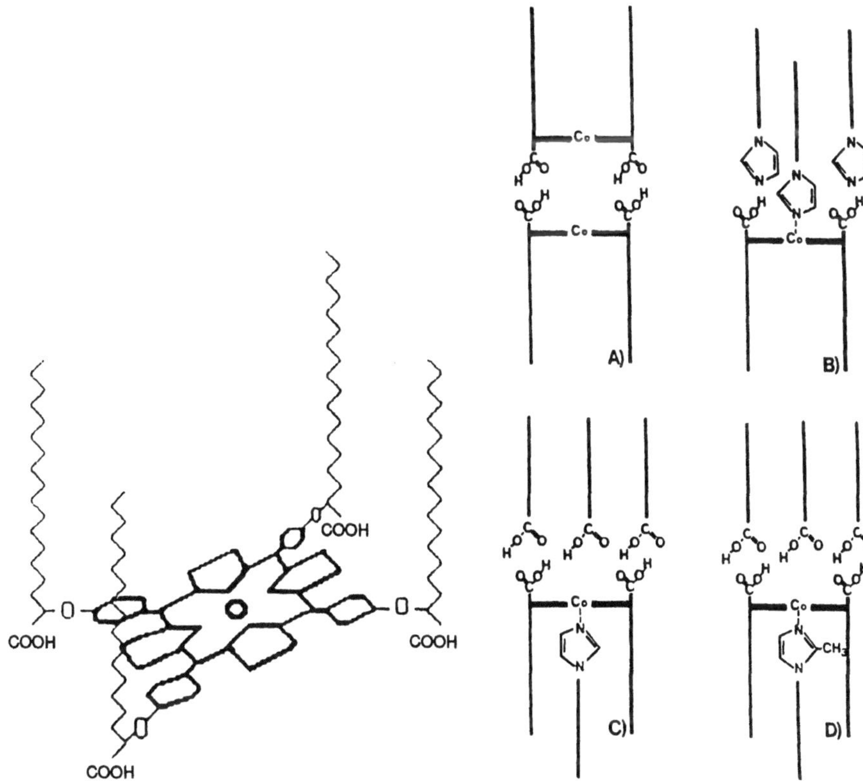

Figure 6 The dioxygen trap. Only molecular edifices which are dissymmetrical (B and C) complex dioxygen. Left: the Co^{II} amphiphilic tetraphenyl-porphyrin seen in profile in the right diagrams.

A. Normal superimposed porphyrin LB layers. This structure is not intented to break the Cobalt axial bond symmetry and as expected it does not complex dioxygen.

B. Alternate LB layers of amphiphilic imidazole and porphyrin. In this technique, imidazole can only contact Cobalt on one side of the macrocycle. As expected, this molecular assembly is very sensitive to dioxygen which is complexed down to 0.1 mbar.

C. Self-organizing equimolar mixture of amphiphilic porphyrin and imidazole. Upon film compression, imidazole, which likes to be complexed by cobalt, "climbs" on top of the macrocycle. Here again this breaks the symmetry of the Cobalt axial bonds. The assembly is complemented by an alternate layer of an inert molecule, stearic acid. As expected, this system complexes dioxygen down to a fraction of a mbar.

D. Same assembly as C, but imidazole has been made bulky by adding a side CH_3 group. Imidazole still climbs on the macrocycle, but, because of the limited room between the four aliphatic chains of the porphyrin, imidazole cannot come close enough to the Cobalt ion to activate the axial bond noticeably. As expected, this bulky system is hindered and does not complex dioxygen, even at air pressure.

These four molecular edifices are clear evidences that assemblies involving several molecules held together only by physical forces can be operated in a texbook way. The efficiency of the method is high because the chemical synthetic work is reduced to a minimum : all these experiments, including syntheses, could be carried out in a few months.

THE MOLECULAR THICK CONDUCTORS.

Two conditions must be fulfilled for an organic molecule, which is by nature insulating, to become conducting:

1. the molecule must be in a mixed valence state, i.e. it must be occupied by a **non integral** number of electrons (for instance its highest occupied orbital bears 0.5 electron) [J. Simon et al., 1985]

2. the molecular orbitals of neighbouring molecules must overlap. This means the molecules are stacked and tightly coupled to one another in a **regular** array. This gives rise to allowed energy bands in which electrons are quasi-free to move if the band is not completely full or empty of electrons. Tetracyanoquinodimethane (TCNQ) is a planar, fully conjugated molecule which stacks easily, especially in the mixed valence state. Although it is not the best molecule for conduction applications, it has been chosen because of the large amount of informations already gathered by physico-chemists on that molecule in bulk.

TCNQ is a fragile molecule : it is best to take it as it is, so that the semi-amphiphilic method is the best fitted method to make TCNQ monolayers, although the molecule is not amphiphilic. Associated to an amphiphilic pyridinium cation and $TCNQ^o$, $TCNQ^-$ gives rise to a semi-amphiphilic compound (fig.3, left) which is conducting in powder. Unfortunately, this mixed valence compound is destroyed at the surface of water, so that the resulting film is insulating ($TCNQ^o$ dissolves in water).

Following the failure of the above direct strategy, we resorted to an indirect one, which circumvented the exposure of the mixed valence compound to water. The semi-amphiphilic compound is again a salt of N-docosyl-pyridinium$^+$ (NDP^+) and $TCNQ^-$, but no $TCNQ^o$ is added (fig.3, right). The compound is not a mixed valence state so that, as expected, both the powder and the LB film are insulating [A. Barraud et al., 1985]. The LB film is then submitted to iodine vapours for a few seconds. Iodine defeats TCNQ in the struggle for the negative charge, and tends to form $TCNQ^o$ while transforming into I_3^-. If the reaction was left to its natural thermodynamics, it would yield only $TCNQ^o$, and no mixed valence state would be obtained. Here come into play the peculiarities of the chemistry in LB films : the reaction is controlled by the lattice of the hydrophobic chains of NDP and stops when all the sites available to iodine in this lattice are filled up with I_3^-. This happens when one TCNQ out of two has been deprived of its electron (fig.7). Thus, when iodine exposure is carried out properly, it automatically yields mixed valence TCNQ.

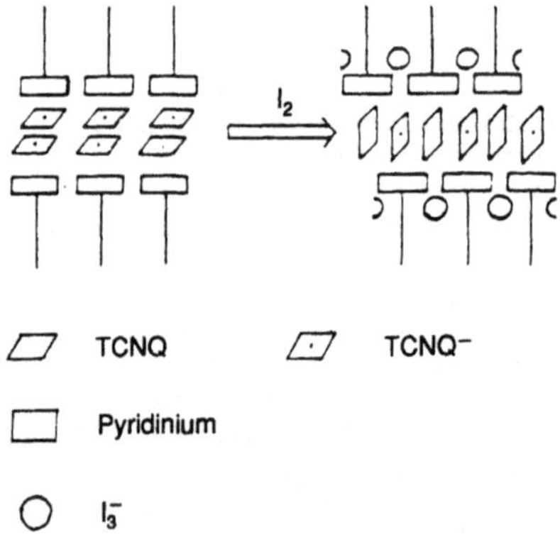

Figure 7 The oxidation of $TCNQ^-$ by iodine is controlled by the lattice of the amphiphilic pyridinium. It stops when all the lattice sites available to iodine are filled with I_3^-. This happens when one TCNQ out of two has lost its electron. This reaction is accompanied by a rotation of the TCNQ's, which form conducting horizontal mixed valence stacks.

A spectacular reorganisation of the TCNQ's takes place spontaneously upon this iodine "doping": the disk-like TCNQ molecules, which were lying flat on the substrate, rotate by a quarter of a revolution and stand up vertical [A. Barraud et al., 1985], in the form of long stacks with a horizontal axis, in which molecules are closely packed and coupled. This structure fulfills the second requirement for conduction. This reorganisation was purposely favoured in the design of the strategy, and the semi-amphiphilic method is quite helpful in this direction, because TCNQ bears no hindering aliphatic chain.

This reorganisation was indeed a crucial point for the strategy to be successful, because we knew that, if a TCNQ reorganisation could take place, it would directly give rise to a conducting structure. This is a direct consequence of band formation: forming an energy band and taking off half its electrons gains an energy roughly equal to half the band-width, i.e. around 0.25 eV in the case of TCNQ. It is the lowest possible energy configuration. This gain in electronic energy is enough to overcome lattice friction in the polar plane of semi-amphiphilic LB films but is not high enough to destroy their lamellar structure.

The resulting films exhibit infrared, high frequency and d.c. semi- conduction, with a d.c. resistivity ranging around 10 ohm.cm [J. Richard et al., 1986]. Their structure could be completely determined owing to the remarkable molecular order observed in the films : the aliphatic chains are tilted and interdigitated, and iodine is located at the end of the aliphatic chains, in the cavity left by the interdigitation [B. Belbeoch et al., 1985].

These films and several variants have found an applied slot which is one of the first outcomes of molecular electronics : gas sensors. In the presence of phosphine (PH_3), they undergo a drastic resistivity increase. Their sensitivity to phosphine is better than 1 part per million (ppm) [L. Henrion et al., 1989], which is the right range for worker protection. They also selectively distinguish phosphine from ammonia (NH_3), a very closely related molecule, with a good selectivity factor (4000).

Their tremendous sensitivity results from their specific molecular architecture. In each stack, band conduction, which is a cooperative phenomenon, requires all the molecules to be exactly identical. When a phosphine molecule comes in contact with a TCNQ, it modifies it, so that the system is no longer periodic (this modified TCNQ behaves as a defect). Because the TCNQ stacks are highly one dimensional in LB films***, charge carriers cannot go round the defect and conduction is killed over the whole stack. Whatever the precise nature and location of the defect, this mechanism, which allies low dimensionality and molecular cooperativity, is likely to explain why the **electrical** sensitivity of these sensors is a hundred to a thousandfold higher than their intrinsic **chemical** sensitivity (seen from spectral modification). This "molecular amplification" has no equivalent in classical electronics, because it takes place in the sensing material itself.

The example of these thin conducting films is typical of the close relationship between the design of the supramolecular architecture and the final properties of a material. Every step in the elaboration strategy must be carefully designed to help tailoring the material and bring as little disorder as possible.

Disorder is indeed one of the basic ennemies. Although the iodine doped conducting LB films exhibit marvellous X-ray diffraction patterns [B. Belbeoch et al., 1985], they show some anomalous properties, which soon appeared to be related to defects arising upon the iodine treatment [A. Barraud et al., 1991]. For instance, a doped mono- or bi-layer is not d.c. conducting, while its infrared spectrum is characteristic of a conducting material ; only above 5 to 6 bilayers is the film conducting (fig.8, left). This anomaly can be explained by insulating grain boundaries in which molecules damaged by side reactions accrete: a few percent of this dead material is enough to make the whole sample long-distance insulating.

In order to obtain a material with less defects so as to render a single bilayer conducting, we imagined a new, doping free, hence direct, strategy [J.P. Bourgoin et al., 1992]. In spite of the heavy synthetic task involved, an aliphatic chain was grafted to $TCNQ^o$ to reduce its solubility and a direct strategy, similar to the one we had used initially, was set up : a semi-amphiphilic salt of amphiphilic octadecyl-sulfonium (ODS^+) and $TCNQ^-$ was mixed with amphiphilic (C_{18}) $TCNQ^o$. The mixture (powder) exhibits a semi-conducting infrared absorption spectrum, indicating that i. a mixed valence state and ii. stacks of mixed valence TCNQ are obtained. After spreading on water and transfer onto an

*** TCNQ stacks in LB films do not exhibit the slight secondary 2D and 3D characters they show in bulk, because of a different arrangement of the stacks.

insulating substrate, this mixture gives rise to infrared and **d.c. conducting single bilayers,** (100 to 1000 ohm.cm, same value as for multiple bi-layers, fig.8, right) and even to d.c.conducting monolayers. This new strategy, which has been given the name "homodoping", is in fact quite general, and applies even to compounds impossible to dope with iodine [J.P. Bourgoin et al., 1991].

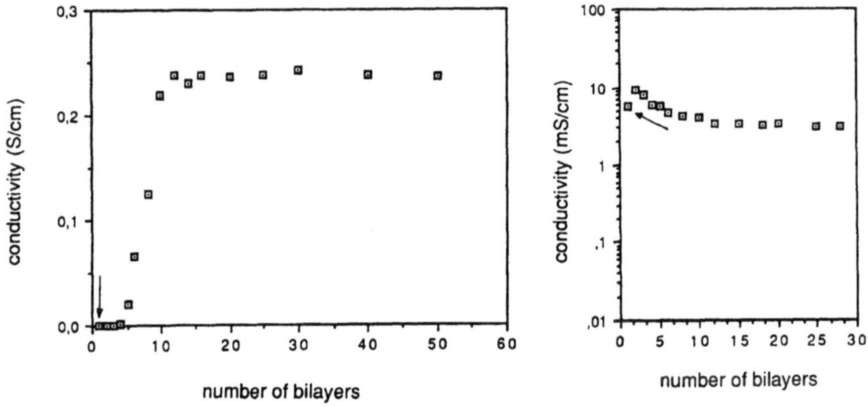

Figure 8 Conductivity (conductance per layer) vs the number of layers in the case of:
- left : iodine doped conducting films
- right : homodoped conducting films
The arrows point to the case of a single bilayer.

The homodoped films are of course not defect-free, but the defect density is low enough to allow charge carriers to percolate in a single polar plane. This opens the way to molecule thick electrodes, flat down to the molecular level; these are a necessary tool to study the effect of high electric fields on molecules, a basic field in molecular electronics.

CONCLUSION

These examples point out the extremely basic aspect of this research: the feasibility of supramolecular architecture is now proved, but it is still stone age and most of the basic concepts of molecular electronics still have to be demonstrated. Nevertheless, it already appears that the very large choice in organic molecules and the quasi-infinite variety in combinations of molecules is a major trump for the design of active materials dedicated to specific functions. Mother nature has already widely exploited supramolecular architecture, but there are many combinations other than those found in biological systems. **Molecular electronics** is expected to be one of the first fields to benefit from this new field of physical chemistry.

REFERENCES

A. Barraud et al., 1985, *Thin Solid Films*, **133**:125.

A. Barraud et al., 1985, *Thin Solid Films*, **134**:195.

A. Barraud, 1987, in "*Proc. 4th Internat. School in Condensed Matter*", World Science publ.

A. Barraud, 1987, *J. Chimie Phys.*, **84**:1105.

A. Barraud and M. Vandevyver, 1987, in "*Nonlinear Optical Properties of Organic Molecules and Crystals*", Academic Press publ., New-York, 1:357.

A. Barraud and M. Vandevyver, 1991, in "*Condensed Systems of Low Dimensionality*", Plenum Press publ., New York.

B. Belbeoch, M. Roulliay and M. Tournarie, 1985, *Thin Solid Films*, **134**:89.

K. Blodgett, 1935, *J. Am. Chem. Soc.*, **57**:1007.

J.P. Bourgoin, 11 Décembre 1991, *Ph. D. Thesis*, Univ. Paris XI.
J.P. Bourgoin et al., 1992, *Thin Solid Films*, **210-211**:250.
G. Gaines Jr., 1966, "*Insoluble Monolayers at Liquid-Gas Interface*", Interscience John Wiley publ., New York.
L. Henrion et al., 1989, *Sensors and Actuators*, **17**:494.
I. Langmuir, 1920, *Trans. Faraday Soc.*, **15**:62.
C. Lecomte et al., 1985, Thin Solid Films, **133**:103.
S. Palacin, 04 March 1988, *Ph. D. Thesis*, Univ. Paris VI.
E. Polymeropoulos, D. Möbius and H. Kuhn, 1980, *Thin Solid Films*, **68**:173.
J. Richard et al., 1986, *J. Phys. D: Appl. Phys.*, 19:2421.
J. Simon and J.-J. André, 1985, "*Molecular Semi-Conductors*", Springer-Verlag publ.

STRUCTURAL PROPERTIES OF MONOLAYERS AND LANGMUIR-BLODGETT FILMS BY X-RAY SCATTERING TECHNIQUES

Franco Rustichelli

Istituto di Scienze Fisiche
Università di Ancona
Ancona, Italy

1. INTRODUCTION

Interdisciplinary research related to molecular electronics is directed towards bringing the miniaturisation of electronic devices, to its extreme limit i.e. to the molecular scale. As a consequence, the different investigations related to chemical synthesis, supramolecular assembly and the characterization of the development units for future electronic devices should be conceived having in mind a spatial resolution of the order of molecular dimensions. In other words, the molecules constitute the building units of supramolecular systems designed within the framework of the so called molecular architecture.

The ultimate goal of the research in this field is to produce stable systems, wich can be used for information storage and processing. The first step is to synthesize molecules containing units suitable for carrying out functions of high specialization. The second step is to mutually arrange different such molecules in a convenient way so that each functional unit contribute in a synergic fashion to achieve the strategic aim of the given new molecular device.

In this context the Langmuir-Blodgett (LB) technique plays a key role, as it enables the creation of arbitrary sequences of monolayers of different chemical compounds in a direction perpendicular to the substrate surface, i.e. supramolecular organizations with molecular scale variations along this direction. The practical feasibility of such superlattices has already been demostrated with the production of LB films of different alternating dielectric monolayers [Yu.M. Lvov et al., 1989].

Furthermore, considerable progress has occurred with the discovery of conductive LB films of different types [A. Ruaudel-Teixier et al., 1985; H.J. Merle et al., 1991]. These advances made it possible to obtain [T.S. Berzina et al., 1992], from mixtures of surfactant donor and acceptor molecules, LB films containing superlattices constituted by alternated bilayers having a dielectric or conducting character, which will be considered in detail below. Finally the important discovery, in view of electronic device development, of the field effect in conducting LB films [J. Paloheimo et al., 1990] should be emphasized. In the context of these investigations, the direction along which compositional or strucutural changes occur is, most frequently, the one perpendicular to the layers, which in turn are generally consisting each one of a single type of molecule and are considered as homogeneous. Structural information related to these supramolecular systems can be obtained by X-ray diffraction, X-ray specular reflection and the X-ray standing wave technique, which will be reviewed in this paper, together with surface X-ray diffraction which provides information on the lateral in-plane molecular organization. Moreover some recent structural investigations related to LB films obtained by solutions of binary mixtures will be considered.

2. X-RAY DIFFRACTION

A review on X-ray and electron diffraction studies of LB films was recently published [L.A. Feigin et al., 1989] and one should refer to it for classical applications and details of the technique. Here only some peculiar and/or recent examples will be considered.

Small-angle diffraction

Fig. 1a shows schematically the geometry of small-angle diffraction by a LB film recorded by a powder diffractometer. The diffracted intensity is recorded by rotating the sample by an angle θ and the detector by an angle 2θ. In a typical diffraction pattern the detected X-ray intensity is recorded as a function of 2θ. Typically from 10 to 15 Bragg reflections are recorded, whose spacing ratios in the order 1:2:3:4:......... clearly indicate a one dimensional lamellar symmetry in correspondence to a stacking of the molecular layers. Fig. 1b reports typical X-ray scattering patterns as a function of the number of bilayers of Tl-behenate [L.A. Feigin et al., 1988]. From the position of the Bragg peaks, by using Bragg equation, one obtains the spatial periodicity of the supramolecular system. Furthermore one can obtain the electron density profile $\rho(z)$ along a direction perpendicular to the layers through:

$$\rho(z) = k \cdot \Sigma F(h) \cos(2\pi hz) \qquad (1)$$

where k is a normalising factor and F(h) is the structure factor of the reflection with Miller index h (the sum concerns all the observed peaks). In eq. (1), which is valid because a LB structure is centrosymmetric, the modulus of each structure factor is connected with the measured intensity I(h) of the corresponding Bragg reflection by:

$$|F(h)|^2 = c \cdot h \cdot I(h) \qquad (2)$$

where c is a constant and the Lorentz factor h takes into account the geometry of the experiment. The crystallographic "phase problem" is reduced in this case to a "sign problem", i.e. to the correct signs of all the observed strucure factors. A new method, based on a pattern recognition approach, was recently proposed, for determining the sign of several LB Bragg reflections and therefore the corresponding electron density profiles [V. Erokhin et al., to be published].

Fig.2 reports the electron density profile $\rho(z)$ and the corresponding structural model of conducting LB film of organic complexes of tetracyanoquinodimethane + N-octadecylpyridine after doping with iodine [Yu. M. Lvov et al., 1988]. A study of the different chain packing possibilities in LB films containing in the head an $MeOF_5$ group (Me indicating a metal) is reported in V.V. Erokhin et al. (1989). Fig.3 reports the electron density profiles for two samples. On the left a classical one for a LB film of a lead salt of stearic acid and the corresponding structural model, on the right the profile for a LB film of stearic acid complex with $TaOF_5$.

The reduced lattice spacing, the lack of minimum in the center of the profile, the presence of two minima in the outer parts of the cell, indicate an interdigitated structure, as shown in figure, for this last case.

Protein LB films were obtained by using as spreading solutions reversed micelles of sodium disooctylsulfosuccinate (AOT) with cytochrome C [V. Erokhin et al., 1994.]. Fig.4 reports the X-ray diffraction pattern from a 10 periods (AOT-cytochrome C-AOT) LB film and fig.5 the corresponding model, where the cytochrome C monolayer (circles) is placed between the head groups of AOT molecules in the adjacent monolayers. The metalloprotein in the film is not denatured and the LB film is more ordered than the corresponding monocomponent LB film.

Polar LB films of p-(p-octadecyloxyphenylazo) benzenesulfonamide films were investigated by X-ray diffraction and did show significant pyroelectric effect [H.J. Merle, et al., 1991]. Recent X-ray diffraction investigations include a study of LB films of Cd soaps in the vicinity of break up point [H.J. Merle, et al., 1991], the structure and epitaxial layer growth of fatty acid LB films [A. Leuthe et al., submitted], the influence of structure change on electrical properties of conducting LB films of amphiphilic liquid crystals [T.S. Berzina, et al., 1991]. X-ray diffraction in a peculiar geometry, namely with a thin substrate perpendicular to the incident beam, was used [P.A. Albouy et al., 1992] [M. Vandevyver et al., in press.] to investigate the liquid crystalline order and phase transition of a disk-shaped heteroaromatic salt. Moreover a review on LB protein films can be found in [Yu. M. Lvov et al., 1991].

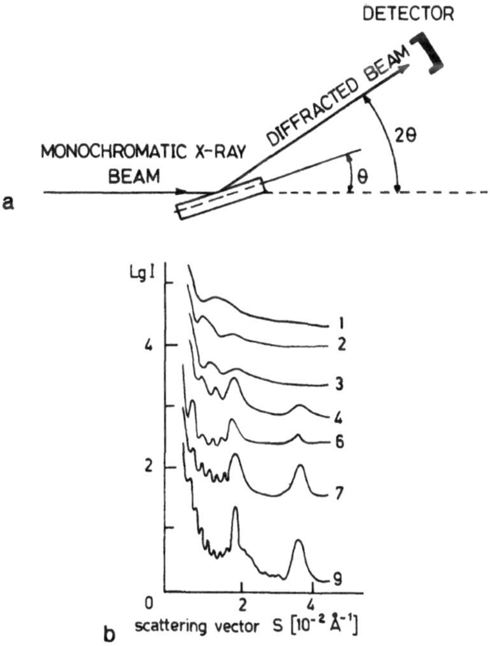

Figure 1 (a) Scheme of X-ray small-angle diffraction by a LB film; (b) typical X-ray scattering patterns as a function of the number (labeling each curve) of bilayers of Tl-behenate [L.A. Feigin et al., 1988].

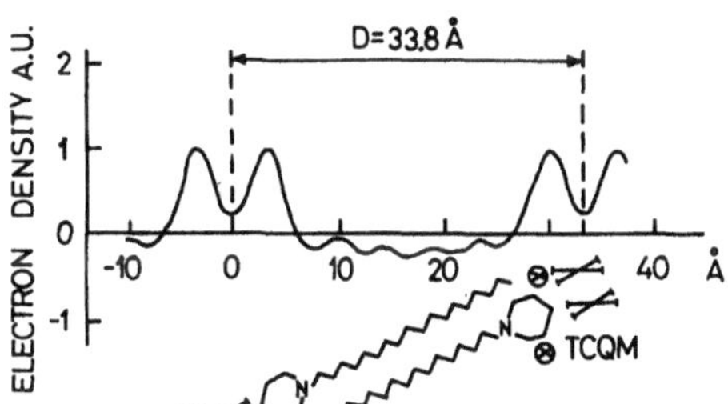

Figure 2 Electron density profile and structural model of a conducting LB film of organic complexes of tetracyanoquinodimethane + N - octadecylpyridine after doping with iodine. The crosses are iodine ions [Yu. M. Lvov et al., 1988].

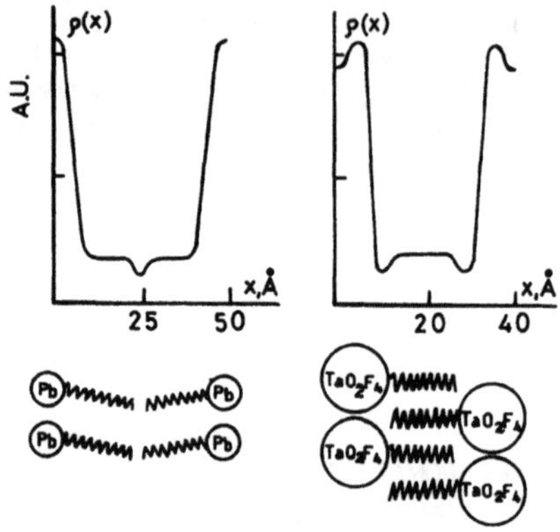

Figure 3 Electron density profile and structural model of a LB film of lead salt of stearic acid (left) and of a stearic acid complex with $TaOF_5$ [V.V. Erokhin et al., 1989].

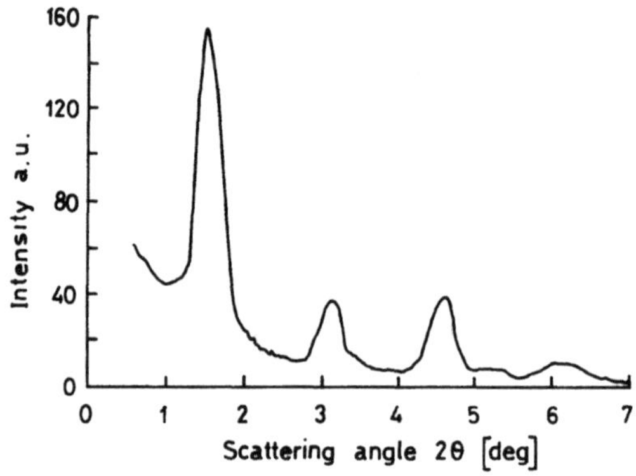

Figure 4 X-ray diffraction pattern from a 10 periods (AOT - cytochrome C - AOT) LB film, obtained using reversed micelles as spreading solution [V. Erokhin et al., in press].

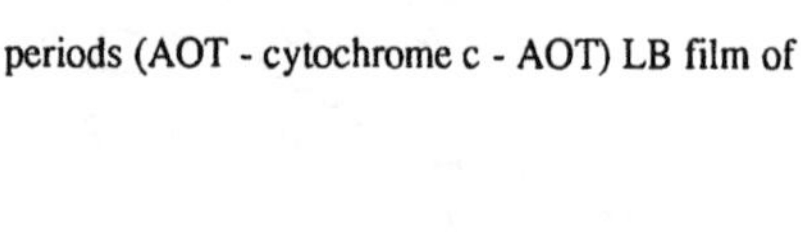

Figure 5 Model for the 10 periods (AOT - cytochrome c - AOT) LB film of fig. 4 [V. Erokhin et al., in press].

Mixed LB films

Some recent investigations concerning the structure of LB films obtained by solutions of binary mixtures will be considered. Such studies are not only of fundamental interest, but also of practical importance in connection with the preparation of LB films including functional molecules (like proteins), embedded, inside a given monolayer, in a matrix of surfactant molecules, like fatty acids. In this context a basic question is wether or not the molecules form a homogeneous mixture inside the monolayer.

Different types of binary mixtures of simple, similar molecules, namely fatty acid salts, were considered in ref. [S. Dante et al., 1992-V. Erokhin et al., 1994]. In spite of the simplicity of the model systems used, unexpectedly puzzling results were obtained. Firstly LB films of mixed behenic and stearic acid salts of Cd were deposited and investigated by small-angle X-ray diffraction. Only when the molecular ratio was 1:1 and the compounds were mixed in a solution before spreading, was a single set of Bragg reflections registered in the diffraction pattern (fig. 6a), indicating the presence of a single phase.

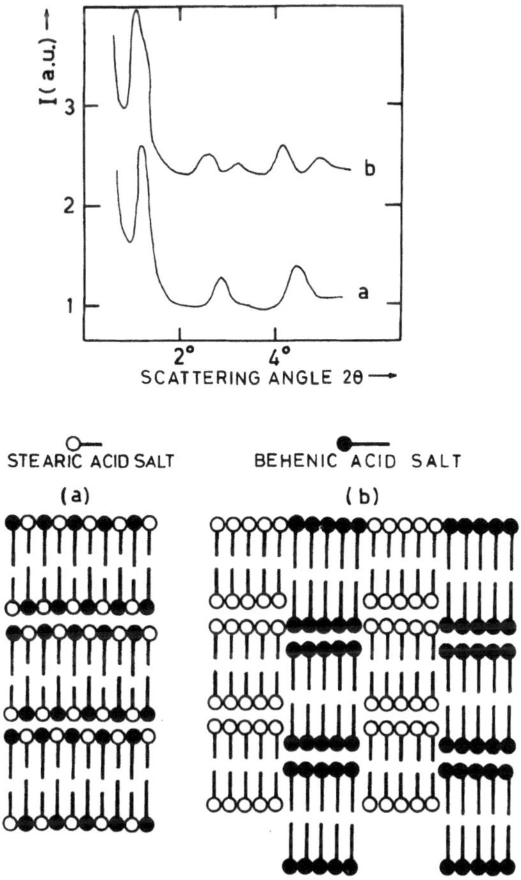

Figure 6 The upper part reports the X-ray diffraction patterns of mixed stearic and behenic acid salts with a molar ratio of 1:1 (a) and with molar ratio different from 1:1 (b). In the lower part the two corresponding probable models for molecular packing are shown [S. Dante et al., 1992; V. Erokhin et al., in press].

The value of the lattice spacing was found to be D=54.0±0.5Å, which is equal to the sum of the total length of the two fatty acid molecules or to twice the chain length of the intermediate fatty acid (arachidic acid). The model of the molecular packing is presented in fig. 6a, according to which the elementary cell in the direction perpendicular to the layers contains one stearic acid molecule facing one behenic acid molecule.

For all the other molecular ratios and even for the 1:1 ratio when the mixture occurred at the water surface, the X-ray patterns of the obtained LB films always contained two sets of Bragg reflections (fig. 6b), indicating the presence of two phases. The spacings obtained,

D_1= 49.0±0.5Å and D_2=59.0±0.5Å, correspond to the ones of pure stearic and behenic acid salts, respectively.

From the analysis of the Bragg peak widths, by using the Scherrer equation, a lower limit of the correlation length x_{min} in the direction perpendicular to the films was obtained. This is approximately the thickness of the sample over which the unidimensional periodic arrangement extends for a given phase. It corresponded to 4-5 bilayers in these cases [V. Erokhin et al., 1994].

The model obtained for the two coexisting phases is reported in fig. 6b: once a segregation of domains of the two different molecules occurs in a given layer, this information is transmitted to the following layers for several layer thickness. This fact indicates that, under these conditions, a selective diffusion involving the monolayer molecules during the deposition process occurs, in order that each one of the two types of molecules is able to deposit on a similar molecule in a kind of epitaxial growth. As a consequence the in-plane size of domains cannot be very large and should depend on the in-plane diffusion coefficient of the molecules and on the deposition speed. One should be able to verify this dependence by detecting the lateral size of segregated domains by atomic force microscopy.

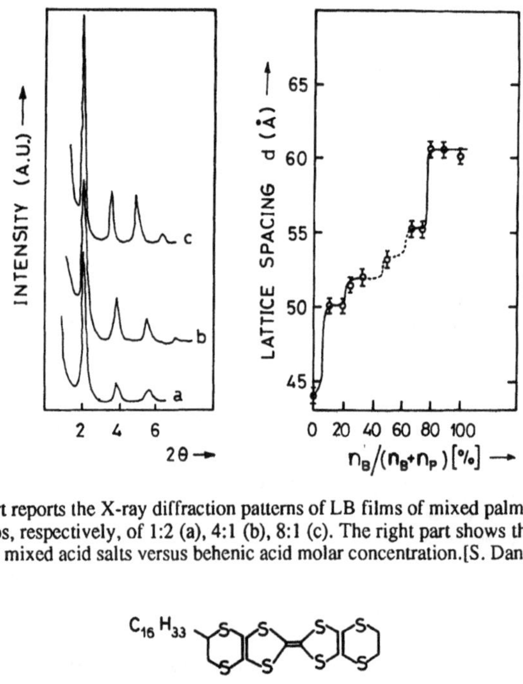

Figure 7 The left part reports the X-ray diffraction patterns of LB films of mixed palmitic and behenic acid salts with molar ratios, respectively, of 1:2 (a), 4:1 (b), 8:1 (c). The right part shows the lattice spacing of LB films of the same mixed acid salts versus behenic acid molar concentration.[S. Dante et al., 1992].

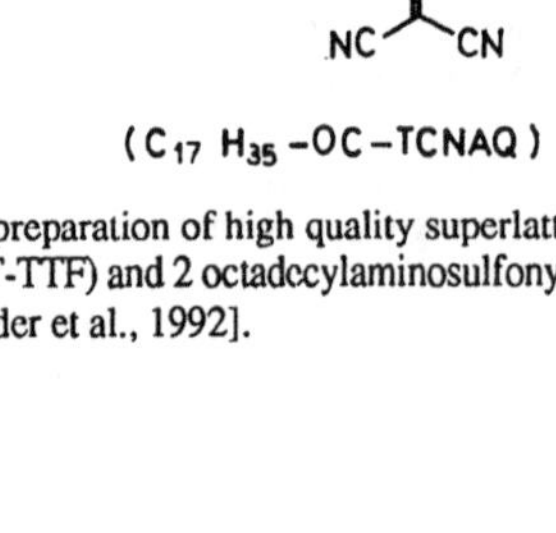

Figure 8 Molecules used for the preparation of high quality superlattices: hexadecylbis (ethylenedithio) tetrathiafulvalene ($C_{16}H_{33}$-BEDT-TTF) and 2 octadecylaminosulfonyl-tetracyanoanthraquinodimethane ($C_{17}H_{35}$-OC-TCNAQ) [V.V. Lider et al., 1992].

The investigation of another binary mixture (namely of behenic and palmitic acid salts of Cd) gave competely different results: a single phase (fig. 7a) was present for all molar ratio of the two compounds and the lattice pacing was found to depend on the molar ratio itself going from a minimum of D=50.0±0.5Å to a maximum of D=60.0±0.5Å (fig. 7b). It is impossible to produce a simple model for this system owing, among other things, to the relatively small widths of the diffraction peaks (fig. 7) which exclude large fluctuations of the lattice spacing [S. Dante et al., 1992].

Good mixing and high ordering was observed [V. Erokhin et al., in press.] in multilayers LB films of the mixture of an aliphatic amine (octadecylamine) and a fatty acid (stearic acid) when the molar ratio was 1:1, and poorer mixing and ordering in the other cases, including sometimes phase segregation.

Superlattices

As it was mentioned in the introduction one of the main interests related to the LB technique is the possibility of creating artificial well organized supramolecular assemblies at molecular scale spatial resolution. In this context, after the discovery of superlattices LB films of different alternating dielectric monolayers [Yu.M. Lvov et al., 1989] the discovery of conductive LB films [A.S. Dhindsa et al., 1987-T.S. Berzina et al., 1989] superlattices, LB films containing alternating sequences of dielectric (D) and two-component conducting (C) layers were recently prepared and investigated by x-ray diffraction.

These alternating sequences lead to electron flow inside the layer plane and provide a much greater flexibility in molecular architecture designs of systems interesting the molecular electronics. The following superlattice periodic sequences were produced: DC, DDC, DCC, DCCC and DDCC. Barium behenate and stearate were used as dielectrics and various surfactant as donor and acceptor molecules.

For instance successful results were obtained by using a sequence of two dielectric bilayers of Ba stearate and one conducting bilayer of a binary mixtures of hexadecylbis (ethylenedithio) tetrathiafulvalene ($C_{16}H_{33}$-BEDT-TTF) with 2-octadecylaminosulfonyl-tetracyanoanthraquinodimethane ($C_{18}H_{37}$-AS-TCNAQ) (fig. 8). The existence of the lamellar order in the so obtained superlattice was demonstrated by X-ray diffraction, as shown in fig. 9 where several higher order Bragg peaks are visible, from which a superlattice periodicity of 148.7 Å associated to the threee bilayers was obtained.

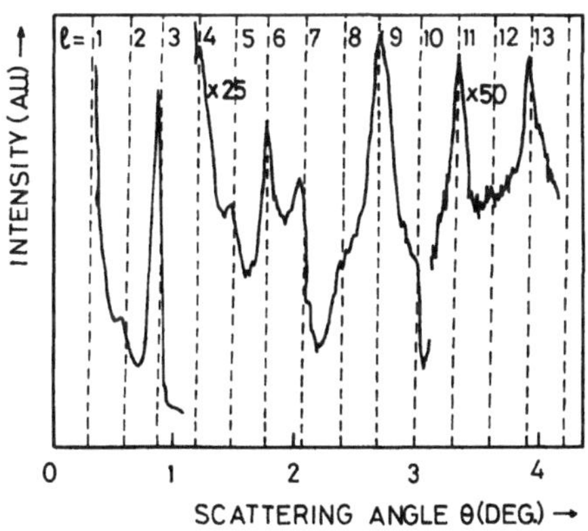

Figure 9 X-ray small angle diffraction pattern of a superlattice LB film constituted by a periodic sequence of two dielectric bilayers of Ba-stearate and one conducting bilayer of a binary mixture of the two molecules reported in fig. 8. Calculated positions of the reflections and their indexes (l) are shown [T.S. Berzina et al., 1992].

3. X-RAY TOPOGRAPHY

Generally, X-ray topography consists of detecting, on a photographic plate, the X-ray intensity diffracted by a single crystal. Whenever an increase of intensity is observed in a nearly perfect single crystal, this should be attributed to a corresponding presence of a defective structure in a region of crystal located along the X-ray path. For instance a single dislocation could be present and visible, or a dislocation cluster, or a fluctuation or gradient in the lattice spacing. Such imperfections, or alternatively the presence or the increase in a given region of the crystal of a certain mosaic spread, produce an increase of the X-ray reflectivity from the value corresponding to the dynamical theory of diffraction towards the one much larger (unless for very thin crystals) of the kinematical theory of diffraction.

When dealing with samples having high periodicities, of the order of few tens of Ångstroms, like LB films, superlattices, liquid crystals, stacked membrane bilayers [M. Hentschel et al., 1991], one could in principle perform x-ray topography. However in practice large difficulty arises, due to the squeezing of the diffracted beam in a narrow surface, as a consequence of the very small Bragg angle. This difficulty was recently overcome [V.V. Lider et al., 1991] by using an asymmetric crystal analyzer in order to expand the recorded image (Fig. 10). Topographs were obtained for a Pb stearate LB film (100 monolayers for a total tickness of 0.25 μm), deposited on Si single crystal with surface orientation (001). The dimensions of the film were 20x20 mm^2. A Ge crystal plate in symmetrical Bragg condition with the surface orientation (112) was used as monochromator and the reflection (113) of an asymmetrically cut Ge single crystal was used in the analyzer. The first order Bragg reflection, corresponding to a Bragg diffraction angle, $\theta_B = 0.16°$, was used. Heterogeneities in the topographs were well visible including a kind of ripple fine structure, which was attributed to the formation of blocks with dimensions of the order of 0.5 mm having a mosaic spread not larger than ±25". A periodic change of contrast along the direction normal to the one of the LB deposition, was also observed and attributed to changes in electronic density or surface roughness. Moreover topographs were obtained by the same geometry in condition of total external reflection of X-ray both from the film and from the substrate with a similar fine contrast. In conclusion it is now possible to perform topographic recording not only from the above mentioned periodic layered structures of large lattice spacings, but also from single layers and interfaces.

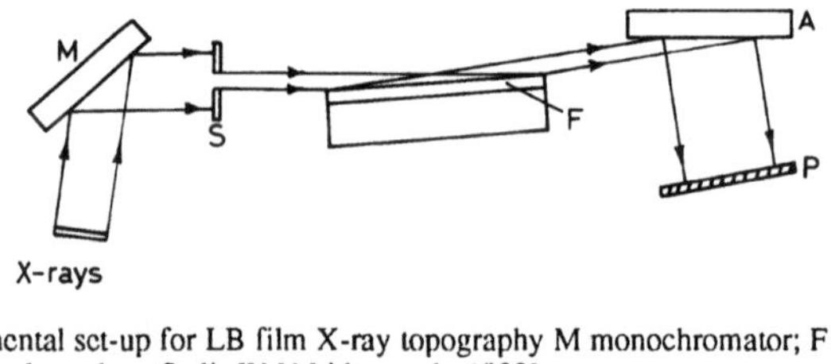

Figure 10 Experimental set-up for LB film X-ray topography M monochromator; F Langmuir-Blodgett film, A analyzer, P photoplate, S slit [V.V.Lider et al., 1992].

4. X-RAY SPECULAR REFLECTION

This technique, which is also called X-ray reflectivity at grazing incidence, provide electron density profiles along the direction normal to the surface, from which interesting structural information can be obtained along this direction, on surfaces, interfaces and layers having thicknesses lower then several towsands Ångstroms. In particular the molecular arrangement along the same direction can be obtained in Langmuir or LB films with a spatial resolution of few Ångstrom. Moreover surface and interface roughess and layer thickness and density can be derived.

Fig. 11 reports schematically the geometry of the technique and of an X-ray diffractometer for liquids [J. Als-Nielsen, 1986] [J.J. Benattar et al., 1991]. An incident monochromatic X-ray beam of wavelength λ, (wavevector $\underline{k}_{in}$), impinges on the sample surface at an angle q and the reflected beam intensity, also at an angle θ, (wavevector $\underline{k}_{out}$), is recorded in the plane of incidence. As a consequence the scattering vector $\underline{q} = \underline{k}_{out} - \underline{k}_{in}$, which represents the exchanged wavevector, is perpendicular to the surface (fig. 11), and its modulus is given by:

$$q = \frac{4\pi}{\lambda}\sin\theta \qquad (3)$$

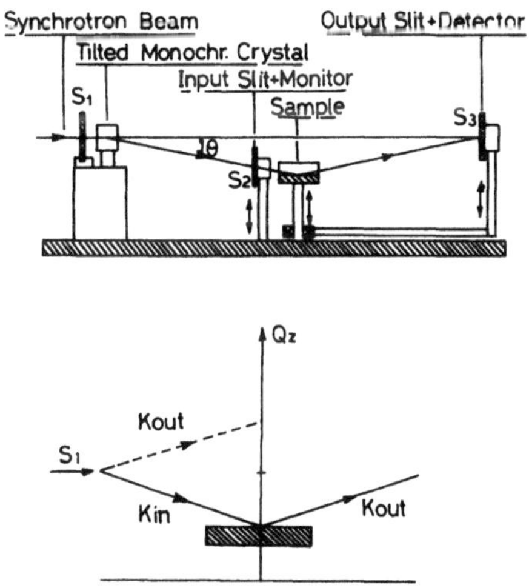

Figure 11 Experimental set-up for X-ray reflectivity from liquids, and reciprocal space representation. [J. Als-Nielsen, 1986; J.J.Benattar et al., 1991].

In a typical experiment a reflectivy profile defined as the ratio between the reflected and impinging X-ray intensity, R(q), is obtained from the measured reflected intensity as a function of the incidence angle θ. The electron density profile ρ(z) is directly related to the refractive index profile n(z). In fact:

$$n = 1 - \delta + i\beta \qquad (4)$$

where:

$$\delta = \frac{\lambda^2}{2\pi} r_e \left(f + \Delta f' \right) \frac{\rho}{M} \qquad (5)$$

$$\beta = \frac{\lambda^2}{2\pi} r_e \left(\Delta f''\right) = \frac{\mu\lambda}{4\pi} \qquad (6)$$

r_e being the classical electron radius $r_e = \dfrac{e^2}{mc^2}$, M the atomic mass, f+Δf'+iDf" the complex atomic scattering factor, μ the linear absorption coefficient.

For waver and the Cu $K\alpha_1$ radiation (λ = 1.5405 Å), β = 0.0126 x 10^{-6} and δ = 3.56 x 10^{-6}. As the real part of the refractive index is smaller than 1, total external reflection occurs for incidence angles smaller than a critical angle $\theta_c = \sqrt{2\delta}$ (θ_c = 2.67 mrad for water). For a single ideal diopter the reflectivy profile follows the Fresnel law:

$$R_F(q) = \left(\frac{q_c}{2q}\right)^4 \qquad \text{for } q \gg q_c. \qquad (7)$$

It can be shown that in general:

$$R(q) = \frac{R_F(q)}{\rho_0^2} \left| \int \frac{d\rho(z)}{dz} \exp[iqz]\, dz \right|^2, \tag{8}$$

where $R_F(q)$ is the Fresnel reflectivity for an ideally sharp interface and $\rho_0{}^2$ is the electron density of the substrate.

A uniform film of thickness L produces interference between X-ray waves reflected at the air/film and film/support interface. As a result one obtains the so-called Kiessig fringes, where the spacing Δq between maxima (or minima) is given by:

$$\Delta q = 2p/L \tag{9}$$

From eq. 9 the thickness can be calculate with high precision. In general one insert in eq. 8 a density profile $\rho(z)$ associated to a given model and containing several parameters, including the roughness σ, and perform a best fit of the experimental reflectivity profile R(q). A first example is reported in fig. 12 [J. Daillant et al., 1989], which shows the reflectivity profiles recorded for a monolayer of behenic acid ($CH_{21}H_{43}COOH$) on water. Curve (a) refers to the liquid condensed phase, (π = 15 mN/m), curve (b) to the solid phase (π = 22 mN/m). Both curves show the Kiessig fringes: the curve (b) is shifted at lower q values, as a consequence of an increase of the film thickness (see eq. 9), due to a decrease of the molecular tilt angle.

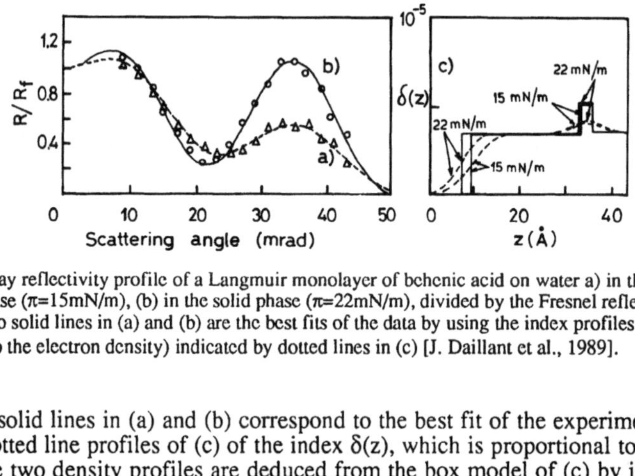

Figure 12 X-ray reflectivity profile of a Langmuir monolayer of behenic acid on water a) in the liquid condensed phase (π=15mN/m), (b) in the solid phase (π=22mN/m), divided by the Fresnel reflectivity of water $R_F(q)$. The two solid lines in (a) and (b) are the best fits of the data by using the index profiles $\delta(z)$ (which is proportional to the electron density) indicated by dotted lines in (c) [J. Daillant et al., 1989].

The solid lines in (a) and (b) correspond to the best fit of the experimental data by using the dotted line profiles of (c) of the index $\delta(z)$, which is proportional to the electron density. The two density profiles are deduced from the box model of (c) by introducing a roughness σ = 3.55 Å at 15 mN/m and s = 3 Å at 22 mN/m. The roughness, which is attributed to thermally induced capillary waves, is measured as a function of the surface pressure and the results discussed using a theory of thermal fluctuations. Moreover the tilt angle is determined as a function of the surface pressure. A strong decrease in roughness is observed at the transition to the solid state, which is attributed to the rigidity of the monolayer in the solid phase.

Successively similar experiments were performed on C_{15}, C_{21} and C_{29} fatty acids [J. Daillant et al., 1991] and the bending rigidity modulus in the solid phase of the Langmuir monolayer was obtained as a function of the aliphatic chain length. The experimental results are reported in fig. 13, where the dotted line was obtained by a simple theoretical treatment of the elastic behaviour of the solid monolayer.

The transition from liquid-expanded (LE) to liquid-condensed (LC) phase was investigated in monolayers of hexadecanoic acid and L-α dypalmitoyl phosphatidilcholine (L-αDPPC) by X-ray specular reflection [J. Daillant et al., 1990]. Fig. 14a reports the phase diagram of hexadecanoic acid and fig. 14b the X-ray reflectivity curves, normalized to

Fresnel reflectivity, for the surface of water and for the hexadecanoic acid at different temperatures and surface pressures. Several structural parameters were extracted from these curves. Fig. 15 reports the roughness (a) and the thickness of the aliphatic medium (b) as a function of surface pressure for the hexadecanoic acid. It appears that the roughness increases as a function of surface pressure, and the thickness of the aliphatic medium increases abruptly from 1.4 nm to 1.7 nm at the first order LE-LC phase transition. The interpretation of this fact, based on molecular-dynamics calculations [J.P. Baremaw et al., 1988], is that in the LE phase, each chain has defects like one or two gauche conformations, leading to a thickness reduction, whereas in the LC phase, the molecules are tilted by approximately 30° with less conformational defects.

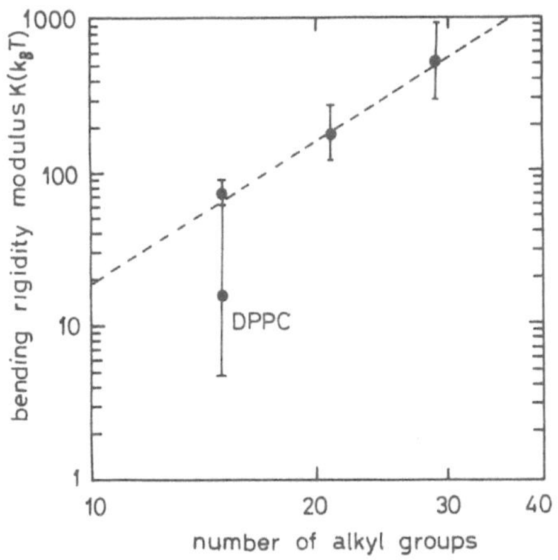

Figure 13 Bending rigidity modulus of a Langmuir monolayer at the water/air interface, in the solid phase, as a function of the number of alkyl groups. The dotted line is obtained by a simple theoretical model [J. Daillant et al., 1991].

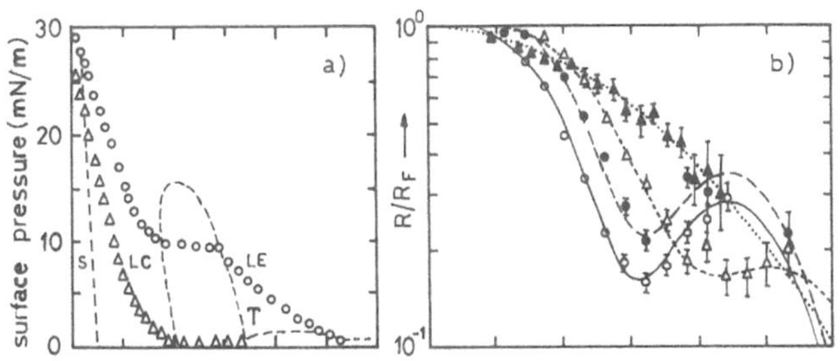

Figure 14 (a) Phase diagram of hexadecanoic acid. Circles, isotherm T=33°C; triangles, isotherm T=22°C. The phase boundaries are suggested by the dotted line. (b) Experimental reflectivity profiles (normalized to Fresnel reflectivity) and best fits. Filled triangles, surface of water; hollow triangles, hexadecanoic acid, T=33°C, π=9mN/m (LE phase); filled circles, T=33°C, π=25mN/m (LC phase); hollow circles, T=22°C, π=5mN/m (LC phase). The curves are best fits [J. Daillant et al., 1990].

Helm et al. [1991] reports an extensive study, by X-ray reflection and in-plane diffraction, of phospholipids (namely two phosphatidyl ethanolamines) monolayers at the air/water interface, from the liquid expanded phase (LE) to the solid phase, through the coexistence range. Fig. 16 reports the normalized X-ray reflectivity profiles R/R_F for the L-α - dimyristoyl - phosphatidylethanolamine (DMPE) at different lateral pressures as indicated in

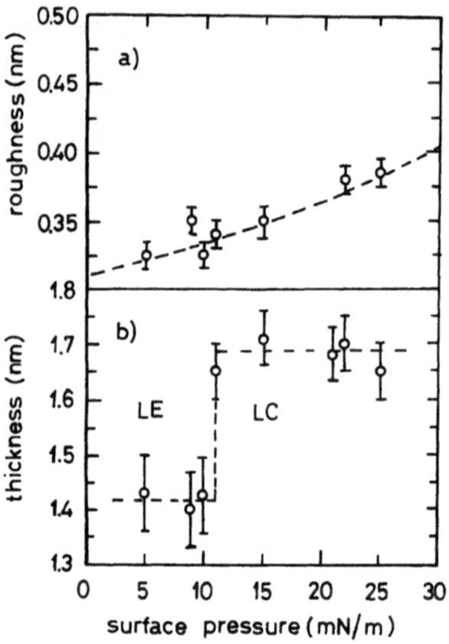

Figure 15 Surface roughness (a) and thickness of the aliphatic medium (b) versus surface pressure for hexadecanoic acid at T=33°C [J. Daillant et al., 1990].

the isotherm of the insert. Region I corresponds to the liquid expanded phase (LE), region II to the phase coexistence, and region III to the solid phase.

The reflectivity profiles a-e including best fits, refer to the liquid expanded phase (LE), the profiles e-k refer to the phase coexistence range (the experimental points were replaced, for clarity, by smooth interpolating curves) and the profiles j-o refer essentially to the region associated to the solid phase.

The electron density profiles and the corresponding models, obtained from the best fits, for the solid phase and the fluid phase (LE) respectively, are reported in fig. 17. For the several other structural informations deduced from the experiment, namely the dependence on the lateral pressure, of the parameters defined in fig. 17, the reader should refer to [A. Helm et al., 1991]. X-ray reflectivity was also used to determine the structure of a Newton black film formed from sodium dodecylsulphate (SDS) solution in presence of salt (NaCl) [O. Belorgey et al., 1991]. Fig. 18 reports the obtained electron density profile along the normal to the film and the structural model. The film resulted much thinner than expected and does not contain any aqueous core but only a hydration layer.

A combined study by both X-ray and neutron reflectivity from spread monolayers of docosanoic acid on an aqueous subphase is reported in [M.J. Grundy et al., 1988] and from polyethylene oxide at air water interface in [E.M. Lee et al., 1990]. The combination of X-ray reflectivity and X-ray fluorescence was used to investigate the interaction of Mn^{2+} cations in the subphase with a behenic acid monolayer [J. Daillant et al., in press]. It was found out that the cations are attracted near the surface (0.47 cations per amphiphilic molecule), in an extremely thin region ($\leq$ 5 Å).

X-ray and neutron reflectivity were used to obtain structural information on a protein, streptavidin, specifically bound at aqueous surface to lipid monolayers functionalized by biotinylated head groups [D. Vaknin et al., 1993]. It was found out that streptavidin forms a homogeneous monomolecular protein layer, of thickness 42 ± 2 Å, just below the lipid monolayer.

Fig. 19 shows schematically a freely-suspended film of liquid crystalline phases. Fig. 20 the X-ray reflected intensity profile from a 26 layer freely-suspended film of 5-(4"-hexyl,3'-fluoro-p-terphenyl-4-oxy)-pentanoicacid ethyl ester at 70°C [G. Decher et al., 1991]. The layer spacing determined from the Bragg peak position is 26.8 Å, and the total film thickness calculated from the Kiessig fringes is 696 Å. Fig. 21 reports the scheme of film transfer on a solid substrate. The preservation of both layer and in-plane structure was confirmed. Fig. 22 reports the model of a silicon substrate covered with a monolayer of phospholipids from lungs, and the corresponding electron density profile as obtained by X-ray specular reflection [A. Bélorgey, in press]. The structure and the phase transitions of a single dimyristoylphosphatidylcholine (DMPC) bilayer absorbed to a planar quartz surface in an aqueous environment was investigated by specular neutron reflection [S.J. Johnson et al., in press].

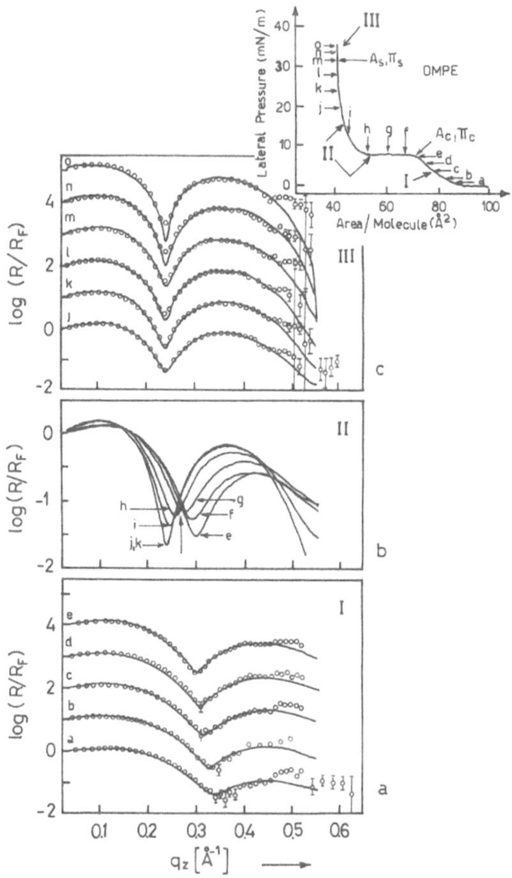

Figure 16 Normalized x-ray reflectivity profiles R/R_F for DMPE monolayers: the insert shows the point of the isotherm corresponding to the different profiles. (a) Liquid expanded (LE) phase range, (b) phase coexistence range, (c) solid phase range [A. Helm et al., 1991].

Several structural parameters of a monomolecular layer of amphiphilic cyclodextrins deposited on silicon wafers were obtained by X-ray reflectivity [A. Schalchli et al., in press] and for alkylsiloxane monolayers on the same substrate as well [I.M. Tidswell et al., 1991]. By the same technique an ultrathin film of alternate layers of DNA and polyallylamine with a periodicity of 40 Å was investigated [Yu. Lvov et al., in press).

The specular X-ray reflection was used to study the structure of several LB films. A beautiful application concerns LB films of oriented hairy rodlike polymers, namely poly(silane)s (fig. 23). Fig. 24a represents the model of an alternate double layer LB film of different hairy rodlike polymers and fig. 24b the corresponding X-ray reflection curve [F.W. Embs et al., 1991] [M. Schaub et al., 1992].

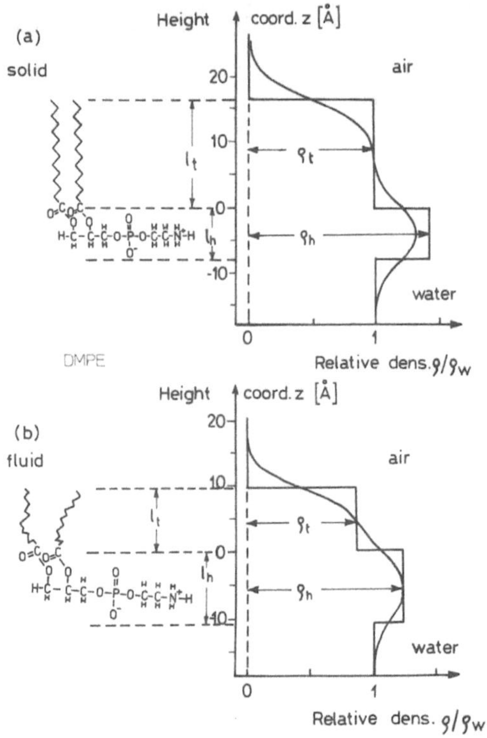

Figure 17 Electron density profiles and molecular conformation obtained from the reflectivity profiles of fig.16 for DMPE monolayers. (a) Solid phase; (b) liquid expanded (LE) phase. The following parameters are also defined: layer thicknesses l_H, l_T, electron densities ρ_H, ρ_T, and the gaussian smearing parameter σ [A. Helm et al., 1991].

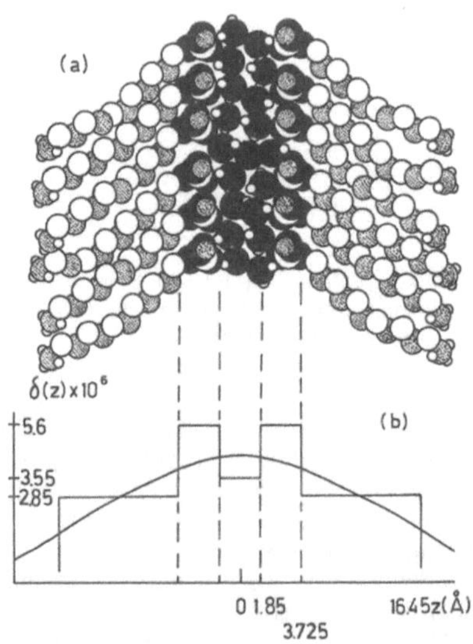

Figure 18 (a) structural model of the film; (b) Electron density profile along the normal to a Newton black film formed from a sodium dodecylsulphate (SDS) solution in presence of NaCl [O. Belorgey et al., 1991].

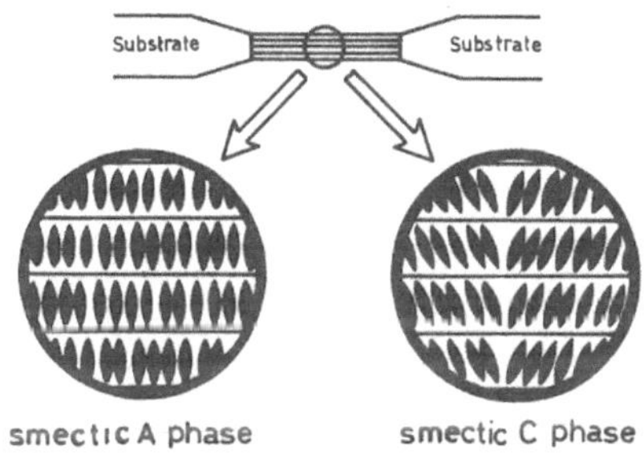

Figure 19 Scheme of a freely suspended film of smectic A or smectic C liquid crystalline phases [G. Decher et al., 1991].

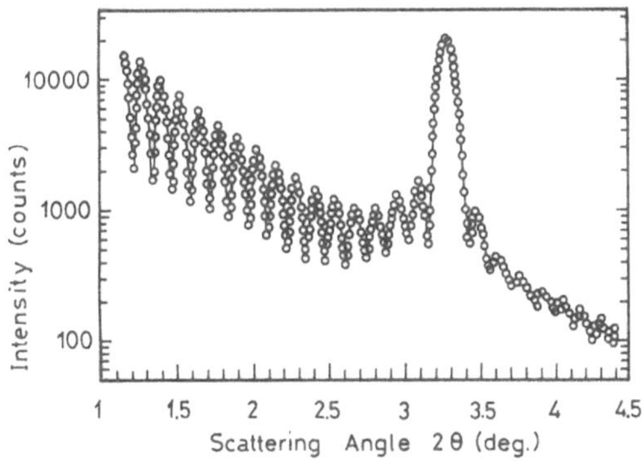

Figure 20 X-ray reflected intensity profile from a 26 layer freely-suspended film of 5-(4"- hexyl, 3' - fluoro - p- terphenyl - 4 - oxy) - pentanoic acid ethyl ester at 70°C [G. Decher et al., 1991].

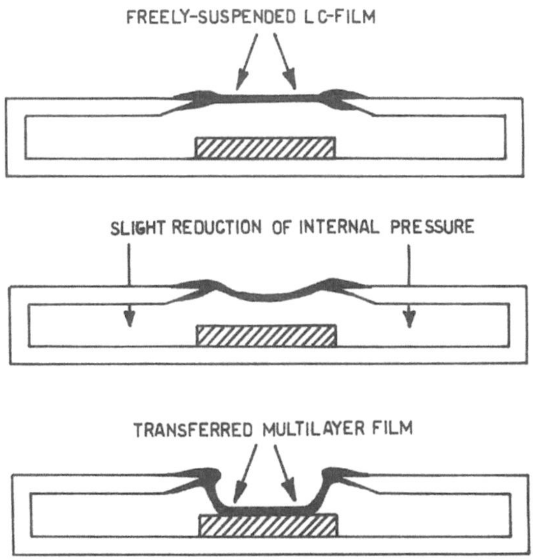

Figure 21 Scheme of the film transfer on a solid substrate [G. Decher et al., 1991].

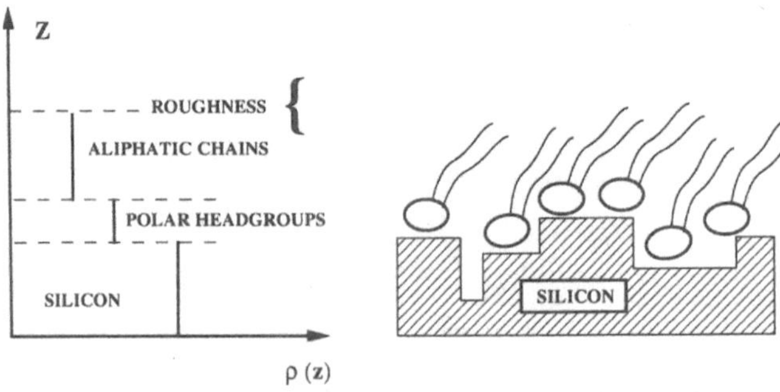

Figure 22 Model of a silicon substrate covered with a monolayer of phospholipids from lungs and the corresponding electron density profile [O. Belorgey et al., in press].

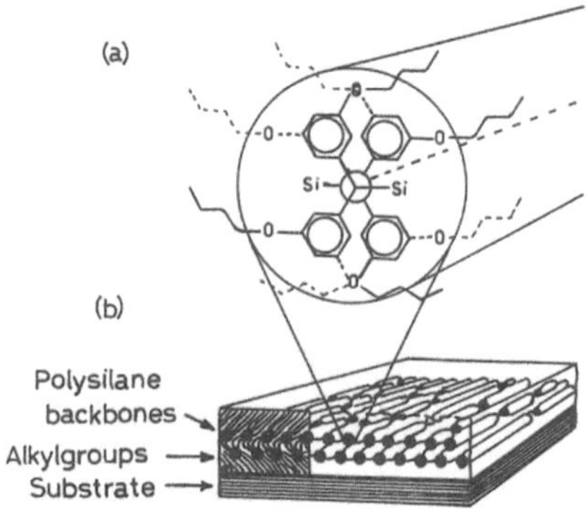

Figure 23 LB film of oriented poly (silane) s. (a) Orientation of the side groups, (b) alignement of the polymer backbones [F.W. Embs et al., 1991].

Another example concerns the study of the desorption of an LB film of 11 cadmium arachidate layers on silicon [P. Tippmann-Krayer et al., 1991]: fig. 25a reports the reflection curves as a function of temperature and fig. 25b the corresponding model for desorption. Alternating multilayer films of polystyrenesulfonate (PSS) and polyallylamine (PAH) were investigated by the same technique [Yu. M. Lvov et al., in press]. In particular fig. 26 reports the obtained film roughness as a function of the number of layers. The "wave"-line is the level of a glass substrate roughness.

5. ANOMALOUS X-RAY REFLECTIVITY

In the conventional x-ray specular reflection technique, as above, is usually assumed a given electron density profile containing several parameters, which is fitted to the observed reflected intensity in order to obtain the values of these parameters. In this way it is not evident that the obtained solution is unique, as the fit concerns only the intensity, and the phase information is lost. This limitation, which is similar to the well known phase problem in crystallography, was overcome very recently and in a very elegant way by using anomalous X-ray reflectivity [M.K. Sanyal et al., 1993].

A LB film consisting of seven Cd arachidate layers deposited on a higly polished Ge single-crystal with a [110] axis normal to the interface, was used for the experiment and is shown schematically in fig. 27. The method is based on the measurement of two X-ray

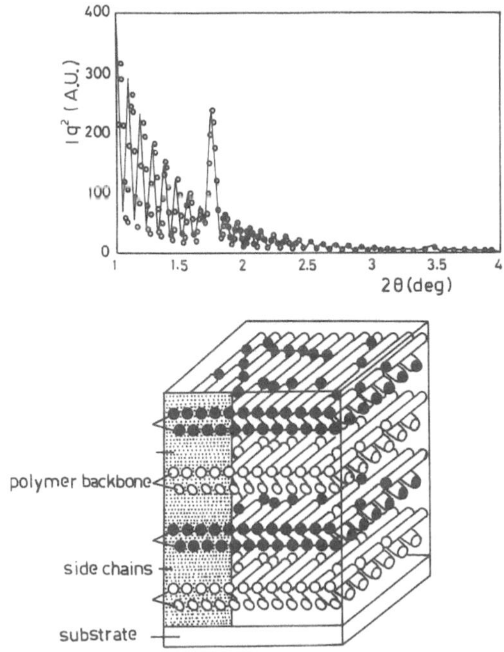

Figure 24 (a) Scheme of an alternate double layer LB film of different hairy rodlike polymers of the type shown in fig. 23; (b) Corresponding X-ray reflection curve (dots, experimental; solid line, fit) [M. Schaub et al., 1992].

reflectivity profiles R(k), one at an X-ray energy close to an absorption edge of Ge and another at an energy away from the edge.

The LB film electron density profile $\rho(z)$ is written as $\rho_0 + \Delta\rho(z)$, where ρ_0 is the average electron density of the film and $\Delta\rho(z)$ is the function which must be determined. The LB film-air interface is taken as z=0, with z positive into the substrate, the other interface being at $z=\delta$ (see fig. 27). The reflecting system is considered as the superposition of two electron densities S_0, S_1:

$$S_0 = \begin{Bmatrix} \rho_0 & for\ \ 0 < z < \delta \\ \rho_3 & for\ \ z > \delta \end{Bmatrix} \tag{10}$$

$$S_1 = \begin{Bmatrix} \Delta\rho(z) & for\ \ 0 < z < \delta \\ 0 & for\ \ z > \delta \end{Bmatrix} \tag{11}$$

The reflection profile R(k) for the system (S_0+S_1) can be calculated approximately by considering S_1 as a small perturbation on S_0 and using the Distorted Wave Born Approximation (DWBA):

$$R(k) = \left| i r_0(k) + \frac{2\pi b}{k}\left[a^2(k)\Delta\rho(q_z) + b^2(k)\Delta\rho(q_z)\right]\right|^2 \tag{12}$$

where b is the Thompson scattering length of the electron, k is the exchanged wavevector, and q_z is given by $2(k^2 - k_c^2)^{1/2}$, k_c being the critical value of k at which total reflection from the film occurs. Moreover $r_0(k)$ is the specular-reflectance coefficient of the sistem S_0 for normal-incidence wave vector k, a(k) and b(k) are the coefficients for the transmitted and reflected amplitudes in the film ρ_0, and $\Delta\rho(q_z)$ is the Fourier transform of $\Delta\rho(z)$.

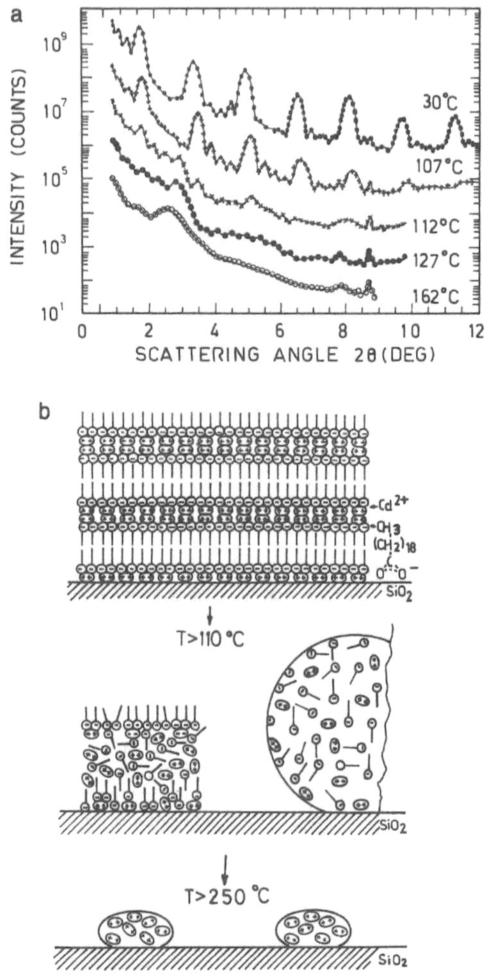

Figure 25 (a) X-ray reflection curves as a function of temperature for an LB film of 11 cadmium arachidate layers on silicon. Each curve is displaced by a factor 10; (b) Model for processes occurring during heat [P. Tippman-Krayer et al., 1991].

By calculating the functions $r_0(k)$, a(k), b(k) and by measuring the reflectivity R(k) at the two X-ray energies mentioned above, it was possible to obtain, through eq. 12, the real and the imaginary parts of $\Delta\rho(q_z)$ corresponding to each measured k. Then $\Delta\rho(z)$, and therefore $\rho(z)=\rho_0-\Delta\rho(z)$, was obtained by conventional Fourier transform of $\Delta\rho(q_z)$, after taking $\rho_0 = 0.37$ electrons/Å^3, as derived from known parameters for Cd arachidate LB films. The electron density profile $\rho(z)$ so obtained trough the overall LB film is reported in fig. 28b, and the reflectivity profiles obtained at the two x-ray energies are reported in fig. 28a. The three peaks correspond to the position of Cd counterions associated with the carboxylic head groups, whereas the plateau on the right side corresponds to the substrate.

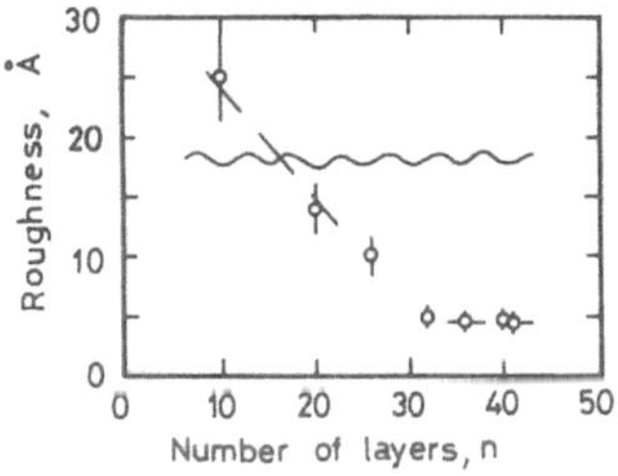

Figure 26 Roughness as a function of the number of layers in a multilayer film of polystyrenesulfonate (PSS) and polyallylamine (PAH). The "wave"-line is the level of a glass substrate roughness [Yu. M.Lvov et al., in press].

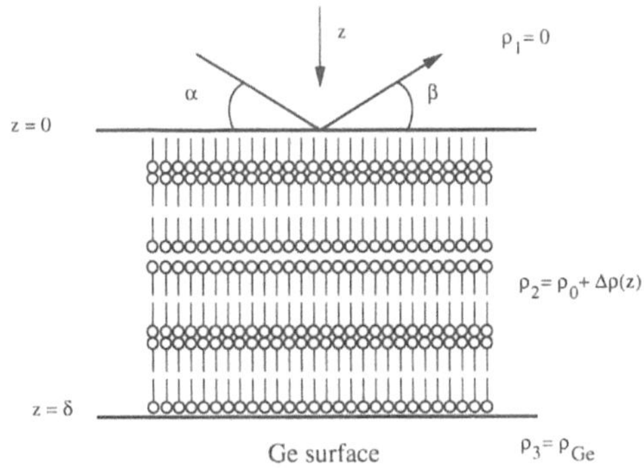

Figure 27 Schematic diagram of the seven layer Cd arachidate LB film on single crystal Ge substrate investigated by anomalous X-ray reflectivity. The X-ray wavectors, the electron densities in the different regions and the z-axis used in the text are also indicated [M.K.Sanyal et al., 1993].

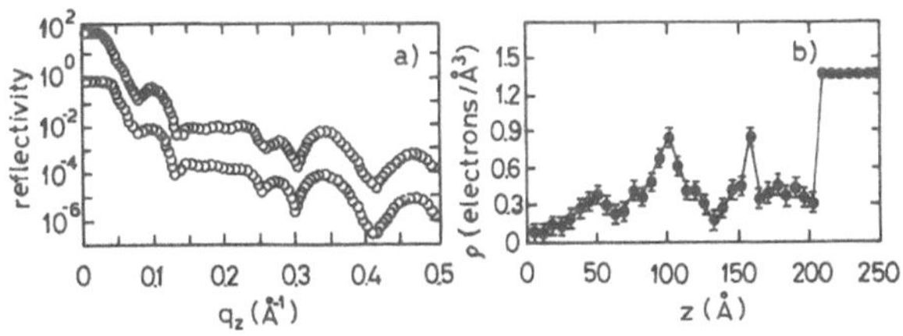

Figure 28 (a) Reflectivity curves for the LB film of fig. 27 at and away from the K-edge of the Ge substrate. o away from the edge; • at the edge (x100). (b) The electron density profile of the seven layer Cd arachidate on single crystal Ge substrate [M.K.Sanyal et al., 1993].

6. X-RAY STANDING WAVES

The technique of X-ray standing waves is the most elegant and sophisticated one among all the X-ray scattering techniques available at present. The standing waves, in the most recent applications, are the result of the interference between the incident, $\underline{E}_0$, and the specular reflected, $\underline{E}_r$, plane waves (fig. 29). The electric field intensity in the vacuum above the reflecting mirror surface can be written [M.J. Bedzyk, 1988; M.J. Bedzyk et al., 1989]

$$I(\theta,z) = |\underline{E}_0|^2\left[1 + R + 2\sqrt{R}\cos(\nu - 2\pi Qz)\right] \tag{13}$$

where $|\underline{E}_0|^2$ is the intensity of the incident beam, R is the reflectivity, ν is the phase of the reflected beam relative to the incident beam, and

$$Q = 2 \sin \theta/\lambda = 1/D. \tag{14}$$

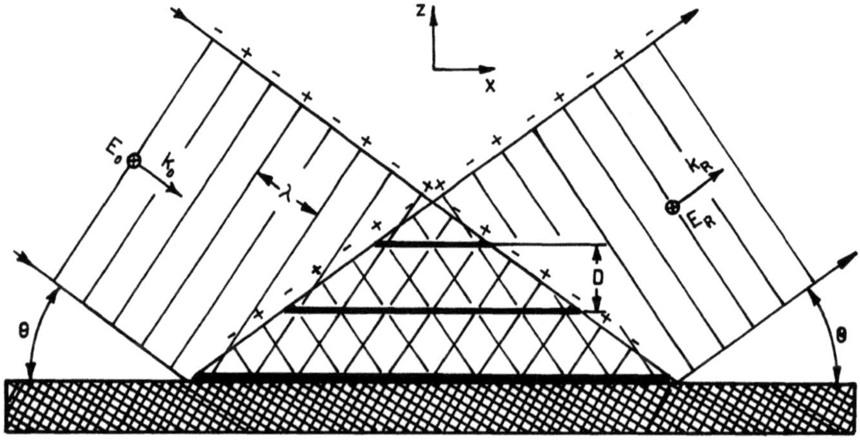

Figure 29 Scheme of the X-ray standing wave field arising from interference between the incident, $\underline{E}_O$, and specular-reflected, $\underline{E}_R$, plane waves above a mirror surface. The antinodes are represented, which have a period D = λ/2 sin θ [J. Wang et al., 1991].

For a fixed value of Q, i.e. of the incidence angle θ, the intensity I(Θ, z) modulates between 0 and $4|\underline{E}_0|^2$, when R = 1 (total reflection region), producing a standing wave, the antinodes of which are parallel to the mirror surface and have a spatial period of

$$D = \lambda/2 \sin \theta \tag{15}$$

as it is indicated in fig. 29. Since the total reflection condition occurs between 0 and θ_C, from eq. 15 one obtains that D varies from ∞ to $D_C = \lambda/2 \sin \theta_C$, i.e. in practice from 1000 Å to 100 Å. At θ = 0 a standing wave node is at the mirror surface and the first antinode at infinity. When q is increased, as D decreases (eq. 15), the first antinode moves toward the mirror surface and practically coincides with it for $\theta = \theta_C$: the other nodes and antinodes of the standing wave move also inward. This technique can be used to evaluate the position of a heavy atom layer above the reflecting surface, by monitoring the fluorescence yield from that atom layer as a function of incidence angle θ, by rotating the sample. In fact the photoelectric-effect cross section is proportional to the E-field intensity at the center of the given atom and is clear that the fluoresce yield strongly depends on the position of the atom layer as compared to the antinode position, i.e. to the reflecting surface position. Heavy atoms can be located with ångstrom or subångstrom resolution.

An LB film of lipid multilayer and a reference sample were investigated by this technique [J. Wang et al., 1991]. The LB film (fig. 30) (sample A) consists of a hydrophobic gold mirror with 14 bilayers of ω-tricosenoic acid (ω-TA) and an inverted bilayer of zinc arachidate, implying a distance of the Zn layer ~ 926 Å above the Au mirror surface. A reference sample (sample B) was prepared without the 14 ω-TA bilayers, implying a distance of the Zn layer ~ 42 Å above the surface. Fig. 31 reports the experimental and theoretical results, namely the fluorescence yield (a) and the X-ray reflectivity profile (b) as a function of the incidence angle, and the obtained zinc distribution above the surface (c) for the sample A. Curves (d), (e) and (f) show the same quantities for sample B. In (a) (b) (d) (e) the continuous curves are the theoretical evaluations.

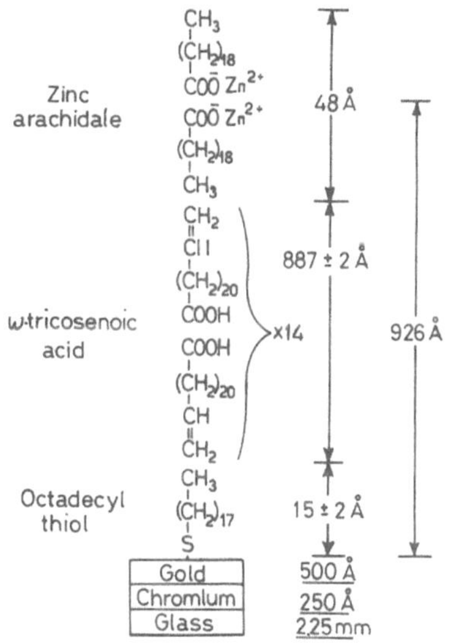

Figure 30 Scheme of the gold mirror and deposited LB film investigated by X-ray standing wave technique [J.Wang et al., 1991].

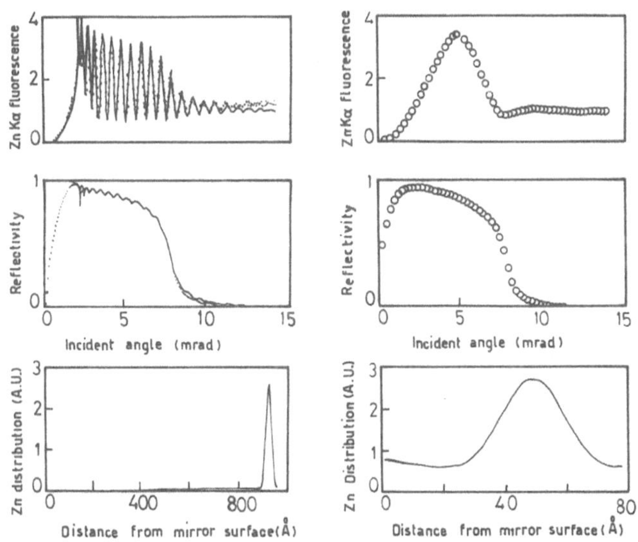

Figure 31 Experimental (.) and theoretical (——) angular dependence of the zinc Kα fluorescence yield (a) specular X-ray reflectivity (b) and calculated zinc distribution above a Au surface for a LB film containing 14 bilayers of ω-tricosenoic acid and an inverted bilayer of zinc arachidate. The same quantities are reported in (d) (e) (f) for a reference sample without the 14 bilayers of ω-tricosenoic acid [J.Wang et al., 1991].

The several peak maxima appearing in (a) represent the crossing of antinodes of X-ray standing wave with the Zn layer, when θ is increased, whereas the minima correspond to the crossing of nodes. In (d) only one antinode is crossing the Zn layer, which is in fact much nearer to the mirror surface. Other investigations of LB films by the same technique can be found in ref. [S.I. Zheludeva et al., 1990; S.I. Zheludeva et al., 1993]. Recently phase transitions in LB films were also investigated [M. Caffrey, private communication.]. In conclusion this technique is very useful in surface structure investigation, including protein films, membrane-membrane and receptor-ligand interactions.

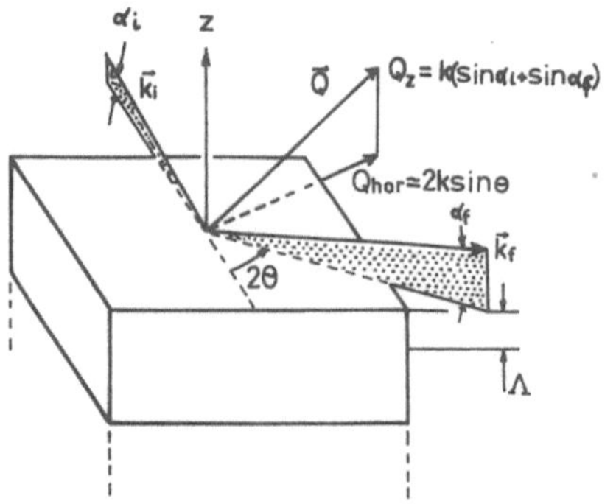

Figure 32 Geometry for surface diffraction with larger horizontal (Q_{hor}) than vertical (Q_z) scattering vector component [A. Helm et al., 1991].

7. SURFACE DIFFRACTION

The surface diffraction (also called horizontal or in-plane diffraction or grazing incidence X-ray diffraction) technique is strictly related to the specular reflection technique and its principle is sketched in fig. 32. A highly collimated and monochromatic X-ray beam impinges on the air-water interface at a vertical angle α_i just under the critical angle of total reflection θ_c, and, if an in-plane periodicity inside the monolayer exists, is reflected at the angle α_f, corresponding to a 2θ deflection in the horizontal plane. In these conditions of total reflection ($\alpha_i < \theta_c$) an evanescent wave travels in a direction parallel to the interface. Its intensity decays exponentially along a direction perpendicular to the interface. The penetration depth (fig. 32), which depends on the incidence angle, is of the order ~ 100 Å. The evanescent wave is therefore diffracted by planes normal to the surface in a very thin region beneath the interface. From the position of the diffracted spot on a bidimensional multidetector on can deduce, through Bragg law, the spacing of the diffracting planes, and from the width of the spot, the correlation length ξ. Fig. 33 reports the X-ray diffracted intensity as a function of the wavevector transfer q for increasing lateral pressure (a-b), corresponding to the arrows of the pressure/area isotherm of the insert, for a phospholipid monolayer at the air/water interface [K. Kjaer et al., 1987].

Qualitatively, one can deduce a decrease in the interplanar distance from the shift to higher q values, and an increase of the correlation length from the narrowing of the peak. Quantitatively, the correlation length ξ, the integrated X-ray intensity X, and the lattice spacing d, are reported in fig. 34, as deduced from the data of fig. 33.

The same technique, using synchrotron X-ray radiation, was employed to detect the complicated polymorphism of insoluble monolayers of several classes of amphiphilic compounds. A review on these studies can be found in ref. [A.M. Bibo, to be published].

Recently a new type of organic monolayers, namely self-assembled monolayers, which form by spontaneous chemisorption of long chain molecules from solution onto many different solid substrates like Au, Ag, Cu, Al, Si, were investigated by grazing incidence X-ray diffraction [P. Fenter et al., 1993]. In particular $CH_3(CH_2)_{n-1}SH$ self-assembled on an Au(111) surface were investigated as a function of the chain length (n) and of the temperature T. A (n,T) phase diagram was explored, in which the individual phases present different tilt direction and two-dimensional periodicity. A bidimensional scan of the diffracted intensity was performed both along the radial ($Q_{||}$) and azimuthal direction (Q_z). Fig. 35a reports the radial scan through the (1,1) diffraction peak (indexed with a rectangular cell) of C_{12} both "as deposited" and after annealing at $Q_z = 0.19$ Å^{-1}, and fig. 35b the scan for the same Bragg peak along the azimuthal direction. From the position of this peak in reciprocal space a

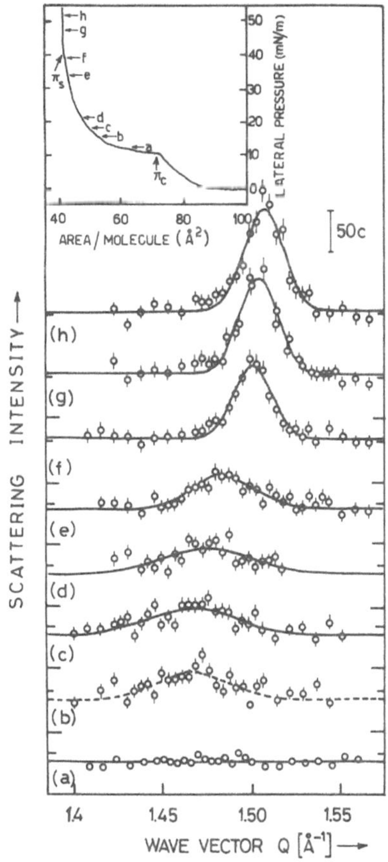

Figure 33 In-plane X-ray diffracted intensity as a function of wave-vector transfer Q for increasing surface pressure (a-h) corresponding to the arrows in the insert. Monolayer of dimyristoyl - phosphatidic acid, pH 5.5, ionic content 10^{-2} M NaCl, 5 x 10^{-5} M ethylene diamine tetroacetic acid, T = 19 ± 2°C [S.I.Zheludeva et al., 1993].

tilt angle of (32.5±1)° in a direction away from the nearest-neighbor direction by (13.8±2)° was derived.

Moreover a hexagonal structure commensurate with the Au lattice was found with an interchain distance perpendicular to the chains equal to 4.5Å. A structural model of the self-assembled layer is given in fig. 36. Fig. 37 reports the temperature dependence of the (1,1) Bragg peak for annealed C_{12} and C_{14} monolayers. Whereas the C_{12} peak does not exhibit any change in position or width but only in intensity, the C_{14} peak splits above 60°C, indicating a two-dimensional phase transition at this temperature, to a nonhexagonal structure which is incommensurate with the underlaying Au lattice.

This fact is related to the observation of two distinct regimes for long (n>14) and short (n<14) chain lengths and reflects the competition of hydrocarbon and interface interactions in the considered self-assembled monolayer. Fig. 38 reports the temperature dependence of the (1,1) peak intensity (at Q_z = 0.19 $Å^{-1}$) for C_{12}. A change of slope is observed at T = 50°C, which is interpreted as a transition to a liquid like phase, i.e. as a melting phase transition. For further interesting structural information obtained in this experiment, the reader should refer to [P. Fenter et al., 1993].

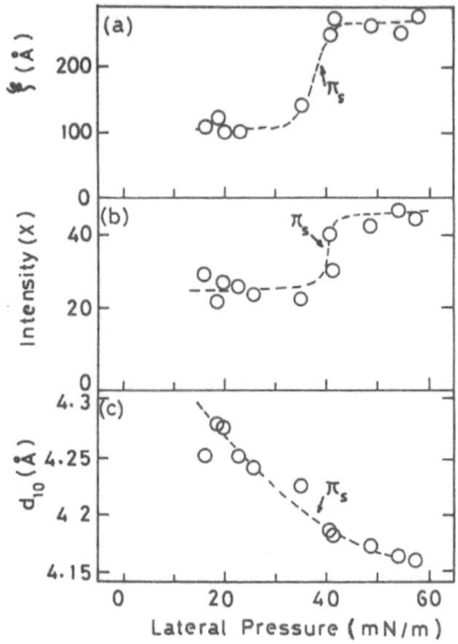

Figure 34 (a) Correlation length x, (b) integrated X-ray intensity, and (c) lattice spacing d, as function of lateral pressure for the dimyristoylphosphatidic acid monolayer of fig. 33 [K. Kjaer et al., 1987].

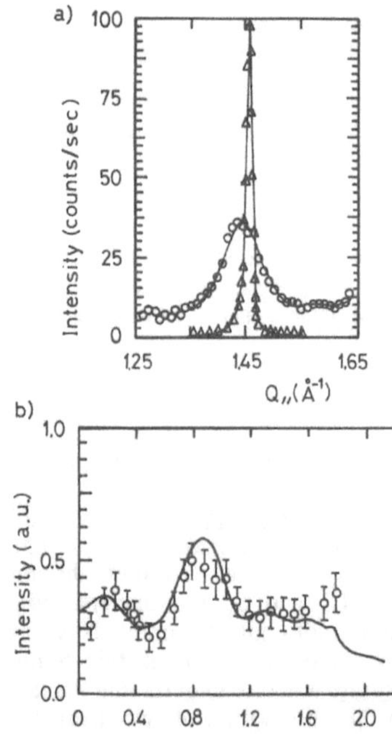

Figure 35 (a) Radial scan of X-ray diffracted intensity by a self-assembled monolayer of $CH_3(CH_2)_{n-1}SH$ on a Au(111) surface. The (1,1) diffraction peaks of C_{12}, as deposited (Q_z=0.40$Å^{-1}$, circles) and after an anneal (Q_z=0.19$Å^{-1}$, triangles) are reported. (b) Scan of the same peak as in (a) along the azimuthal direction [P. Fenter et al., 1993].

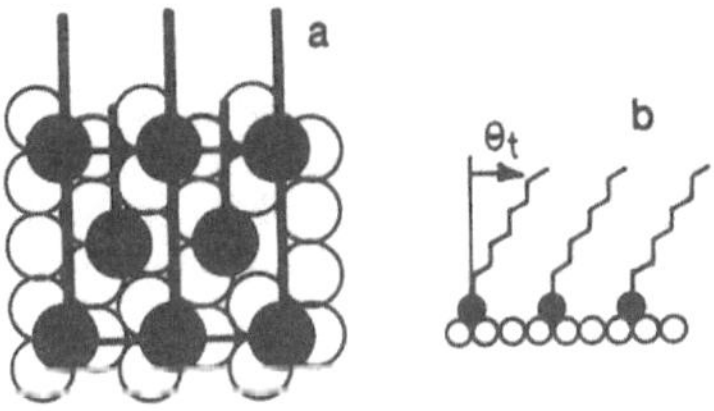

Figure 36 (a) The commensurate hexagonal structure of the $CH_3(CH_2)_{n-1}SH$ monolayer on the Au (111) lattice. The shaded circles represent S atoms and the open circles Au atoms. (b) Side view of the interface, with Au and S atoms, and the tilted hydrocarbon chains [P. Fenter et al., 1993].

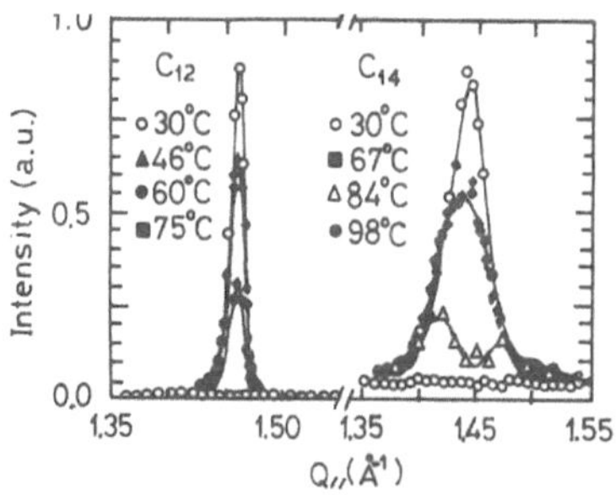

Figure 37 Temperature dependence of the (1,1) Bragg peak as a function of temperature for annealed C_{12} and C_{14} monolayers of $CH_3(CH_2)_{n-1}SH$ on a Au (111) surface [P. Fenter et al., 1993].

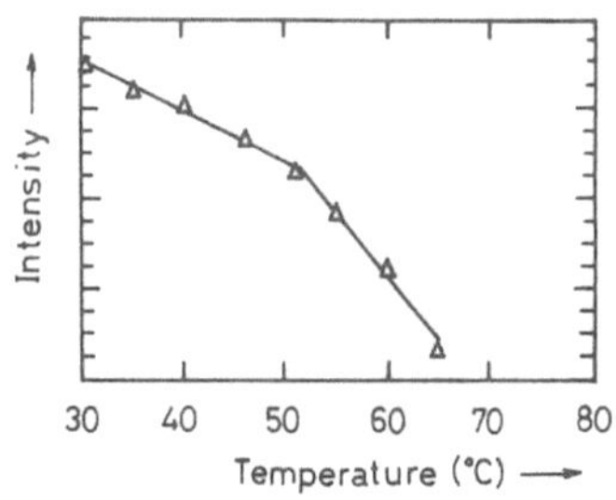

Figure 38 Temperature dependence of the (1,1) Bragg peak intensity (at Q_z=0.19 A^{-1}) for C_{12} as deduced from fig. 37 [P. Fenter et al., 1993].

8. CONCLUSION

The several examples reported above show that the available X-ray scattering techniques can produce fundamental information on the structure of supramolecular layers interesting the molecular electronics research. In particular the available very intense synchrotron radiation sources allow the development of more and more sophysticated experiments.

ACKNOWLEDGEMENTS

It is a pleasure to thank Prof. L. Feigin and Dr.s V. Erokhin, M. Kovalchuk, V. Troitsky for very useful discussions in the recent years on the subjects of this paper. Moreover, I acknowledge Miss S. Franguelli, Mr. Emanuele Mosca for typing the manuscript and Mr. Enzo Mosca, Mr. S. Polenta for the drawings preparation.

REFERENCES

Albouy P.A., Vandevyver M., Perez X., Ecoffet C., Markovitsi D., Veber M., Jallabert C., Strzelecka H., 1992, *Langmuir* , 8:2262

Als-Nielsen J., 1986, *Physica*, 140A:376

Baremaw J.P., Cardini G., Klein M.L., 1988, *Phys. Rev. Lett.*, 60:2152

Barraud A., Lequan M., Lequan R.M., Lesieur P., Richard J., Ruaudel-Teixier A. and Vandevyver M., 1987, *J. Chem. Soc., Chem. Commun.*, 797

Bedzyk M.J., 1988, *Nucl. Instr. and Meth.*, , A266:679

Bedzyk M.J., Bilderback D.H., Bommarito G.M., Caffrey M., Schildkraut J.S., 1988, *Science*, 241:1788

Bedzyk M.J., Bommarito G.M., Schildkraut J.S., 1989, *Phys. Rev. Lett.*, 62:1376

Belorgey O., Benattar J.J., 1991, *Phys. Rev. Lett.*, 66:313

Belorgey O., Tchoreloff P., Benattar J.J., Proust J.E., *Journal of Colloid and Interface Science*, in press

Benattar J.J., Daillant J., Belorgey O., Bosio L., 1991, *Physica A.*, 172:225

Berzina T.S., Shikin S.A., Troitsky V.I., 1991, *Makromol. Chem. Macromol. Symp.*, 46:223

Berzina T.S., Troitsky V.I., Vorobyova S.L., Feigin L.A., Yasunova L.G., Micheletto R., Rustichelli F., 1992, *Thin Solid Films*, 210/211:309

Bibo A.M., Knobler C.M., Peterson I.R., to be published.

Caffrey M., private communication.

Daillant J., Bosio L., Benattar J.J., 1990, *Europhys. Lett.*, 12:715

Daillant J., Bosio L., Benattar J.J., Blot C., *Langmuir letters*, in press

Daillant J., Bosio L., Benattar J.J., Meunier J., 1989, *Europhys. Lett.*, 8:453

Daillant J., Bosio L., Harzallah B., Benattar J.J., 1991, *J. Phys. II*, 1:149

Dante S., Erokhin V., Feigin L., Rustichelli F., 1992, *Thin Solid Films.*, 210/211:637

Dante S., Rustichelli F., Erokhin V., 1992, *Mol. Cryst. Liq. Cryst.*, 215:205

Decher G., Maclennan J., Reibel J., Sohling U., 1991, *Adv. Mater.*, 3:617

Decher G., Sohling U., 1991, *Ber. Bunsenges. Phys. Chem.*, 95:1538

Dhindsa A.S., Bryce M.R., Lloyd J.P. and Petty M.C., 1987, *Synth. Met.*, 22:185

Dhindsa A.S., Bryce M.R., Lloyd J.P. and Petty M.C., 1988, *Thin Solid Films*, 165:97

Embs F.W., Wegner G., Neher D., Albouy P., Miller R.D., Willson C.G., Schrepp W., 1991, *Macromolecules*, 24:5068

Erokhin V., Dante S., Rustichelli F., *Mol. Cryst. Liq. Cryst.*, in press

Erokhin V., Feigin L., Kayushina R., Lvov Yu., Mariani P., Rustichelli F., to be published.

Erokhin V.V., Lvov Yu.M., Mogilevsky L.Tu., Zozulin A.N., Ilyin E.G., 1989, *Thin Solid Films*, 178

Erokhin V., Vakula S., Nicolini C., 1994, *Thin Solid Films*, 238:88-94

Feigin L.A., Lvov Yu.M., Troitsky V.I., 1989, *Sov. Sci. Rev. A. Phys.*, 11:285

Feigin L.A., Lvov Yu.M., 1988, *Makromol. Chem. Macromol. Symp.*, 15:259

Fenter P., Eisenberger P., Liang K.S., 1993, *Phys. Rev. Lett.*, 70:2447

Grundy M.J., Richardson R.M., Roser S.J., Penfold J., Ward R.C., 1988, *Thin Solid Films.*, 159:43

Helm A., Tippman-Krayer P., Möhwald H., Als-Nielsen J., Kyaer K., 1991, *Biophys. J.*, 60:1457

Hentschel M., Rustichelli F., 1991, *Phys. Rev. Lett.*, 66:903

Johnson S.J., Bayerl T.M., Mc Dermott D.C., Adam G.W., Rennie A.R., Thomas R.K., Sackmann E., *Biophysical Journal*, in press.

Kjaer K., Als-Nielsen J., Helm C.A., Laxhuber L.A., Möhwald H., 1987, *Phys. Rev. Lett.*, 58:2224

Lalanne C., Delhaes P., Dupart E., Garrigou-Lagrange Ch., Amiell J., Morrand J.P. and Desbat B., 1989, *Thin Solid Films*, 179:171

Lee E.M., Simister E.A., Thomas R.K., 1990, *Mol. Cryst. Liq. Cryst.*, 179:151

Leuthe A., Li Feng Chi, Riegler H., *Thin Solid Films*, submitted

Lider V.V., Kovalchuk M.V., 1992, *Phys. Stat. Sol.*, 130:K1

Lvov Yu.M., Decher G., Möhwald H., *Langmuir*, in press

Lvov Yu.M., Decher G., Sukhorukov G., *Macromolecules*, in press
Lvov Yu.M., Erokhin V.V., Zaitsev S.Yu., 1991, *Biol. Mem.*, 4:1477
Lvov Yu.M., Haas H., Decher G., Möhwald H., *J. Phys. Chem.*, in press.
Lvov Yu.M., Kalinin V.I., Petty M.C., Dhindsa A., 1988, *JETP Lett.*, 48:576
Lvov Yu.M., Troitsky V.I., Feigin L.A., 1989, *Mol. Cryst. Liq. Cryst.*, 172:89
Merle H.J., Lvov Y.M., Peterson I.R., 1991, *Makromol, Chem. Macromol. Symp.*, 46:271
Morand J.P., Lapouyade R., Delhaes P., Vandevyver M., Richard J. and Barraud A., 1988, *Synth. Met.*, 27:B569
Nakamura T., Takei F., Tanaka M., Matsumoto M., Sekiguchi T., Manda E., Kawabata Y., and Saito G., 1986, *Chem. Lett.*, 323
Novak V.R., Myagkov I.V., Lvov Yu.M., Feigin L.A., 1992, *Thin Solid Films*, 210/211:631
Paloheimo J., Kuivalainen P., Stubb H., Vyuorimaa E., Yli-Lahti P., 1990, *Appl. Phys. Lett.*, 56:1157
Pearson C., Dhindsa A.S., Bryce M.R. and Petty M.C., 1989, *Synth. Met.*, 31:275
Ruaudel-Teixier A., Vandevyver M., Barraud A., 1985, *Mol. Cryst. Liq. Cryst.*, 120:319
Sanyal M.K., Sinha S.K., Gibaud A., Huang K.G., Carvalho B.L., Rafailovich M., Sokolov J., Zhao X., Zhao W., 1993, *Europhys. Lett.*, 21:691
Schalchli A., Tchoreloff P., Zhang P., Coleman A.W., Benattar J.J., *Langmuir letters*, in press
Schaub M., Mathaner K., Schwiegk S., Albouy P.A., Wenz G., Wegner G., 1992, *Thin Solid Films*, 210/211:397
Tidswell I.M., Rabedeau T.A., Pershan P.S., Kosowsky S.D., Folkers J.P., Whitesides G.M., 1991, *J. Chem. Phys.*, 95:2854
Tippmann-Krayer P., Möhwald H., Lvov Yu.M., 1991, *Langmuir*, 7:2298
Vaknin D., Kjaer K., Ringsdorf H., Blankenburg R., Piepenstock M., Diederich A., Losche M., 1993, *Langmuir*, 9:1171
Vandevyver M., Albouy P.A., Mingotaud C., Perez J., Barraud A., *Langmuir*, in press
Wang J., Bedzyk M.J., Penner T.L., Caffrey M., 1991, *Nature*, 354:377
Zheludeva S.I., Kovalchuk M.V., Novikova N.N., Sosphenov A.N., 1993, *Phys. D. Appl. Phys.*, 26:A206
Zheludeva S.I., Kovalchuk M.V., Novikova N.N., Sosphenov A.N., Erokhin V., Feigin L., 1993, *J. Phys. D. Appl. Phys.*, 26:A202
Zheludeva S.I., Lagomarsino S., Novikova N.N., Kovalchuk M.V., Scarinci F., 1990, *Thin Solid Films*, 193/194:395

LIPIDS OF THE ARCHAEA DOMAIN: STATE OF THE ART AND COMPUTER SIMULATION STUDIES OF THE ARCHAEAL MEMBRANE

Mario De Rosa and Alessandra Morana

Istituto di Biochimica delle Macromolecole, Facoltà di Medicina, II Università di Napoli - Via Costantinopoli 16 - 80138 Napoli - Italy

The Archaea domain, formed exclusively of extremophilic prokaryotic organisms, represents a third evolutive line distinct from the well-known Bacteria and Eukarya domains. When comparing molecular components of Archaea with the other forms of life, one is particularly struck by the extensive chemical differences observed in their lipids, which can be considered specific and useful taxonomic markers of this new evolutive line.

Despite their great structural variations, all the lipids of the organisms known so far, from the simplest bacterium to the human being, are almost always esters formed by fatty acids and alcohols. Lipids of the Archaea are an exception to this general rule. They are based on ether links, originating in the condensation between two alcohols. Furthermore, the aliphatic part of these lipids is isoprenic in nature, and thus has nothing in common with fatty acids, which in Archaea are lacking. In Figure 1 is reported an overview of the archaeal core lipids, which can be obtained by acid hydrolysis of complex lipids of these microorganisms. These compounds originate in the formation of two or four ether links between two vicinal hydroxyl groups of a glycerol or more complex polyols and C20, C25, or C40 isoprenoidic alcohols. While diethers are monopolar amphipathic molecules, characterized by one polar head to which a hydrophobic core is linked, tetraether lipids are bipolar amphipatic molecules with two equivalent or nonequivalent polar heads linked by two C40 alkyl components, which are about twice the average length of the aliphatic components of classic ester lipids. The cyclopentane rings on C40 isoprenoids of tetraether lipids depends on environmental parameters such as temperature, and acts as a molecular control for the homeoviscous adaptation of membrane fluidity. In *Sulfolobus solfataricus* and *Thermoplasma acidophila* it has been shown that more rigid C40 chains, with a higher number of flat cyclopentane rings, occur in the lipids of the microorganisms when the growth temperature increases.

The occurrence of isoprenoid ether lipids, the relative ratio of diethers and tetraethers and the nature of the isoprenoid chains have proven of value both in the identification and in the taxonomy of Archaea. The isoprenoid nature of the aliphatic chain, the 2,3 glycerol stereochemistry, and the ether linkage are structural elements which characterize all archaeal lipids and can be considered specific markers of this evolutive line. In contrast, there are some structural details, such as the bipolar architecture of the macrocyclized tetraethers, the cyclopentane ring formation, and the C40 isoprenoid chains, which seem to be limited to specific groups of microorganisms and can be considered phenotypic responses to specific environmental stress such as high temperatures and pressures.

The different structures of the Archaea lipids examined so far give rise to the problem of how these molecules are organized in the membrane. Figure 2 proposes five hypotheses of lipid organization in archaeal membranes. In Figure 2a is schematized the double layer of the membrane of halophilic Archaea, characterized by unusual intermolecular interactions, owing to the isoprenoidic nature of the lipid hydrophobic chains. In Figure 2b is sketched a hypothetical model of lipid organization found mainly in the membrane of haloalkaliphilic

From Neural Networks and Biomolecular Engineering to Bioelectronics
Edited by C. Nicolini, Plenum Press, New York, 1995

Archaea. Strong and unusual interactions are expected to occur between alkyl chains localized on opposite sides of the lipid bilayer, owing to major penetration of the longer C25 chains into the opposite lipid layer, effectively causing a "zip" effect in the middle of the membrane. Figure 2c proposes the model of lipid organization that could be operative in some methanogens. In this membrane monopolar diether lipids are organized in a classic double layer, and the relative ratio between the more flexible diphytanyl diethers and the more rigid C36 macrocyclized diethers is the molecular strategy responsible for the homeoviscous adaptation, which keeps membrane fluidity relatively constant, in spite of environmental temperature changes. The organization of the lipids occurring in the "zip" membrane of alkaliphilic halophiles may be similar in stability to the mixed C20 and C40 membrane structure that could operate in certain methanogens, as reported in Figure 2d. This structure, named "rivetted" membrane, is characterized by the presence of a partly covalently-linked double layer. Figure 2e proposes the monolayer structure of extreme thermoacidophiles, based on C40 macrocyclic tetraethers, in which only covalent bonds are operating in the middle of the lipid layer. The uniqueness of this membrane model is remarkable, considering that the organization of the lipids in a double layer in the biological membranes have so far appeared to be a universally repeated element. In a natural environment in which water is close to the boiling point, a biological membrane based on a lipidic monolayer represents one of the possible strategies to limit the lipid mobility, maintaining the fluidity of the system at values compatible with the biological processes.

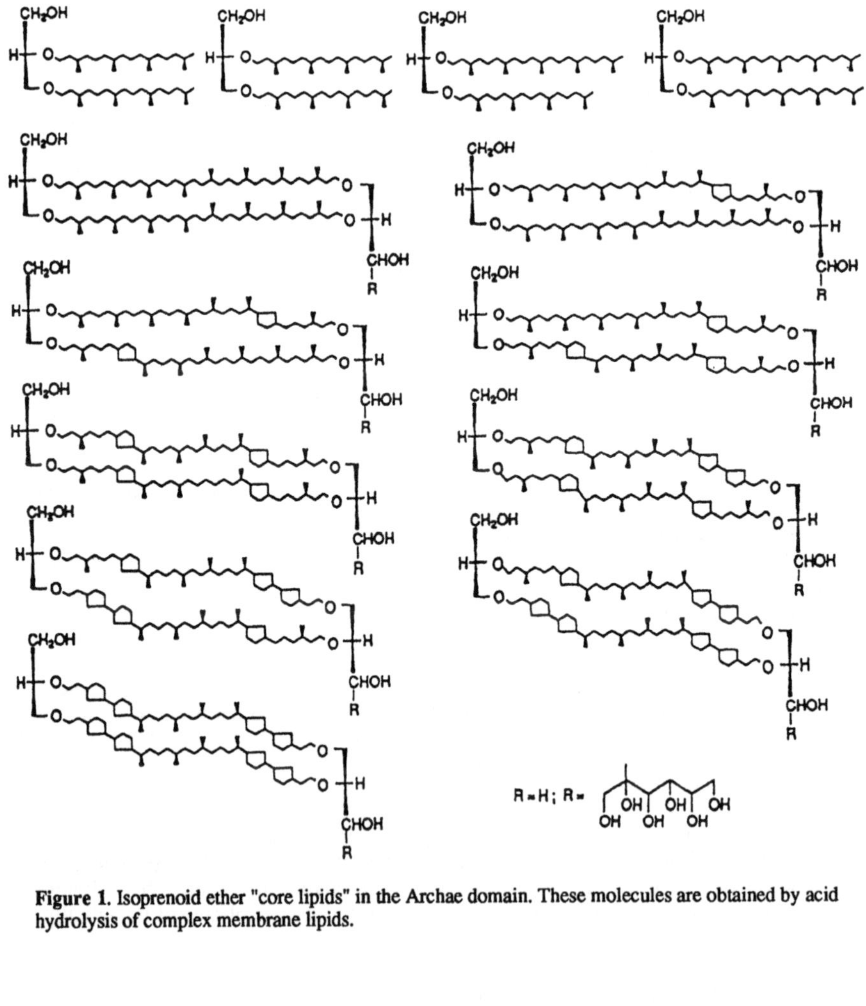

Figure 1. Isoprenoid ether "core lipids" in the Archae domain. These molecules are obtained by acid hydrolysis of complex membrane lipids.

To obtain an innovative approach to the problem of the structural organization and dynamics of bipolar lipids in the monolayer membrane of thermophilic Archaea, we have developed, in collaboration with Dr. Ottaviano Incani of Tecnofarmaci, Italy, a study of computer simulation of this structure at the molecular level using a Convex C120, a mini-super computer with vectorial architecture, and Biosym molecular modeling software for macromolecular computer simulation, named Insight and Discover.

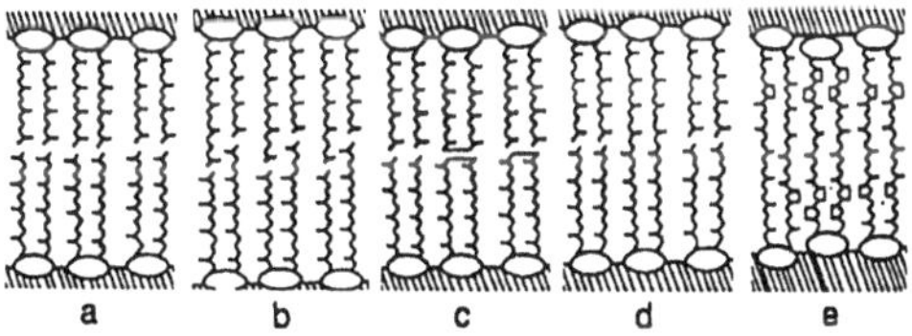

Figure 2. Membrane models in the Archae domain.

In our studies we applied methods of molecular mechanics, molecular dynamics, and periodic boundary simulation to three different lipid structures: the glycerol dialkyl glycerol tetraether (GDGT), obtained by acid hydrolysis of complex membrane lipids of thermophilic Archaea, and the two complex lipids present in *S. solfataricus*, that are derivatives of GDGT having as polar heads -glucopyranosyl-(1-4)--galactopyranosyl and phosphomioinositol.

Figure 3 shows the results of the energy minimization process for an isolated GDGT molecule with C40 isoprenoid chains without cyclopentane rings and for the corresponding complex lipid that has phosphomioinositol and a galactosyl-glucosyl disaccharide as polar heads. These conformations, with isoprenoidic methyls regularly oriented on one side of the bent aliphatic chains, represents the minimum of energy for the isolated molecules in the vacuum. In fact, the same results can be obtained starting the minimization energy process from different extended structures. These arched conformations, mainly originated by intra- and inter-isoprenoid chain interactions inside the same molecule, are far from the true lipid organization in the archaeal membrane of thermophiles, in which the conformation of each lipid strongly depends on intermolecular interactions with neighbouring molecules at the level both of isoprenoid moieties and polar heads.

Adequate computer simulations of biological membranes can be obtained with the periodic boundary conditions technique, where explicit periodic images of the real molecules are generated. These images, replicated to as great a distance as necessary, interact with the real molecule. To improve the rigor and realism of the model, in all our computer simulations two water layers of 13 angstroms interact with the two faces of the membrane originated by the polar heads of the lipids, thus mimicking the aqueous environment in which biological membranes operate. Periodic boundary computer simulation of the GDGT gives rise to a bent conformation, not consistent either with the compact organization of lipid molecules in a biological membrane, nor with the thickness of this structure in Archaea. This behaviour indicates that intramolecular interactions at the level of isoprenoid chains, responsible for the arched structure in the isolated molecule, in this case also prevails on intermolecular interactions, originating this non-natural and non-stable membrane structure. These data suggest that intermolecular interactions at the polar heads level, which are present in the intact complex lipids of Archaea, could play a key role in the lipid organization in the archaeal membrane. This hypothesis is confirmed by the periodic boundary computer simulation of the glycolipid and phosphoglycolipid derivatives of GDGT, which have respectively one and two polar heads attached to the glycerol(s). In this case both molecules assume an extended conformation, originating a membrane monolayer of bipolar lipids 40 angstroms thick, a value that agrees with the estimated hydrophobic membrane thickness of the archaeal membrane. These data indicate that only one polar head on the GDGT core lipid is sufficient to obtain an extended structure. These structures, that resemble the *S. solfataricus* membrane with all the disaccharide heads regularly oriented outside the cell, are strongly stabilized by the optimization of intermolecular polar interactions and hydrogen bonds among the polar heads of surrounding molecules, that prevents isoprenoid chain bending. The molecules show a cylindrical geometry, with a diameter of about 11 angstroms, that allows compact lipid assembly in the archaeal membrane, with a surface area per lipid molecule of about 110 square angstroms. The structures are characterized by an aliphatic moiety 40 angstroms long, and two polar heads, of about 10 angstroms for the phosphomioinositol residue and 11 angstroms for the glucosyl-galactosyl disaccharide end respectively. The presence of one to four cyclopentane rings on the isoprenoid C40 chains of GDGT moiety cause the shortening of isoprenoid hydrophobic core to 39 - 37 angstroms and a small increase in the surface area of the lipid molecules.

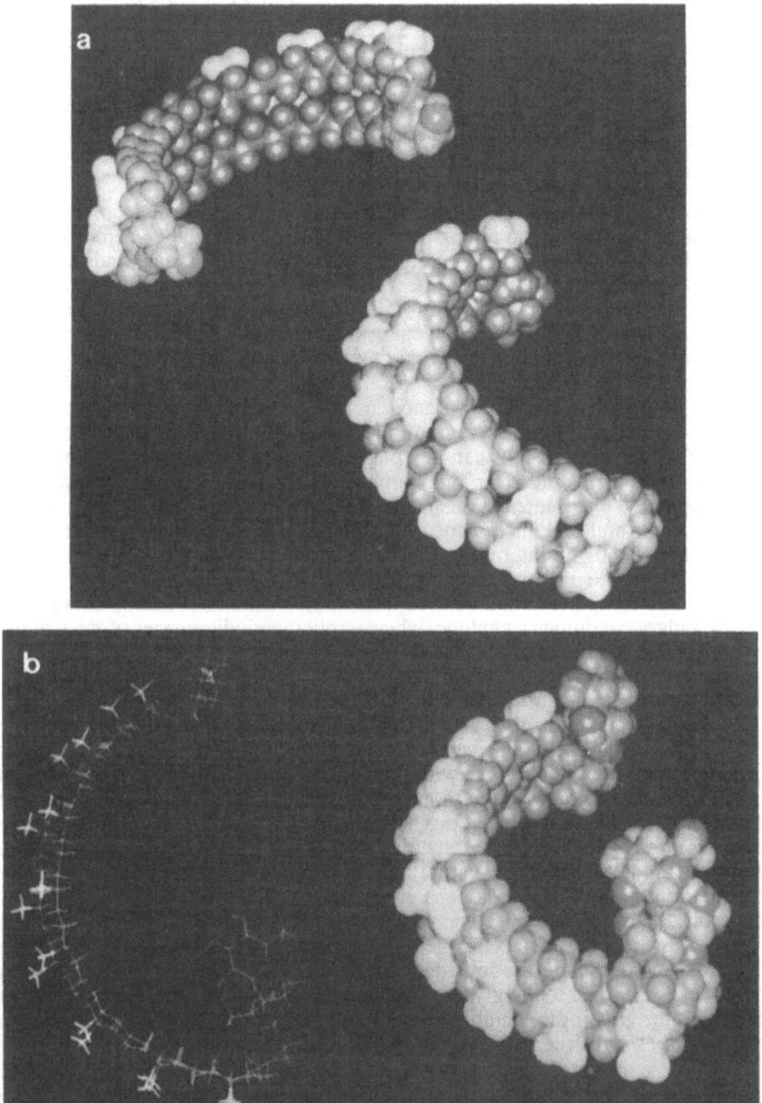

Figure 3. Computer simulation of archaeal lipids. Energy minimized conformations of the glycerol dialkyl glycerol tetraether (GDGT) (a) and the glycophospholipid of *S. solfataricus* (b) as isolated molecules.

Further information on lipid organization in archaeal membrane can be obtained with the molecular dynamics approach, which solves the equations of motions for the system, giving a dynamic vision of the membrane structure. In our computer simulation we explore 25 picosecond intervals, a time sufficiently long to see the most important molecular movements. The molecular dynamics with the periodic boundary simulation, at 80°C, of *S. solfataricus* phosphoglycolipid shows that the isoprenoid moiety fluctuates between 37 and 40 angstroms. These data suggest that at physiological temperatures for the *S. solfataricus* the dynamics of the lipid molecules, fully stretched and organized in a true monolayer, originates a membrane "respiration" based on small variations of the lipid layer thickness. The same analysis, performed at 100°C, shows that at this too high, non-physiological temperature, the molecules assume a bent conformation, causing the collapse of the monolayer membrane organization. The data of molecular dynamics performed on polar lipids with cyclized C40 isoprenoids indicate that cyclopentane rings, strategically closed on the aliphatic chains of *S. solfataricus* lipids, control chain motion and increase the break-down temperature of the simulated membrane structure, that collapses at higher temperatures, confirming that cyclopentane rings could represent the structural element for the homeoviscous control of the membrane lipid layer to environmental temperature changes.

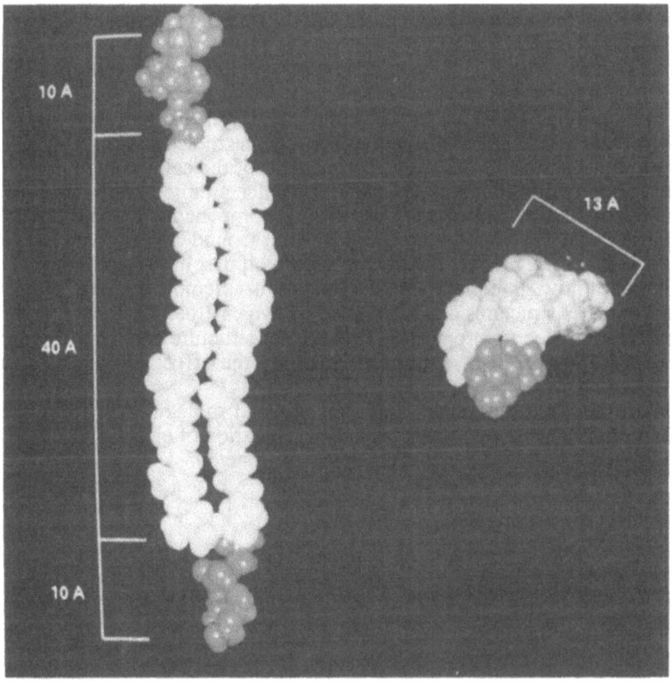

Figure 4. Computer simulation of archaeal lipids. Periodic boundary simulation of the phosphoglycolipid of *S. solfataricus*.

I'm going to briefly illustrate potentialities of archaebacteria and their lipids in future development of bioelectronics, an emerging discipline in which the development of electronic components containing biological elements is the primary objective.

The spectacular evolution of electronic devices of the last few years will be limited in the future by the fundamental laws of physics, that prevent further increases in complexity and miniaturization of electronic devices.

In this respect molecular electronics and in particular the capacity of some biochemical structures to transport electrons on a molecular scale, together with the inherent capability of self assembly and organization of such compounds, realistically represent the new frontier toward obtaining denser and faster circuits.

Biosensor and molecular electronics research are now the starting points of bioelectronics toward more futuristic targets such as biochips, biocomputers and the utilization of biological systems to achieve artificial intelligence that may even approach that of the human.

The relative chemical instability of many biological components, compatible with physiological conditions and metabolic turnover, but far from the silicium inertness, represent the most severe limit to obtaining realistic, pratical bioelectronic devices.

Therefore the development of bioelectronics is, by necessity, highly dependent on the search for stable biological systems, useful for the assembly of bioelectronic devices.

In this respect the use of biochemical components of extremophilic microorganisms, naturally selected during bilions of years in the severe conditions of ecosystems with temperatures of up to 120 °C, pH ranging from 0.5 to 12 and ionic strength near saturation, is one of the most promising options in the development of strategies for bioelectronics.

In view of the interest of biolectronics for stable Langmuir - Blodgett lipid films, I will summarize some characteristics of isoprenoid ether lipids of archaebacteria, that are of interest for this type of technique.

These unique molecules are more chemically stable than conventional ester lipids, as demonstrated by the presence of undegraded molecules of archaebacterial lipids in ancient sediments and petroleums.

Physical properties of these molecules are unusual, because their aliphatic components have less structural variety than fatty acid chains and have a comb structure originated by a regular presence of isoprenoid methyls on the aliphatic chains, that strategically controls hydrophobic interactions between the molecules in the lipid layer.

Some archaebacterial lipids, basic components of the cellular membrane of the more thermophilic forms of life, have an unusual molecular architecture, characterized by the presence of two polar heads, linked by two C40 alkyl components. These molecules originate stable monolayers, that can by considered as covalently-bound conventional bilayers.

Black-membranes, prepared with asymmetric bipolar lipids, are the most stable lipid films prepared until now. In these layers the lipids self-assemble with the same orientation, originating asymmetric structures characterizated by two external polar surfaces with different chemico-physical and electric properties. This type of structure could be of great interest in preparing protein engineered layers and in studying the effect of electric field on these surfaces.

In this respect extremophiles and their unusual molecules probably represent an interesting opportunity to shorten times for a new generation of electronic components based on an efficient integration of biological and electronic systems.

A CIDS-ACTIVATED CELL SENSOR FOR MONITORING DNA SUPERSTRUCTURES

Paolo Facci, Alberto Diaspro and Claudio Nicolini

Institute of Biophysics, University of Genova Via Giotto 2, 16153 Genova, Italy

ABSTRACT

An instrument for sensing DNA superstructure modifications by performing CIDS (Circular Intensity Differential Scattering) measurements of single biological specimens has been developed. Sample drops have been suspended in the laser beam interaction area and were impinged by left and right circular polarized light. The differentially scattered radiation was acquired as a function of the azimuthal scattering angle. The prototype was tested by means of a monodispersed micro sphere sample. Preliminary results on rat liver nuclei fixed with glutaraldehyde and ethanol point out that this instrument is sensitive to the structural alterations caused by the action of these fixatives, suggesting the possibility of implementing such kind of CIDS module in usual flow cytometers.

CIDS (acronym for Circular Intensity Differential Scattering), is an optical parameter defined as follows:

$$\mathrm{CIDS}(\vartheta,\varphi)=\frac{I_L(\vartheta,\varphi)-I_R(\vartheta,\varphi)}{I_L(\vartheta,\varphi)+I_R(\vartheta,\varphi)}, \quad (1)$$

where:
$I_L(\vartheta,\varphi)$:= light intensity scattered by the sample when excited by left circular polarized light;
$I_R(\vartheta,\varphi)$:= light intensity scattered by the sample when excited by right circular polarized light;
ϑ:= azimuthal angle;
φ:= zenith angle.

This parameter has been shown to be sensitive to the chiral structure of the observed biopolymer (e.g. for chromatin it depends upon pitch and radius of DNA quaternary structure) (Diaspro and Nicolini, 1987; Diaspro et al., 1991; Facci, 1990; Facci et al., 1990; Nicolini and Kendall, 1977).

It has been successfully used to characterize chromatin structures (Diaspro et al., 1991; Nicolini et al., 1988), both in vitro and in situ, the degree of supercoiling in -DNA (Nicolini et al., 1991) and micro-organisms as bacteria or viruses (Salzman et al., 1982). Each different sample exhibits a characteristic angular dependency of the CIDS signal that represents a "finger print" of the sample itself. Therefore, even if a complete theoretical understanding of the CIDS signal from biological complex structures (e.g. micro-organisms or cells) doesn't exist so far, it should be possible to identify a particular sample by means of its CIDS angular behaviour (Nicolini, 1986; Salzman et al., 1982; Bickel and Stafford, 1981).

From Neural Networks and Biomolecular Engineering to Bioelectronics
Edited by C. Nicolini, Plenum Press, New York, 1995

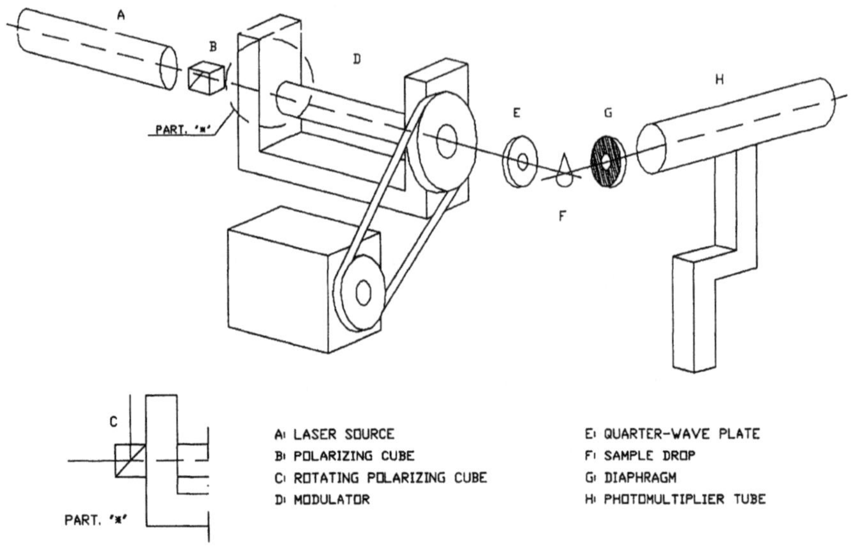

Figure 1 Experimental set-up of the sensor prototype for CIDS measurements.

All these facts point out the charming possibility of realizing a sensor based on CIDS measurements for the detection, identification and sorting of viable cells or nuclei (Facci et al., 1990).

The instrument set-up, schematically sketched in fig.1, is made of a HeNe unpolarized LASER source radiating at 632.8 nm (Diaspro et al., 1991), followed by a linear polarizer which provides linear polarization in vertical direction. Then, the polarized radiation passes through a modulator which generates the suitable polarization by means of a rotating beam splitter cube allowing light modulation at 184 Hz and providing the trigger signal for syncronizing data acquisition. The ray emerging from the modulator is linearly polarized along the current angular orientation. This modulator is followed by a quarter wave retardation plate with the fast axis in vertical position. Circular left (or right) polarized light is produced every time the beam impinging on the phase shift plate is polarized to +(-) 45° with respect to the plate's fast axis. This polarized beam impinges on the small sample drop and the differential scattering (circular right versus left polarized light) is collected at the various azimuthal angles. This last operation is performed by means of a photomultiplier tube (PMT) mounted on a goniometric mechanical arm which is computer driven and which performs the angular scanning (Facci et al., 1990).

To test the sensitivity of our instrument to the chirality of the sample (Facci et al., 1990; Diaspro and Nicolini, 1987; Diaspro et al., 1991) we performed initially a calibration with monodispersed latex micro spheres, expected to yield consistently zero CIDS at any angle (Bhoren and Huffman, 1983).

CIDS measurements, carried out on micro spheres of 0.091 m in diameter (Sigma Chemicals Co.) are shown in fig.2.

Micro spheres, in a concentration of $3.9\ 10^{11}$/ml, have been suspended in tris HCl 10 mM (pH 8) in order to avoid their aggregation.

After this check, we applied the instrument to the detection of the structural changes induced by different fixation techniques on rat liver nuclei. Fixing the sample with two different fixatives, glutaraldehyde and EtOH, we expected to alter in a detectable way the structure of the biopolymers inside each nucleus. Glutaraldehyde, providing cross-links between amino and hydroxil groups, builds up a rather rigid network which should "freeze" the structures at issue. On the other hand, ethanol is known to cause a shrinking effect on the samples. Therefore, treating the samples with these two different fixatives could give basic answers to the sensitivity of our sensor prototype in monitoring changes in the sample chirality. Data obtained on nuclei fixed with glutaraldheyde (0.5%) and ethanol-acetic acid (1:4) are reported in fig.2.

The nuclei have been analyzed at a very low concentration (10^3/ml) in order to have only few nuclei (ideally one) in each drop. We measured the number of nuclei contained in each drop by a phase contrast microscope finding a number ranging between 2 and 5 nuclei per drop.

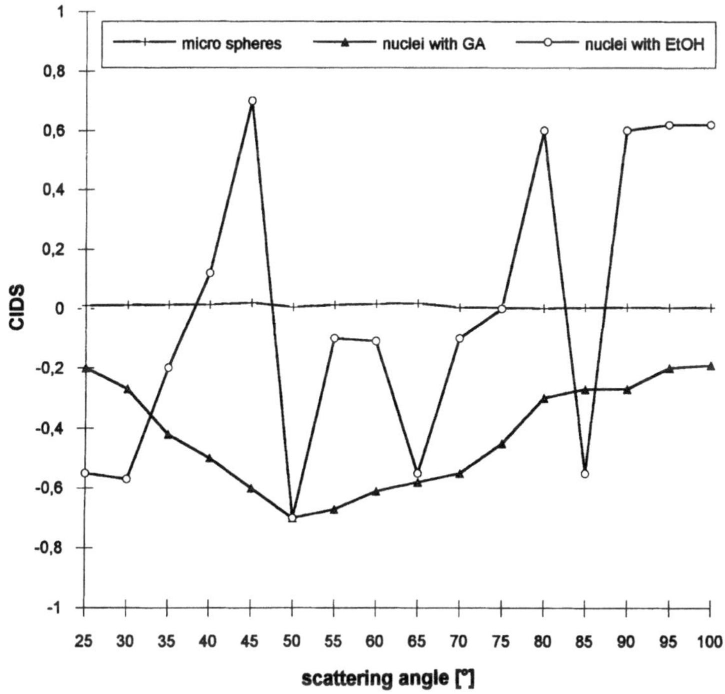

Figure 2 Behaviour of the CIDS parameter as a function of the scattering angle for a solution of latex microspheres of 0.091 mm in diameter and rat liver nuclei fixed with GA and EtOH. Solid lines just facilitate to follow the trend of the data points.

We can see a marked difference in these two samples. Particularly, the nuclei fixed with glutaraldehyde show a behaviour of the CIDS signal which is smoother than the one displayed by the nuclei fixed with EtOH. This fact correlates well with the knowledge about the effect of these different fixatives on chromatin, as the shrinking of the structure can cause an enhancement in the chirality of the samples which results in a different trend of the CIDS signal.

These drastic differences were found to be reliable in several analyzed samples. Namely, even if the curve for different samples could slightly differ in the absolute value of the CIDS parameter, it was always easy to discriminate between nuclei treated with the two different fixatives.

The prototype here described represents the first step towards the realization of a sensor, perhaps operating in flow conditions, based on the differential scattering of polarized light from single cells or nuclei.

Chiral and unchiral structures in single nuclei have been discriminated even without the use of a nozzle vibrator for the generation of small drops, which would minimize the background signal due to the scattering by the drop itself.

Even if several improvements can be thought (e.g. two lasers emitting respectively circular left and right polarized radiation), we may conclude that the data here presented, both on standard and biological specimens, indicate the feasibility of a flow instrument based on the described electronics and optics.

Our prototype can be used both as an additional module for a classic flow cytometer and as an independent instrument.

Its utilization may range from basic to clinical or industrial research.

ACKNOWLEDGEMENTS

This work has been supported by the C.N.R. target project "Biotechnology and Bioinstumentation".

REFERENCES

Bickel, W.S., and Stafford M.E., 1981, *J. Biol. Phys.* 9:53.

Bohren, C.F. Huffman, D.R., 1983, "*Absorption and Scattering of Light by Small Particles*", Wiley, New York.

Diaspro, A., and Nicolini, C., 1987, *Cell Biophys.* 10:45.

Diaspro, A., Bertolotto, M., Vergani, L., and Nicolini, C., 1991, *IEEE Trans. Biomed. Eng.* BME-38(7):670.

Facci, P., 1990, Thesis dissertation, Institute of Biophysics, University of Genova.

Facci, P., Diaspro, A., and Nicolini, C., 1990, *Bas. Appl. Histochem.* 34(4):321.

Nicolini, C., Kendall, F., 1977, *Physiol. Chem. Phys.* 9:265.

Nicolini, C. (1986) "*Biophysics and Cancer*", Plenum Publ., New York.

Nicolini, C., Vergani, L., Diaspro, A., and Scelza, P., 1988, *Biochem. Biophys. Res. Comm.* 155:1396.

Nicolini, C., Diaspro, A., Bertolotto, M., Facci, P., and Vergani, L., 1991, *Biochem. Biophys. Res. Comm.* 177(3):1313.

Salzman G.C., Griffit, J.K., and Gregg C.T., 1982 , *Appl. Envirom. Microbiol.* 44:1081.

BIOINTERFACING FOR ELECTRON TRANSFER OF REDOX ENZYMES

Masuo Aizawa

Department of Bioengineering, Tokio Institute of Technology
Nagatsuta, Midori-ku, Yokohama 227, Japan

INTRODUCTION

Most bioelectronic and biophotonic devices are designed by mimicking the biological systems which undergo signal transduction and information processing [A. Aviram, 1989, 1991; F. T. Hong, 1989]. Although the ultimate goal is to realize a neurodevice based on the brain, some other novel functional devices have also been pursued by many researchers. Of these functional devices protein molecular devices have gained an increasing attention due to the characteristic molecular functions of proteins.

Enzymes, antibodies, and bacteriorhodopsin have successfully been incorporated in electronic and optical devices to form biosensors and biophotonic devices. The molecular recognition function of these proteins have, however, been coupled with the electronic function of inorganic electronic materials as metal and semiconductors in an indirect manner. Electronic proteins as redox enzymes can confine electrons at their active centers where the insulated layer of protein is surrounded, and find the difficulty in linking the electron transfer site of these proteins with the electronic materials. Even though a protein molecule works as a minimum element of electronic function, the current technologies have been far behind in utilizing the electronic function of a single molecule. That is primarily due to a lack of technology in fabricating "biointerface", which is a molecular level of interface to facilitate electron transfer between a protein molecule and a conventional electronic material. The electronic communication has currently been available between protein molecules and an electrode through the biointerface. This accomplishment has opened a door for us to fabricate not only a new type of biosensor but a novel biodevice for an intelligent bioreactor. This paper will focus on the concept and design of the biointerface for proteins, specifically redox enzymes, and their applications in fabricating biodevices.

BIOINTERFACING TECHNOLOGY FOR ELECTRONIC PROTEINS

The protein molecular electric devices are constructed in such a manner that an electronic protein molecule is electronically linked with an electrode to make an electric communication. A bionsesor comprising of a redox enzyme may be fabricated by forming an enzyme layer on the working electrode with a counter electrode in the vicinity an a reference electrode if necessary. The one surface of the enzyme layer faces a solution containing a determinant, and the other is attached to the electrode surface. The determinant is captured and recognized by the enzyme. The event is followed by electron transfer from/to the redox center of the enzyme. The electron transfer should be monitored through electronic communication with the electrode. A very limited number of redox enzymes is inhibited primarily due to steric hindrance against an access of the active centers of redox enzymes to the electrode surface.

From Neural Networks and Biomolecular Engineering to Bioelectronics
Edited by C. Nicolini, Plenum Press, New York, 1995

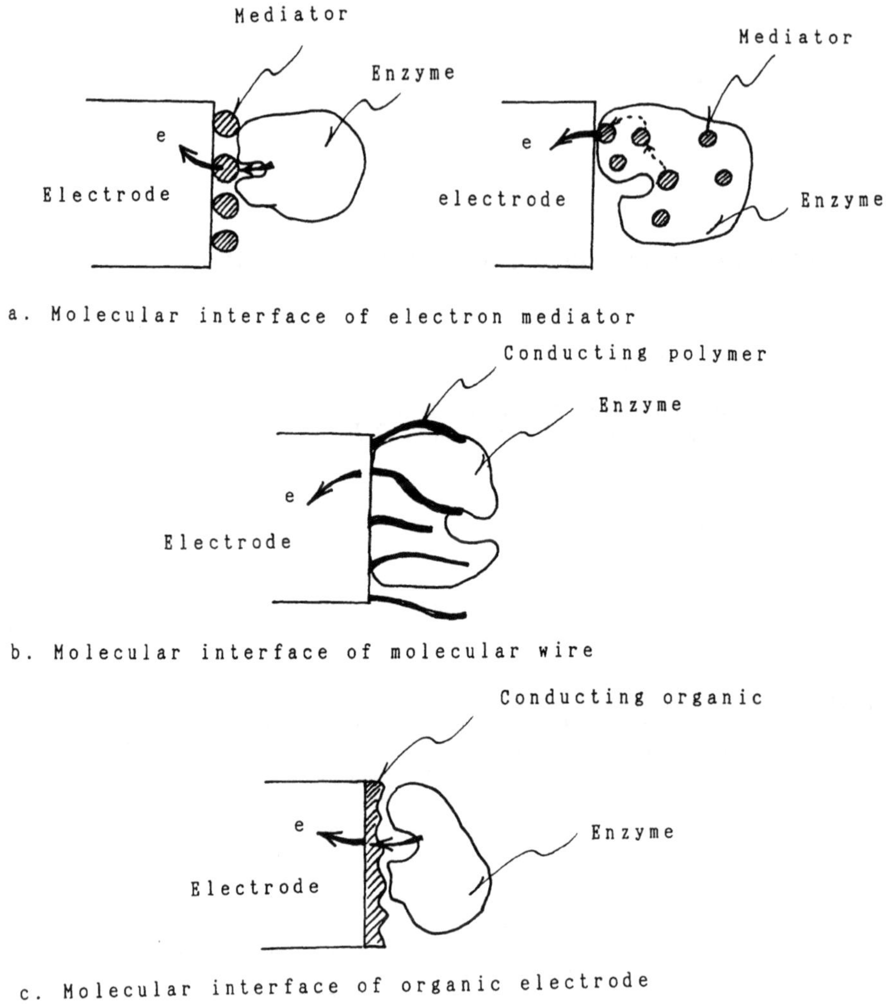

Figure 1 Molecular interfaces

Pioneering works of Hill et al. have opened the way to realize fast and efficient electron transfer between an enzyme molecule to the electrode surface [A. E. Cass et al., 1989]. They modified a gold electrode with 4,4'-bipyrydyl, an electron promoter, not a mediator since it does not take part in electron transfer in the potential region of interest, to accomplish rapid electron transfer of cytochrome c. Their work has triggered intensive investigation of electron transfer of enzymes using modified electrodes.

Apart from electron promoters, a large number of electron mediators have long been investigated to make redox enzymes electrochemically active on the electrode surface. In this line of research electron mediators such as ferrocene and its derivatives have successfully been incorporated into an enzyme sensor for glucose. The mediator was easily accessible to both glucose oxidase and an electrode to transfer an electron in an enzyme sensor. Heller et al. have chemically modified glucose oxidase with an electron mediator [Y. Degani and A. Heller, 1989]. They presumed that an electron tunneling pathway could be formed within the enzyme molecule, although the mediator-modified enzymes have not been immobilized on the electrode surface. The authors and Lowe used a conducting polymer as a molecular wire to connect an redox enzyme molecule to the electrode surface [S. Yabuki et al., 1990; N. C.

Foulds and C. R. Lowe, 1986]. These progresses in electron transfer of enzymes have led us to conclude that a molecular level of assembly should be designed to facilitate electron transfer may be termed specifically "Molecular Interface" defined as Biointerface.

There are several molecular interfaces for redox enzymes to promote electron transfer at the electrode surface:

1) Electron mediator: either the electrode or the enzyme is modified with an electron mediator in various manners.

2) Molecular wire: the redox center of an enzyme molecule is connected to an electrode with such a molecular wire as a conductng polymer chain.

3) Organic salt electrode and conducting polymer electrode; the surface of an organic electrode may provide enzymes with smooth electron transfer.

DIRECT UNMEDIATED ELECTRON TRANSFER OF REDOX ENZYMES ON THE ELECTRODE SURFACE

Until very recently, direct electron transfer of redox enzymes has not been statisfactorily reported. Guo et al. [1989] have achieved the direct unmediated electron transfer of p-cresolmethyl-hydroxylase (PCMH) at an edge-plane graphite (EPG) electrode. This enzyme is a tetramer of molecular weight 115,000 Da and catalyzes the specific oxygenation of p-cresol to p-hydroxybenzaldehyde. The larger of the two subunits contains a covalently bound FAD, which is the site of substrate oxygenation. The other subunits, c-type cytochromes, serve to relay electrons to an acceptor molecule which is thought to be azurin. They showed that the direct electrochemistry of PCMH at EPG electrodes could be observed both through the electrocatalytic oxidation of p-cresol with a low concentration of PCMH in solution or through the reversible response of enzyme in the absence of the substrate as presented in Fig. 2. Considerable selectivity was presented among various cations as electroinactive promoters of the electrochemistry. The aminoglycoside, neomycin, was shown to be the most potent reaget. They also showed that the catalytic response could be achieved with gold electrodes modified by the adsorption of peptides.

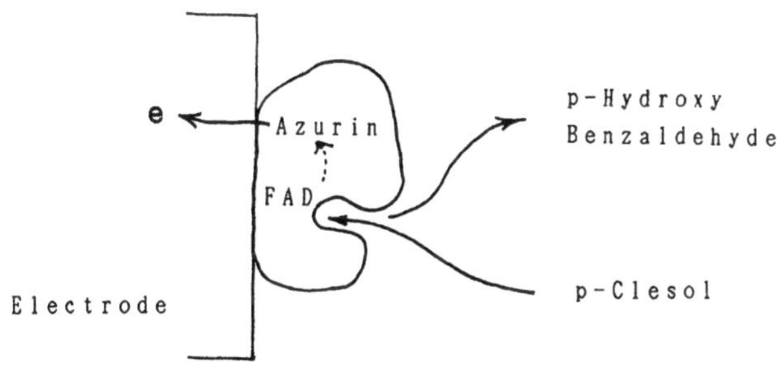

Figure 2 Direct electron transfer of p-cresolmethyl hydroxylase on the electrode surface

Burrows et al. [1991] reported that the direct unmodified electron transfer of methylamine dehydrogenase (MADH) was achieved at polished EPG or negatively charged modified gold electrodes without the need for promoters. The enzyme is a tetramer compound of two identical larger subunits and two identical smaller subunits which possess a covalently-bound form of a quinine cofactor. MADH catalyzes the oxydation of methylamine to formaldehyde and ammonia.

Although direct unmediated electron transfer of a number of redox enzymes has been achieved, there is still a long way to go to make it work for a wide range of enzymes of interest. An increasing interest has arisen in designing and modifying enzymes to be electrochemically active.

ELECTRON MEDIATORS FOR MOLECULAR INTERFACE

Mediator-Modified Electrodes

The most successful approach to construct more practical devices has been to use an electron transfer mediator used as a shuttle to carry charge between the electrode and the active site of an enzyme. Cass et al. [1984] found that ferrocene and its derivatives are very efficient electron acceptors for glucose oxidase in their oxidized form. Cyclic voltammetric studies showed good, quasi-reversible electrochemistry of ferrocene in the presence of glucose alone. On the addition of enzyme, however, a large catalytic current flows at oxidizing potentials. The ferrocene-mediated approach offered the change of developing a biosensor for glucose. It was 1987 that a disposable biosensor for glucose has been commercialized for self-monitoring by diabetic patients.

Electron mediators successfully used with oxidases include 2,6-dichlorophenolindophonol, hexacyanoferrate-(III), tetrathiafulvalene, tetracyano-p-quinodimethane, various quinones and ferrocene derivates. From Marcus' theory it is evident that for long-range electron transfer the reorganization energies of the redox compound have to be low. Additionally, the redox potential of the mediator should be about 0 to 100 mV vs. standard calomel electrode (SCE) for a flavoprotein (formal potential of glucose oxidase is about - 450 mV vs SCE) in order to attain rapid vectorial electron transfer from the active site of the enzyme to the oxidized form of the redox species. The formal potentials of ferrocene derivatives are listed in Fig. 3 [W. Schumann, H. L. Schmidt, 1992].

Ferrocene-mediated enzyme sensors have usually been obtained by adsoprtion of the mediator onto the electrode surface because of this insolubility in aqueous solution. Due to the good solubility of ferricinium cations in aqueous solution complications arise from leakage of the mediator from the electrode surface.

To overcome the poor stability of ferrocene-mediated enzyme sensors, mediator-modified electrodes have been used. In the case of glucose oxidase, the cofactor FAD is deeply buried within the protein matrix. The depth of the active center is estimated to be 0.87 nm. Therefore, one cannot expect that the mediator covalently attached to the electrode surface via a short spacer retain the possibility of closely approaching the cofactor of the enzyme.

Mizutani et al. [1988] have demonstrated that ferrocene derivatives, attached by means of covalent bonds to the surface of bovine serum albumin, have been able to mediate the electron transfer between the glucose oxidase and the electrode through the osemium complex.

Mediator-modified Enzymes

In contrast to the mediator-modified electrodes, Degani et al. modified glucose oxidase itself by means of covalently bound ferrocene [Y. Degani and A. Heller, 1987]. After modifying enzymes with ferrocene carboxylic acid, they observed direct electron transfer from the active site of the enzyme to a gold or platinum electrode at the potential given by the ferrocene redox couple. It has been claimed that the tunneling distance for electron transfer from the active site to the surface of the macromolecule could be drastically shortened after incorporation of about 12 ferrocene carboxylic acid moieties between the two subunits of the enzyme, and that this might be possible through an electron-hopping mechanism between adjacent mediator molecules.

Schuman et al. have synthesized ferrocene-modified glucose oxidase with the ferrocene derivatives bound via long and flexible chains directly to the outer surface of the enzyme [1991]. A peripherally attached redox mediator may accept electrons through either an intramolecular or through an intermolecular process.

Aizawa et al. [1991] have immobilized mediator-modified glucose oxidase within micropores of a gold black electrode by self-assembling via the thiol-gold interaction as schematically presented in Fig. 4.

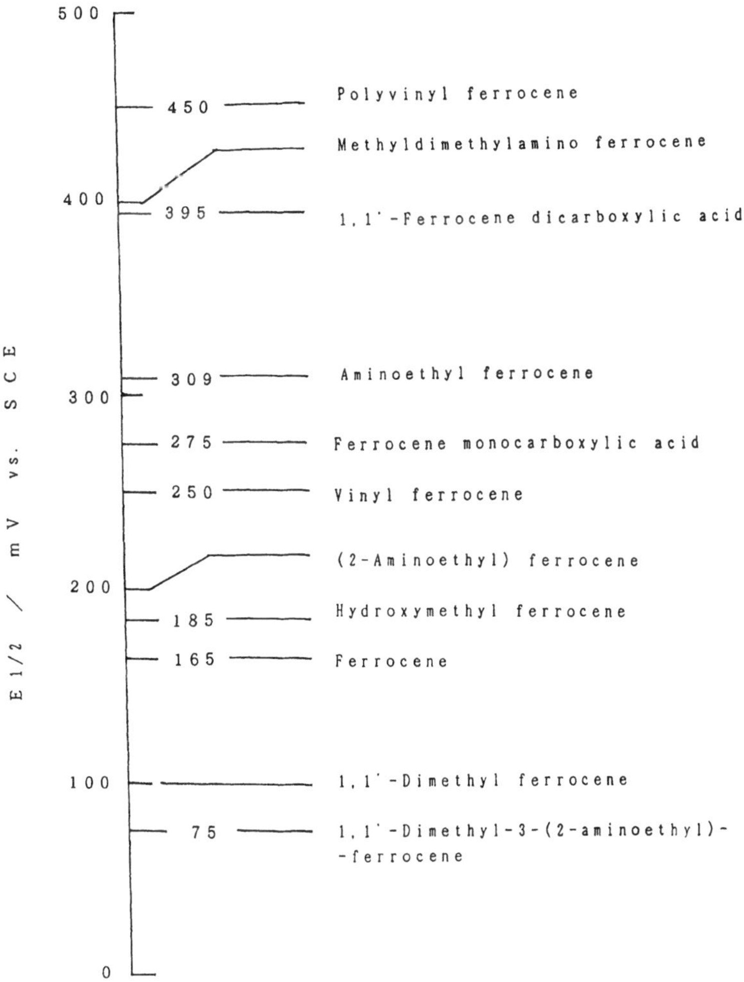

Figure 3 Formal potentials of ferrocene derivatives

CONDUCTING POLYMER MOLECULAR WIRE FOR MOLECULAR INTERFACE

Preparation of the Conducting Molecular Interface

Conducting polymers or molecular wires can be feasibly used as the molecular interface for redox enzymes on the electrode surface. Some of conducting polymers may be electrochemically synthesized in aqueous solutions and be deposited on the electrode surface, which allow us to regulate the membrane thickness, i.e. the length of the molecular wire, by adjusting the total electric charge passed during electrolytic polymerization. The immobilization of enzyme activity into or onto the polymer could be obtained, either by electrochemical polymerization of pyrrole in the presence of the enzyme [N. C. Foulds and

C. R. Lowe, 1986; M. Umana and J. Waller, 1986; P. N. Bartlett and R. G. Whitaker, 1987a, b, 1988; E. A. Hall, 1988; G. Fortier et al., 1988; P. C. Pandey, 1988; De Taxis du Poet et al., 1990] or by the adsorption of the enzyme to polypyrrole films grown from aqueous solution [J. M. Dicko et al., 1989; E. Tamiya et al., 1989]. Polypyrrole is one of the conducting polymers which are suitable for use as the molecular wire of interface for redox proteins, and works as a molecular wire through the polaron mechanism.

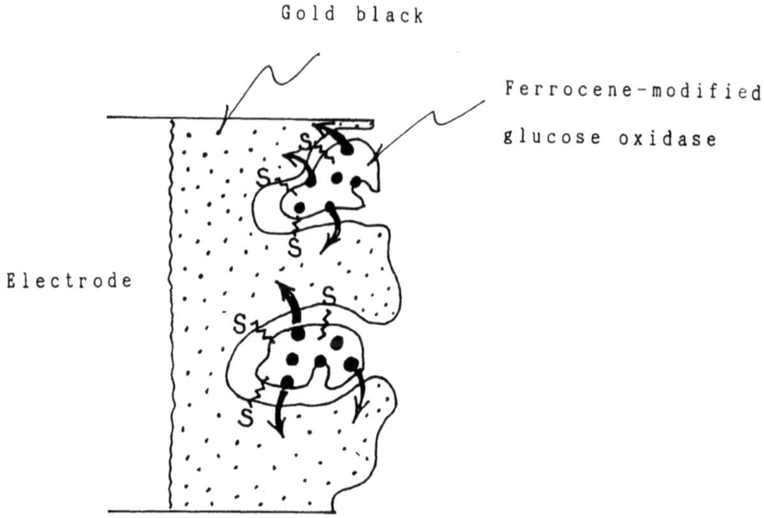

Figure 4 Ferrocene-modified glucose oxidase self-assembled within micropores of an gold black electrode

Glucose oxidase is one of the few redox enzymes that can be molecularly interfaced with an electrode through the molecular wire of polypyrrole. The molecular-interfaced glucose oxidase was prepared by the following process.

Glucose oxidase was first adsorbed onto the platinum surface at a controlled potential. The protein adsorption varied depending on several factors such as electrode potential, glucose oxidase concentration, pH and temperature. The effects of these factors on protein adsorption were carefully investigated.

Glucose oxidase gradually adsorbed onto the platinum electrode at a potential of 0.5V vs. Ag/AgCl and pH5.5, reaching a saturated state within 60 min. Adsorbed glucose oxidase was assayed for its enzyme activity and protein content. These results indicate that a monolayer of adsorbed glucose oxidase at a different surface coverage may be prepared under a controlled electrode potential. According to this characterization, a glucose oxidase monolayer was prepared at a surface coverage of 60-70% on the platinum electrode surface.

The adsorbed glucose oxidase was brought in contact with a solution containing pyrrole after rinsing. The electrode potential was immediately controlled at 0.7V vs. Ag/AgCl to initiate polymerization of pyrrole. Total charge passed during polymerization was controlled with a coulombstat. Polypyrrole deposited on the electrode surface supposedly in such a manner that the glucose oxidase intermolecular space could be filled. Polymerization was terminated at an estimated membrane thickness of 2 nm. On the assumption that glucose oxidase is a globular protein with a diameter of 3.5 nm, the monolayer of glucose/polypyrrole on the electrode is schematically illustrated in Fig. 5. Polypyrrole is very strongly adhered onto the electrode surface to prevent glucose oxidase from detaching.

Some other redox enzymes including fructose dehydrogenase (FDH) and alcohol dehydrogenase (ADH) have been molecularly interfaced with an electrode through polypyrrole conducting polymer according to processes similar to those described above.

The above process was found applicable to any size and shape of electrodes on which a monolayer of redox enzymes could be formed.

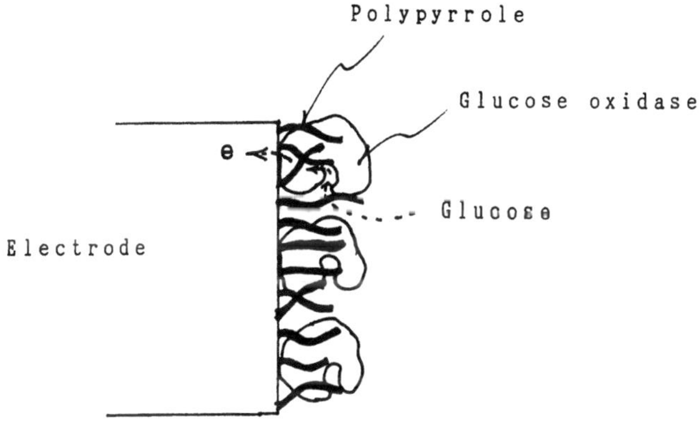

Figure 5 Postulated scheme of the molecular-interfaced glucose oxidase on the electrode surface

Electron Transfer of Glucose Oxidase through Polypyrrole

Electron transfer involving glucose oxidase (GOD) is difficult when GOD is adsorbed on the surface of a platinum electrode, although several papers suggest that electron transfer on the electrode surface is possible under uncertain conditions. The molecular interface of a conducting polymer presumably accesses the active site of GOD where flavinadenine dinucleotide (FAD) is bound as prosthetic group, while the other terminal end of the molecular interface of conducting polymer adheres to the electrode surface. Glucose diffuses into the active site of GOD to be oxidized to gluconolactone, which is followed by reduction of GOD-bound FAD. In case no oxygen is dissolved in the solution, the reduced form of GOD-bound FAD has no pathway to yield an electron. An electron or polaron may pass from the reduced form of GOD-bound FAD to the electrode through the molecular interface of conducting polymer, if an appropriate potential gradient could be created between the reduced form of GOD-bound FAD and the electrode.

Electron transfer of the molecularly interfaced glucose oxidase on the electrode surface was confirmed by differential pulse voltammetry and cyclic voltammetry. The glucose oxidase clearly exhibited both reductive and oxidative current peaks in the absence of dissolved oxygen in these voltammograms. These results indicate that electron transfer takes place from the electrode to the oxidized form of glucose oxidase and the reduced form is oxidized by electron transfer to the elctrode through polypyrrole. It may be concluded that polypyrrole works as a molecular wire between the adsorbed glucose oxidase and the platinum electrode.

ELECTRON TRANSFER OF POLYPYRROLE-INTERFACED FRUCTOSE DEHYDROGENASE

Fructose dehydorogenase (FDH) has a prosthetic group of pyrroloquinoline quinone (PQQ) at the active center where the substrate molecule of D-fructose transfers an electron to be oxidized. The authors showed that PQQ could undergo reversible electron transfer on a polypyrrole-coated electrode and in a polypyrrole membrane-bound form. Theşe indicate a possible electron transfer of the molecular-interfaced FDH on the the electrode surface [G. F. Khan, H. Shinohara et al., 1991; G. F. Khan, E. Kobatake et al., 1992].

Reversible electron transfer of the molecular-interfaced FDH was confirmed by differential pulse voltammmmetry. The differential pulse voltammograms of the polyoyrrole-interfaced FDH on the Pt electrode surface and FDH adsorbed on the Pt electrode surface are presented in Fig. 6. In both cases a pair of anodic and cathodic peaks were obtained which were attributed to the electrochemical oxidation and reduction of the quinoprotein at redox potentials of 0.08 and 0.07 V vs. Ag/AgCl, respectively. We reported elsewhere that the redox potential of PQQ in a solution of pH 4.5 is 0.06 V vs. Ag/AgCl with the polypyrrole-coated electrode. Therefore, the redox peaks of the voltammogram should be related directly to the electrochemical process of the prosthetic PQQ. Due to the incorporation of conductive polymer molecules in the vicinity of adsorbed FDH on the electrode surface, the prosthetic group PQQ electronically communicates with the electrode. These results clearly indicate that conductive polymer of polypyrrole works as a molecular wire of molecular interface.

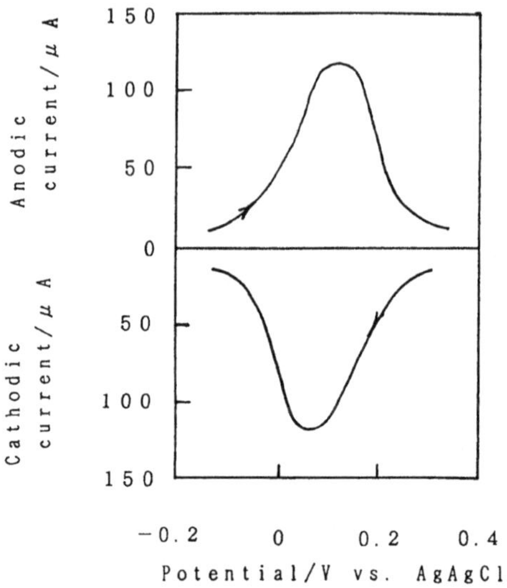

Figure 6 Differential pulse voltammogram of the molecular interfaced fructose dehydorogenase

ELECTRICALLY MODULATED ACTIVITY OF MOLECULAR-INTERFACED ENZYMES

Enzymes possess two characteristic functions: molecular recognition and selective biocatalytic functions. The molecular recognition function is essentilally important for constructing biomolecualr devices, especially molecular sensing devices, because their selectivity depends on the molecular recognition of enzymes. The biocatalytic function has potentially to create novel devices. In the molecular sensing devices, the molecular information to be determined is transduced in an amplified manner into the output signal. However, further investigation on implementation of the biocatalytic function is highly demanded or urgently needed.

Investigation on the molecular interfacing of redox enzymes yields the following important findings. The molecular-interfaced redox enzymes showed the potential dependency of enzyme activity. The enzyme was inactive when the electrode potential was set below a certain threshold. In contrast, the enzyme activity increased with an increase in the electrode potential above the threshold. The activity of the molecular-interfaced enzyme is reversibly modulated by changing the electrode potential.

The molecular-interfaced glucose oxidase (GOD) catalyzes the cycling reaction. The substrate molecule of glucose transfers an electron to the active site of GOD and is oxidized to gluconolactone. Since the prosthetic FAD is located at the active site of GOD, it is reduced to $FADH_2$. It is necessary to regenerate the oxidized form of the prosthetic FAD to enable continuous catalytic reaction. If the electrode potential is lower (which means less negative) than the redox potential of the prosthetic FAD, an electron should be transferred from the reduced form of FAD to the electrode through the molecular interface. The catalytic process could be cycled. Unless the energy balance is satisfied, the catalytic process could be inhibited.

The voltage dependency of the molecular-interfaced GOD activity is presented in Fig. 7.

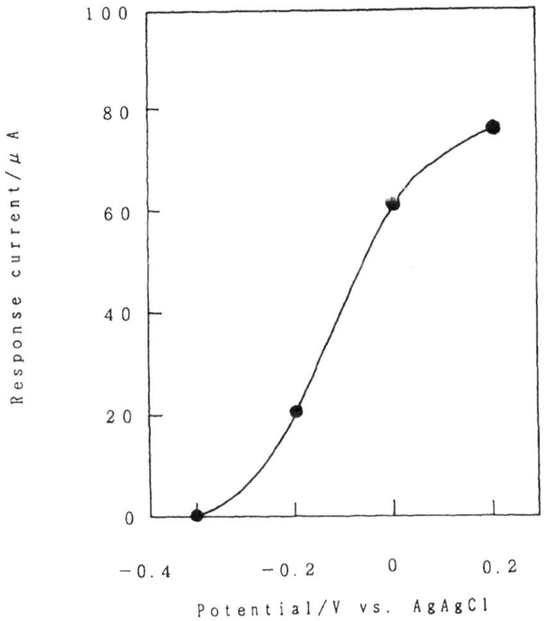

Figure 7 Potential dependency of the enzyme activity of the molecular interfaced glucose oxidase

The glucose oxidase/polypyrrole membrane electrode was immersed in a pH 5.5 citrate buffer solution containing potassium choride with a counter electrode and an Ag/AgCl reference electrode. The solution was bubbled with N_2 gas for deoxygenation prior to the experiment. The electrode potential was set to 0.0 V vs. Ag/AgCl, which might be sufficient to oxidize the reduced form of FAD to the oxidized form. After a steady anodic current was measured, glucose was added to the solution. The anodic currrent immediately increased with addition of glucose, which resulted from the oxidation of the reduced form of FAD due to the enzymatic oxidation of glucose in solution. These results show that the molecular function of glucose oxidase is transduced to the electron flow to the electrode with the aid of polypyrrole.

The anodic current increase is caused by the electrochemical oxidation of the enzymatically generated reduced form of glucose oxidase, which corresponds to the enzyme acitivity. As shown in Fig. 7, the enzyme activity of the molecular-interfaced glucose oxidase changed depending on the electrode potential. The enzyme activity instantly changed when the electrode potential was varied in the range from 0 V vs. Ag/AgCl. Such a change was completely reversible. These results lead us to conclude that the enzyme activity of molecular-interfaced glucose oxidase can be controlled by the electrode potential.

The potential dependency of the biocatalytic activity of the molecular-interface FDH is also shown in Fig. 8. The response current was measured in a similar manner as in the case of the molecular-interfaced glucose oxidase. Therefore the response current corresponds to the enzyme activity of the molecular-interfaced FDH.

The FDH remained inactive in the potential range below the redox potential of the prosthetic PQQ of the FDH. When the electrode potential exceeded the redox potential of the prosthetic PQQ, the enzyme activity increased sharply depending on the potential.

In Fig. 9, the energy correlation is schematically presented. The pontential-controlled modulation of the molecular-interface enzymes may be interpreted by Fig. 9. The enzyme and its substrate molecule have their intrinsic redox potentials. The redox potentials of oxidases and dehydrogenases are determined by an electron transferring molecule, i.e. a cofactor such as FAD, which is localized at the active site of the enzyme. Due to potential gradient, an electron can be transferred from the substrate molecule to the active site of the enzyme,if the substrate molecule is accepted by the molecular space of the enzyme active site. However, the electron transfer between the active site of the enzyme and the electrode is regulated by the electrode potential, even if the molecular wire could be completed. It should be reasonable that the enzyme activity is electrically modulated at a threshold of the redox potential of the enzyme.

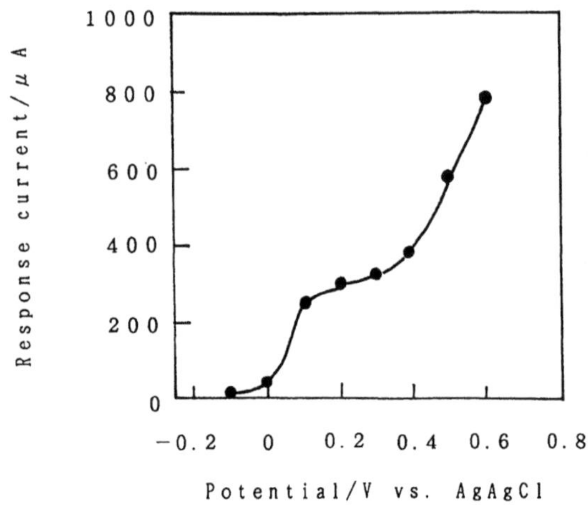

Figure 8 Potential dependency of the enzyme activity of the molecular interfaced fructose dehydorogenase

ELECTRON TRANSFER TYPE OF BIOSENSORS

A biosensor consists, in general, of two functional parts for molecular recognition and signal transduction. Either biocatalyst of bioaffinity substance is used as major material for molecular recognition to attain extremely high selectivity [A.E.G. Cass et al., 1984]. The signal transducing part involves typically an electrochemical, an optical, a thermal, and a piezoacoustic devices. To design a highly sensitive biosensor, it is important to link efficiently, in function, a biological substance for molecular recognition with a signal transducing device.

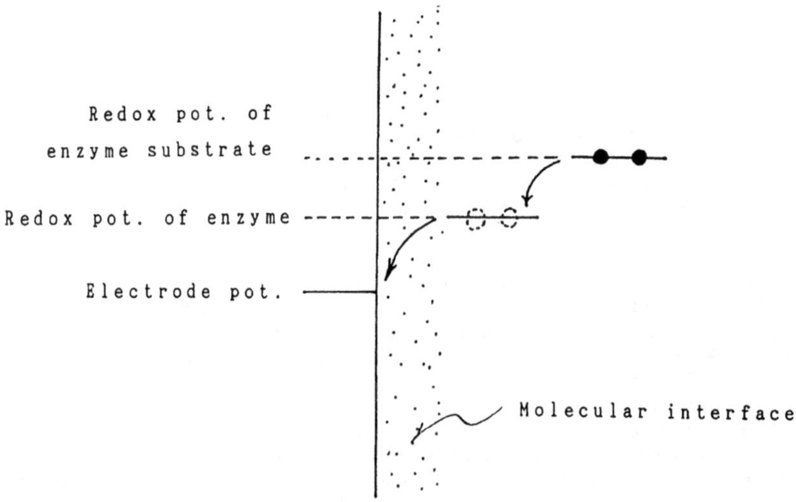

Figure 9 Postulated potential profile of the molecular interfaced enzyme on the electrode surface

Redox enzymes are recognized as major materials in constructing both biocatalytic sensors and bioaffinity sensors. Biocatalytic sensors for glucose, lactate, and alcohol utilize glucose oxidase, lactate dehydrogenase (lactate oxidase), and alcohol dehydrogenase (alcohol oxidase) and molecularly recognizable material. Since these redox enzymes are mostly associated with the generation of electrochemically active substances, many electrochemical enzyme sensors have been developed by linking redox enzymes for molecular recognition with electrochemical devices for signal transduction. These enzymes, however, have been linked in an indirect manner with electrochemical devices, resulting in a loss in sensitivity.

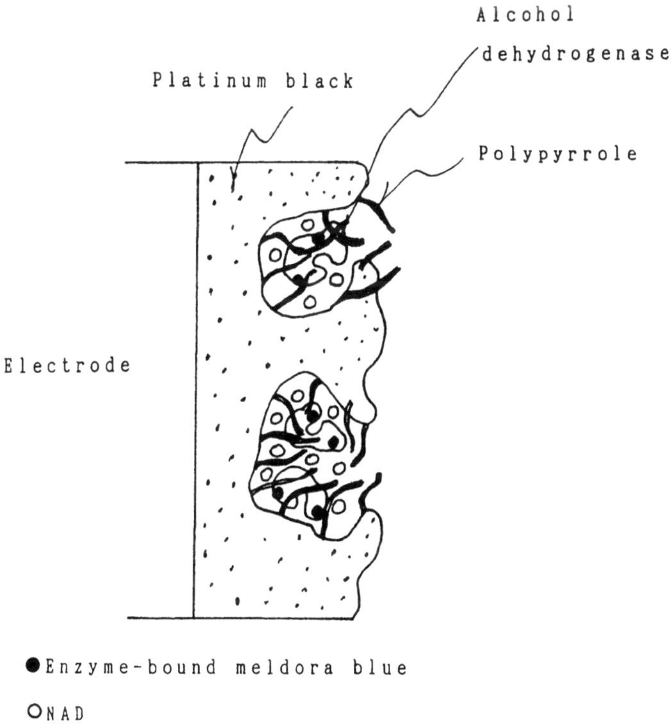

Figure 10 Electron transfer system constucted in a biosensor for methanol

The molecular-interfaced redox enzymes have offered a new design principle for realizing an electron transfer type of enzyme sensor. It is difficult, however, to incorporate dehydrogenases that are coupled with $NAD(P)^+$ into amperometric enzyme sensors owing to the irreversible electrochemical reaction of NAD^+. We have developed an amperometric dehydrogenase sensor for ethanol in which NAD^+ is electrochemically regenerated within a membrane matrix [T. Ishizuka et al., 1991].

The amperometric dehydorgenase sensor for ethanol consists of a platinum black electrode on the surface of which alcohol dehydrogenase (ADH), Meldola blue (MB) and NAD^+ are immobilized with a conductive polypyrrole membrane as schematically illustrated in Fig. 10.

A platinum disc electrode ($0.2cm^2$) was electrolytically platinized in a platinum chloride solution to increase the surface area and enhance the adsorption power. The platinized platinum electrode was then immersed in a solution containing 10 mg ml^{-1} ADH, 0.75 mM MB and 6.2 mM NAD^+. After sufficient adsorption of these molecules on the elctrode surface, the electrode was transferred into a solution containing 0.1 M pyrrole and 1 M KC1. Electrochemical polymerization of pyrrole was conducted at +0.7 V vs. Ag/AgCl.

The electrolysis was stopped at a total charge of 1 C cm^{-2}. An enzyme-entrapped polypyrrole membrane was deposited on the electrode surface.

Cyclic voltammetry was performed with the ADH-NAD-MB/polypyrrole electrode in 0.1 M phosphate buffer (pH8.5) at a scan rate of 5 mVs^{-1} as presented in Fig. 11. The substrate od ADH caused the anodic current to increase. These results suggest a possible electron transfer from membrane-bound ADH to the electrode through membrane-bound NAD^+ and MB with the help of a conductive polymer of pyrrole.

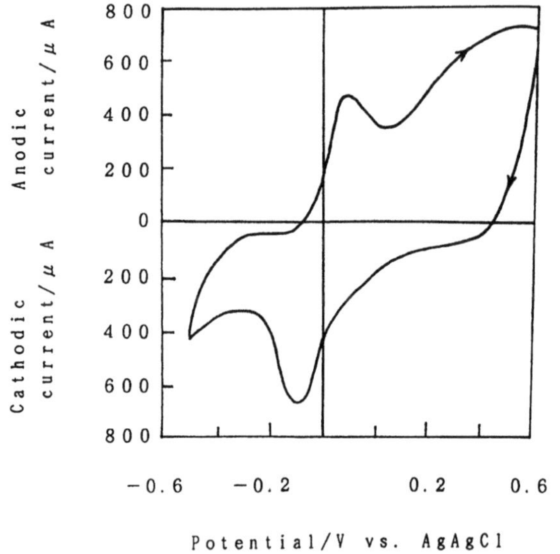

Figure 11 Cyclic voltammogram of the ADH-NAD-MB-polypyrrole electrode

The electrode gave a steady-state current when the electrode potential was maintained at 0.35V vs AG:AgCl in 0.1M phosphate buffer. Addition of ethanol to the buffer solution resulted in an increase in the anodic current, which was attributed to the oxidation of membrane-bound NADH. A steady response was obtained within 40 s. The increase in the anodic current was linearly correlated with the concentration of ethanol. An electron transfer type of enzyme sensor was thus fabricated by an electrochemical process. Although no appreciable leakage of ADH and MB from the membrane matrix was detected, NAD^+ leaked slightly. To prevent this leakage, the ADH-MB-NAD/polypyrrole electrode was coated with Naflon. Ethanol was selectively and sensitively determined in an aqueous solution.

In a further development, and ADH-MB-NAD/polypyrrole electrode, a platinum counter electrode and an Ag/AgCl reference electrode were assembled and covered with a gas-permeable polymer membrane to form a gaseous ethanol sensor. This appears to be the first time that a complete enzyme sensor for gaseous ethanol has been fabricated in such a manner with NAD^+ incorporated in immobilized form.

CONCLUDING REMARKS

Several protein molecular biodevices have been realized by the molecular interfacing technology that can provide proteins with electronic communication. The molecular interfacing technology, however, should be extensively improved to be precisely designed in a single molecular level. The molecular-scale of protein molecular biodevices may be accomplished by the innovative molecular interfacing technology. It is also important to pursue a possibility to utilize the self-organization of biomolecules for information processing, which should make it possible to create a molecular network-based biodevice without any wiring among molecules.

REFERENCES

Aizawa M., 1991, *Anal. Chim. Acta*, 250:249

Aviram A., ed. "*Molecular Electronics - Science and Technology*", 1989, Eng. Found., New York

Aviram A., ed. "*Molecular Electronics - Science and Technology*", 1991, Am. Inst. Phys., New York

Bartlett P.N. and Whitaker R.G., 1987a, *J. Electroanal. Chem.*, 224:27-35

Bartlett P.N. and Whitaker R.G., 1987b, *J. Electroanal. Chem.*, 224:37-48

Bartlett P.N. and Whitaker R.G., 1988, *Biosensors*, 3:359-379

Burrows A.L., Hill H.A.O., Leese T.A., McIntire W.S., Nakayama H. and Sanghera G.S., 1991, *Europ. J. Biochem.*, 199:73-78

Cass A.E.G., Francis D.F., Hill H.A.O., Aston W.J., Higgins I.J., Plotkin E.V., Scott L.D.L., and Turner A.P.F., 1984, *Anal. Chem.*, 56:667-671

Degani Y. and Heller A., 1987, *J. Phys. Chem.*, 71:1285-1289

De Taxis du Poet, Miyamoto S., Murakami T., Kimura J., and Karube I., 1990, *Anal. Chim. Acta*, 23:255-263

Dicko J.M., Hattori S., Karube I., Turner A.P.F. and Yokozawa T., 1989, , *Ann. Biol. Clin.*, 47:607-619

Fortier G., Brassard E. and Belanger D., 1988, *Biotechnol. Tech.*, 2:177-182

Foulds N.C. and Lowe C.R., 1986, *J. Chem. Soc., Faraday Trans. I*, 82:1259-1264

Gregg B.A. and Hiller A., 1990, *Anal. Chem.*, 62:258-263

Guo L.H., Hill H.A.O., Lawrence G.A. and Sanghera G.S., 1989, *J. Electroanal. Chem.*, 266:379-396

Hall E.A., 1988, *Int. J. Biochem.*, 20:357-362

Hong, F.T., ed. "*Molecular Electronics; Biosensors and Biocomputers*", 1989, Plenum Press, New York

Ishizuka T., Kobatake E., Ikariyama Y., and Aizawa M., 1991, *Tech. Digest 10th Sensor Symp.*, 73

Khan G.F., Shinohara H., Ikariyama Y., and Aizawa M., 1991, *J. Electroanal. Chem.*, 315:263

Khan G.F., Kobatake E., Shinohara H., Ikariyama Y., and Aizawa M., 1992, *Anal. Chem.*, 64:1254

Mizutani F. and Asai M., 1988, *Denki Kagaku*, 56:1100-1101

Pandey P.C., 1988, *J. Chem. Soc. Faraday Trans. I*, 84:2259-2265

Schumann W., Ohara T.J., Schmist H.L., and Heller A., 1991, *J. Am. Chem. Soc.*, 113:1394-1397

Schumann W., Schmidt H.L., 1992, in "*Advanced in Biosensors*", Vol. 2 (A. P. F. Turner, ed.) JAI Press, London, p.82

Tamiya E., Karube I., Hattori S., and Suzuki M., 1989, *Sensors and Actuators*, 18:297-307

Umana M. and Waller J., 1986, *Anal Chem.*, 58:2679-2983

Yabuki S., Shinohara H., Ikariyama Y. and Aizawa M., 1990, *J. Electroanal. Chem.*, 277:199

processing, which should make it possible to create a molecular-scale-based device without any wiring among molecules.

REFERENCES

BIOLOGICAL INFORMATION PROCESSING AND MOLECULAR NANOCOMPUTING

J. J. Hopfield

Californian Institute of Technology, Pasadena, USA

ABSTRACT

The scale on which computing devices are made with current technologies will grow ever smaller. At some small scale of devices it will become more effective to make nano-devices through biochemistry than to continue to manipulate the most microscopic scales with macroscopic apparatus. Microelectronics, premade macromolecular nano-components, and biological ideas of self-assembly can merge in future. Biological examples as the genetic code, the brain and the photosynthesis show us the possibility of computing at the molecular level. Their examinations give us insight into how to engineer computers at the molecular level. Fundamental issues involved in biological computing are physical scales, power (supplied energy), clock (timing structure), error (error rate and the restoration process). This talk provides a review of several biological examples and some ideas, problems, and prospects for the development of a molecular devices technology and a nano-materials approach.

INTRODUCTION

When doing engineering, it is always useful to have what the mathematicians would call an "existence theorem" at hand. Mathematicians use this term in reference to a proof that some function which they want to construct actually does exist, and is not impossible. In this sense, the observation of birds flying provides an "existence theorem" that an engineer should be able to design a flying machine. A study of bird flight then provides useful input about the principles on which the engineering could be based. When we ask about the possibility of computing at the molecular level, biology again provides us, by example, with "existence theorems". I will begin with a brief examination of some of these biological examples for the insight which they give us into how to engineer computers at the molecular level. There are two different physical scales on which biology computes. One involves truly molecular computation at the nuclear and cellular level. The other involves using aggregates of cleverly designed molecules as the fundamental computing elements in a larger and more flexible system, the nervous system and the brain. In each case, there are both some general lessons and some particular cautions and problems which the examination of biology brings to our engineering problem.

BIOLOGICAL COMPUTING AT THE CELLULAR LEVEL

The fundamental information about the structure of biological molecules and the structure of organisms is carried in DNA. Each cell has a copy of such information, which is

From Neural Networks and Biomolecular Engineering to Bioelectronics
Edited by C. Nicolini, Plenum Press, New York, 1995

written as a string of symbols G, C, A, T taken from a four letter alphabet. The physical representation of this information is a polymer of DNA, in which the successive polymer units are chosen as either guanine, cytosine, adenine, or thymine. The physical molecule of DNA is in simple isomorphism to the information contained. The DNA of a bacterium thus represents a string of about 5 symbols in the base 4. A simple cell has only a single copy of this essential information. Cellular reproduction involves among other things copying the DNA, and thus copying this bit string. The molecules responsible for the "copy" operation are called DNA polymerase. An individual molecule of this polymerase (in conjunction with appropriate supplies of precursor molecules of the four monomer types) is capable of generating a copy of the information in the DNA polymer, a complementary molecule. (In the case of DNA, the copy generated would be termed the "1's complement" in computer science). A single DNA polymerase molecule can copy at a rate of about 1000 symbols/second. In the cell, copying of several parts of the strand takes place in parallel in order to obtain a higher rate. This is particularly important for higher animals, when the total length of DNA can excede 10^9 symbols.

Protein synthesis is a string copy with symbol conversion operation. A typical protein is, in its essential chemical structure, a one-dimensional polymer with units taken from a 20-symbol alphabet. Protein synthesis symbolically consists of translating a string of information from the 4-symbol DNA into a string of symbols for the 20-symbol proteins. In the case of proteins, the different symbols are physically represented by the 20 different amino acids used in protein synthesis. The major apparatus to do this translation operation is a complex of about 50 molecules, and has a total molecular weight of about 500,000. When thinking about nanocomputation at the molecular level, three of the major issues are:
POWER How is the necessary energy supplied to the computing system?
CLOCK What is the general timing structure?
ERRORS What is the fundamental error rate and the restoration process for dealing with errors?

POWER Within molecular biology, the power supply is chemical. By a process both chemically and spatially separated from the copy and translation mechanisms, the chemical reaction

P + ADP —> ATP

is kept out of equilibrium by about a factor of 10^8, corresponding to a driving energy of 0.5 eV/molecule. An ATP splitting reaction is coupled to steps of the copy or translation processes when they require energy. Thus the ATP system is equivalent to a ubiquitous power supply, available everywhere without the need of wires, and used through coupling in such processes as

polymer of n units + next unit + ATP —> polymer of n+1 units + ADP

CLOCK Protein synthesis and DNA replication do not truly have clocks. These processes are enabled and proceed in a self-timed fashion until completed or interrupted by another control molecule. As a result, the fundamental operations can procede in a self-timed fashion. The addition of each further monomer unit does not have to be synchronized with anything else in the case of protein synthesis (for example) since nothing is going on in parallel. The precise time that a molecule is finished does not have to be synchronized with anything else. Protein synthesis is fundamentally a self-timed computation.

ERROR Both protein and DNA synthesis depend on the accurate choice of monomer units to add to a growing polymer strand. In the case of protein synthesis, the average rate of making errors is probably no worse than 1 in 10^4 units added. Typical RNA synthesis is no more accurate than this. DNA synthesis is typically carried out at a level of accuracy of about 1 in 10^6 - 10^7. In the case of protein synthesis, it is known that the particular errors (symbol substitutions) which are most difficult to prevent (from a chemical standpoint) introduce errors about 1% of the time, and that a restoration process of error checking or proofreading is used (with a similar accuracy) to reduce the total error rate for these worst errors to about 10^{-4}. Both in the case of DNA synthesis and of RNA synthesis, the restoration process used is simple redundancy. The probability that a typical protein molecule is correctly made is only about 0.95. This would be a disaster if the information system relied on typical molecules. But since the usual result of errors in protein synthesis is merely non-functional molecules, and since the number of copies of typical proteins is at a minimum of tens or hundreds per cell (and usually much larger), this gives excellent accuracy. One mode of operation of many proteins is simply through their concentration. If 1000 copies of a protein are made, and 5 % of them at random are defective, the uncertainty in the concentration would only be 7/1000. When we examine whether this will have an appreciable effect on biological accuracy, the intuitive and correct answer is "no". In short, compared to digital machines, the rate of making errors in protein synthesis is quite high, but the restoration process of using many molecules to represent the same piece of information is quite adequate.

NERVOUS SYSTEM BIOLOGICAL COMPUTATION

A nervous system or brain uses molecular devices in a totally different fashion. This system dominantly works through exchanging messages and non-linear computation using the rate of production of pulses of activity by a nerve cell as the signal or information being sent to another cell. There are about 10^{11} nerve cells in the human brain, and each is typically connected to about 1000 others. The volume per cell is about 10^{-8} cell (including all wiring and connections), or about 10^{-11} cc/connection (including wiring). The computational nature of the network is rather like the a large and highly non-linear analog computer, and the understanding of these systems is rapidly advancing through the study of the theory of models or "artificial neural network". In the nervous system, the most fundamental active unit is perhaps the synapse, the cell-to-cell connection in which all memories and algorithms are written. The molecular element essential to the synapse is a chemically controlled conductivity gate through a membrane. (Similar channels, but gated by voltage, are responsible for the generation of the pulses necessary for the expression and transmission of information within a cell). Each of these channels is a complex of a few protein molecules, with a total molecular weight on the scale of 100,000. The binding of a particular neurotransmitter molecule, such as glutamate, controls a conductivity path for a specific kind of ion. Each conductivity path, caused by a single such molecular complex, has an ionic conductance on the scale of 10 picoseimens when open, and many orders of magnitude smaller when closed. The channels are located in a membrane which has an effective chemical potential difference of about 0.1 volts across it, so a binding event of a single neurotransmitter molecule at this channel complex controls a readily measurable current. The molecules in the synapse are used in a way which is conceptually simpler than the information processing in molecular biology. The network of neurons is itself the information processing system, rather like a network of connected amplifiers, and the information being controlled is much more macroscopic. While in molecular biology, the control, reading, and manipulation of a single DNA molecule is the essential information act, the nervous system is intrinsically more macroscopic (and probably more collective) in its operation. And the algorithms used by the nervous system seem to be such that the precise function of an individual synapse is not essential. The level of accuracy of individual synaptic function during a particular step of an elementary computation cannot be more than about 30%. So a synapse need not be made with exactly the correct number of channels in it. In addition, "learning mechanisms" are available which change the channel numbers or conductivity in an adaptive fashion, reducing the demands on fabrication accuracy.

Of the three basic questions for molecular computing in general, the POWER question is provided by having the molecular channels, which fundamentally are a controlled resistance, connected across a potential difference which is supplied elsewhere. The ACCURACY question is taken care of by the use of many identical channels at a synapse, brute force parallelism. In addition, the system algorithms used, although analog in nature, are explicitly insensitive to small variations in the connection parameters. The CLOCK is taken care of at the systems level (although, indeed, there is no true clock). Neurobiology thus uses molecular devices in a rather more collective fashion than does neurobiology. The most fundamental device, the synapse, is part of a much more macroscopic system of devices and communication which provides the connectivity between the fundamental molecular devices, which are (conceptually) simple two terminal devices. (A regional "enable learning" signal based on a diffuse chemical effector may also be present.)

One very important lesson from the existence theorem of neurobiology is that for some computations architecture is all-important. Neurobiology has no true clock, but the effective time constants of the system mean that its equivalent clock rate cannot be faster than about 100 Hz. Nevertheless, a pigeon, a bird with about 10^9 brain cells and 10^{12} connections, can visually recognize a particular known person in about a second. On the pigeon architecture, this computation takes no more than 10^{14} device-cycles. By contrast, this problem would be extremely difficult to program on a typical modern workstation, although they compute at a rate of about 4×10^{16} device-cycles/second.

ELECTRONIC DEVICES

Electronic motions allow much more rapid changes and faster computational possibilities. Neither molecular nor neurobiology make use of electronic devices, but biology does so in the photosynthetic apparatus. The early stages of this process are carried out by rapid electron transfers, and a molecular assembly that functions like a pico-solar cell operates in this system at a speed of about 10^{-6} seconds. This photosynthetic reaction center is my last existence theorem for useful molecular devices for nanocomputing.

DEVICES AT THE MOLECULAR LEVEL

I do not believe that anyone has yet described a molecular computational device which is truly likely to function, capable of being fabricated, and is even remotely competitive with ways of performing the same task with existing semiconductor technology. And one must also remember that if we are to speak about a technological competition, semiconductor technology will advance greatly during the time it takes to develop a molecular technology. But in order not to offend anyone else by attacking their pet ideas, I will procede by reviewing one particular molecular design which I have been involved with, as a means of illustrating ideas, problems, and prospects.

The molecularly based shift register of HBO was based on the same kinds of charge-transfer reactions which are present in the photosynthetic apparatus. In the bacterial photosynthetic reaction center, the absorption of light results in the transfer of an electron from the outside surface of a lipid bilayer to the inside surface about 40 Å distant. The biomolecular system has the energy levels and spatial positions sketched in Fig. 1. When site A is photoexcited, it results in the rapid transfer of an electron from A to B to C. The hole on chromophore A is refilled by the oxidation of site D. As a result, an electron has been effectively moved from site D to site A, across the membrane. The biological system does this with a quantum efficiency greater than 97%. The molecular shift register is made as a parallel set of defined-length polymers, of total length perhaps 1000 units.

DABCDABCDABCDABCDABCDABCDABCDABCDABC

The following five points are the dominating ideas of this design.

1) INFORMATION is represented by the state of oxidation of the D elements of the polymer.
2) ENERGY The power for operation is supplied by pulses of light.
3) CLOCK The clock is also supplied by the pulses of light.
4) ALGORITHM The shift-register algorithm is supplied by the molecular architecture. If an electron is present on site D(n) before a pulse of light, that electron is moved to site D(n+1) after the pulse.
5) COMMUNICATION with the outside world of macro-electronics is provided by tethering the initial ends of polymers at a sending electrode, whose electrical potential during a flash of light can is set low if no electron is to be initiated along a chain for that pulse (writing a 0) or is set high if an electron is to be sent during that pulse (writing a 1).

The other end of each polymer molecule is attached to a gate. The arrival of many electrons (from many parallel polymer molecules) at the gate after a particular light pulse signifies the arrival of a "1". If no electrons arrive, the bit was a "0". This design solves most of the conceptual problems of making a molecular device with a useful function and compatible with VLSI technology. Its fabrication is a problem of synthesis, engineering and technology, not one of principle. It would potentially have a bit storage capacity of perhaps

100 times that of conventional lithography at the 1 micron level, due to the fact that use can be made of the large number of monomer units in a polymer 1 micron long. The real reason for describing this device in detail is to point out its weaknesses and drawbacks, as an indication of the directions in which molecular designs must go if they are to be of actual use, rather than being only imaginative hypothetical systems. The following are major issues:

DESIGN. The shift register algorithm is too special a design. Massive memory might easily be a role for molecular technology, but it should be a memory by address, whether ROM or rewritable.

POWER. Like all shift registers, this one consumes a lot of power because all bits must be shifted each cycle. If 1000 polymer strands in parallel are used to represent the data, it would take about 10000 eV to transfer one bit. Unfortunately, to assure the transfer of one bit, about 4 photons must be absorbed. In addition, because the absorbtion cross sections of molecules are not as large as would be liked, much more energy must flow through the system than would be desired. Power issues would probably limit the clock rate of such devices to a maximum of 10^3 to 10^6 Hz.

RESTORATION. There is no restoration at the molecular level. The occasional failure of electrons to transfer must be taken care of at the VLSI level. Many parallel strands must be used to assure the information is correct. Worse, the maximum strand length is strictly limited, since the strand length determines the number of clock cycles before restoration takes place. The system must therefore go from the molecular to the VLSI level too often.

COMMUNICATION with the VLSI world. This requires that more polymer strands be used than otherwise necessary, for the minimum charge arriving at a gate which can be conveniently sensed is on the scale of several hundred electrons if it must turn the transistor from off to strongly "on", and the transistor is made on the 1 micron scale. Also, because of the restoration problem, the amount of VLSI hardware is larger than one would like compared to the molecular hardware

FABRICATION. As originally described, the fundamental electronic polymer ABCDABCDABCD... was to be of defined length, and semi-rigid. The addition of such polymers to VLSI pads by electrochemistry would require polymer solubility at high concentrations. This is a very difficult problem for conventional organic chemistry polymers. The following are brief descriptions of how to improve some of these situations.

FABRICATION. It has been pointed out by Heller and others that the highly developed state of DNA chemical technology should be a prime candidate for a chemical means of implementing nano-molecular technology. For example, applied to the shift register, it would conceptually solve some of the major fabrication problems. Defined-length DNA polymers having lengths of 1000 monomer units are easily made. Double-stranded DNA is appropriately semi-rigid. DNA is soluble in water in high concentrations, and strands tend to stay away from each other and not precipitate. "DNA" can be synthesized with base analogs which have edges containing activatable groups, or even appreciable side groups. Thus a polymer of the form ABCDABCDABCD..... can be easily made with present DNA technology. The desired side chains can either be constructed directly on the bases or linked afterward on specific reactive groups. The research program necessary would be to identify the appropriate side groups for the photo-excitation and electron transfer characteristics. amplification of bits. The use of DNA technology would greatly simplify the problem of making an appropriate polymer for the shift register, and is an example of the potential power of this technology for molecular electronics.

RESTORATION. Consider three parallel polymer strands which are close together. All have an electron on D(0), and a flash of light arrives which is to result in each electron going to D(1) on its particular chain. Suppose that one electron fails to be photo-excited. If the chains are independent, this electron will not become shifted. However, if there is interchain cooperation, then the fact that two of the electrons become shifted could move the third along. In order for this to happen, an effective attractive interaction between three electrons at site D(0) or D(1) is needed.

Cooperative oxidations are common in chemistry at the two electron level. Cooperation in the oxidation of three separate sites is seen in the hemoglobin molecule. Cooperation of a few electrons is involved in photographic film, and is responsible for the fact that a very long, very low level light exposure has much less effect than a short, high intensity exposure involving the same number of photons.

Thus there are mechanisms for promoting cooperativity at the molecular or nanoscale. To harness one of these, permitting restoration by 3, 5, or 7 chains, would immensely improve the shift register. Such an advance would be of fundamental significance to the general problem of restoration and error correction in molecular computational devices.

ARCHITECTURE. To go to more useful architectures, means must be developed to have bits, which might be represented by single electrons, interact as logical operations. (Restoration is also an example of a particular such operation.) Cellular automata can be examples of architectures in which the architecture itself leads to restoration or error correction, as described by Huberman and others. The layout of a two-dimensional periodic structure of cellular automaton is a natural way to use molecular level electronics in a computing structure which is complex enough to be significant and feasible to build. However, cellular automata require logical operations.

LOGICAL OPERATIONS. From a conceptual viewpoint, it would be simple in principle to allow one electron to "gate" the possibility of electron transfer between two sites. Such gating can either be direct, through a an electrostatic influence, or indirect, by means of producing a local change of molecular configuration which alters the degree of contact between the two electron transfer sites. A more macroscopic effect of this can be readily observed in CMOS structures, where the presence of a single electron at a trap in the oxide can generate a measurable current in the underlying channel.

At present, the absence of a single successful demonstration of this control in the context of molecular electronic devices is a major stumbling block to molecular electronics based on electron transfer. A stumbling block is the strong desirability for a "duplication" operation in information processing. The most elementary use of such an operation would be to convert, through a cascade of duplication operations, a single electron on a site into 1000 electrons to arrive simultaneously at an electrode. Such a duplication operation demands either that one give up the use of an electron to carry a bit of information, or that an additional source of electrons be inserted into the system in the form of oxidizing or reducing agents. This latter is fraught with peril. The most direct way of dealing with the former, without giving up the idea of using electrons as the fundamental bit, is to work with a system which always uses electrons in electron-hole pairs, with the two components generally well separated.

A NANOMATERIALS APPROACH

The largest single source of difficulty with the approach to molecularly based computation described above is the necessity to solve two problems at once; the making of elementary computational devices, and the problem of organizing them into appropriate computational structures. I will now try to simplify the problem by considering the possibility of making computationally useful molecular structures whose organization into computing structures is either carried out by using these materials in conjunction with conventional lithography or as optical elements in in optics-based computing. In many ways, this is not a new subject. Indeed, photographic film and zerographic drums are in fact sheets of small microelectronic devices with some clear computing aspects. In the case of film, for example, the requirement to have a good shelf life before exposure, and that development need not be done immediately after exposure necessitates a kind of "restoration process" at the molecular or microcrystalline level. A grain of silver halide must not become developable if reducing electrons are transferred to it from photoexcited sensitizer molecules at too slow a rate, while the same number of electrons should result in a developable grain if they arrive within a fraction of a second of each other. At low levels, light should be a switch, and an intensity below a certain threshold intensity should be regarded as equivalent to "no light at all". While film does not have such a strong computational switch action as this, nonetheless the engineering of microstructures and molecular components to result in an appropriate threshold characteristic is an important aspect of film electronics. And of course, the sophistication in the layer structure of modern color film and its processing is extraordinary.

One problem which can be attack in such a fashion is the development of an appropriate "synapse material" for artificial neural networks. We have already noted that a synapse in neurobiology really consists of a multimolecular composite of many identical channels operating in parallel. In a semiconductor-based artificial neural network, the wires (the equivalent of axons and dendrites) could be made on conventional VLSI technology, while the appropriate synapses could be generated by a materials technology. The general idea of such a structure is represented by a structure where one array of parallel wires represents a set of axons, and the orthogonal parallel array represents a set of dendrites. A "synapse material" is placed as a layer between the two of them. This material needs to have properties such that:

1) It is a weakly conductive material based on electronic (not ionic) conductivity.
2) In the presence of a large electric field in one direction, its electrical conductivity increases with time, allowing a "conductivity write" operation.
3) In the presence of a large electric field in the other direction, or some other well-defined large-field situation, its electrical conductivity decreases, allowing a "conductivity erase" operation.
4) At small electric fields, it is merely a conducting material. Conductivity changes occur only for large fields.
5) A global enabling electrode (or light) might be included.

These are not impossible characteristics. Indeed, work by Potember and collaborators on Cu TCNQ shows many of the desired effects, particularly a reversible conductance change with a strong hysteresis. Calchogenide glasses also have many of the relevant characteristics. The tungsten bronzes in work by the group at Jet Propulsion Laboratory have also shown some possibilities for applications in such directions. Unfortunately, there has been little work specifically aimed at neural network applications in such materials. And it may be necessary to more carefully engineer the molecular andelectronic nature of such materials, to build more sophisticated molecular devices, in order to find a simultaneous solution to all the above essential characteristics. This molecular materials approach to making synapses for integrated circuit neural networks has the following conceptual advantages.

1) In neural networks, the by far the most numerous devices are the synapses. The approach thus concentrates on the devices whose reduction in size is most essential.
2) The problem of architecture is left to more macroscopic lithography. Arrays of parallel wires are all that would be required, which are in turn connected to much larger "neuron amplifiers".
3) The synapse material could be macroscopically homogeneous; the synapse connections themselves develope at wire intersections in a self-registering fashion.
4) This arrangement is only suitable for artificial neural network circuits, and is not useful for making conventional computer hardware. However, such circuits and algorithms are understood to be fault tolerant. The neural net architecture is known (again, through the "existence theorem" of biology) to be adequate and appropriate for the solution of major problems of language, robotics, and pattern recognition.

Designed molecular (or defined microcrystal) components as elements in materials for optical elements of computers are another way in which the materials aspects of macromolecular materials can be readily used. For example, consider the question of optical storage of information on disks. The use of rhodopsin-like material as the active material for optical disks, with the storage of information taking place through conformation change, has been worked on since Lamola first described the possibility 20 years ago. Birge and Lewis have also been involved in advancing these ideas. Another idea in optical storage has been the use of multiple wavelengths to store many bits at the same point in space. Experiments on this have been done at low temperatures, using the inhomogenous broadening of a narrow absorbtion line to permit "hole burning" and simultaneous storage at several different wavelengths. But the materials and techniques used have been restricted to low temperatures. These two approaches very much need to be combined. A series of related molecules could be designed, whose optical absorbtions have little overlap and which could be used at room temperature to produce the necessary multiple wavelengths. The storage of information by them could be through photo-isomerization or photo-charge transfer. [1]

These molecules would need to be kept physically separated so that dimerization does not destroy the carefully engineered optical properties. The absorbtion lines should be intense and narrow. (For reasons of depth-of-focus, the total layer to be written should be only a few

[1] One should note that the storage of information requires the motion of something physical to represent the stored bit. The question then is whether to store and move electrons or nuclear motions. Because of the comparitive quantum mechanical wavelengths of nuclei and electrons, it is easier to store nuclear motions in verysmall volumes than it is to store electrons.

microns.) Such a device is almost mundane. Yet its success could raise the density of information stored on optical disks by at least a factor of 10, and perhaps much more.

CONCLUSIONS AND SUMMARY.

In selecting the material for such a brief talk on a subject with such a variety of possible goal, approaches, and uses, I was guided in greatest part by an attempt to be responsible. The origins of this field lie in the creative imaginations of its founders. One should remember Forrest Carter in this respect. But in my view, the field now suffers from an excess of imagination and a deficiency of accomplishment. So I deliberately chose not to present an imaginative viewpoint.

It seems to me important that workers in the field think about particular problems. It does no good to be able to make elaborate three dimensional molecular computing assemblies if there is no way to get the energy necessary to compute delivered throughout the array. Molecular architecture and algorithms or computer structure must be planned together. The problem of restoration must be carefully thought about, for we are inevitably not going to be able to insist on very high precision of bit-operations in molecularly based computation. I believe that the materials uses of molecular (or microcrystalline) design will be useful much sooner than the ambitious schemes which more nearly use molecules as complete circuits.

Finally, the scale on which computing devices are made will grow ever smaller. At some small scale of devices, perhaps 100 Å, it will become more effective to make nano-devices through chemistry and cause them to self-assemble on a substrate than to continue to manipulate the most microscopic scales with macroscopic apparatus. Contemporary lithography, STEM, and molecular beam epitaxy approaches to nanoelectronics seem increasingly like making a suspension bridge by carving it out of a large block of steel. I think there is an inevitable coming together of microelectronics with the use of premade macromolecular nano-components, whether these be active or passive, and with biological ideas of self-assembly.

ACKNOWLEDGMENT

Thanks to Dr.S.Magrì, who helped to revise the manuscript.

INDEX

MIX
Papier aus verantwortungsvollen Quellen
Paper from responsible sources
FSC® C105338

If you have any concerns about our products,
you can contact us on
ProductSafety@springernature.com

In case Publisher is established outside the EU,
the EU authorized representative is:
Springer Nature Customer Service Center GmbH
Europaplatz 3, 69115 Heidelberg, Germany

Printed by Libri Plureos GmbH
in Hamburg, Germany